Proceedings of the 13th International Symposium of
The Princess Takamatsu Cancer Research Fund, Tokyo, 1982

# ADP-RIBOSYLATION, DNA REPAIR AND CANCER

Edited by
MASANAO MIWA, OSAMU HAYAISHI,
SYDNEY SHALL, MARK SMULSON,
and TAKASHI SUGIMURA

CRC Press
Taylor & Francis Group
Boca Raton London New York

CRC Press is an imprint of the
Taylor & Francis Group, an **informa** business

First published 1983 by Japan Scientific Press
and VNU Scientific Societies Press

CRC Press
Taylor & Francis Group
6000 Broken Sound Parkway NW, Suite 300
Boca Raton, FL 33487-2742



ISBN-13: 978-90-6764-003-9 (hbk)



**Visit the Taylor & Francis Web site at**
**http://www.taylorandfrancis.com**

**and the CRC Press Web site at**
**http://www.crcpress.com**

Organizing Committee of the 13th International Symposium

Masanao MIWA
Virology Division, National Cancer Center Research Institute, Tokyo 104, Japan
Osamu HAYAISHI
Department of Medical Chemistry, Kyoto University, Faculty of Medicine, Kyoto 606, Japan
Sydney SHALL
Cell and Molecular Biology Laboratory, University of Sussex, Brighton, East Sussex BN1 9QG, England
Mark SMULSON
Department of Biochemistry, Georgetown University, School of Medicine and Dentistry, Washington, D.C. 20007, U.S.A.
Takashi SUGIMURA
Biochemistry Division, National Cancer Center Research Institute, Tokyo 104, Japan

# Preface

Cancer is currently a major human disease although several tropical diseases such as malaria, bilharzia and trypanosomiasis cause more ill-health and deaths. The diagnosis and treatment of human cancer has steadily improved over the last half-centry, but we still do not understand the essential features of the disease; while cancer prevention is available (but rarely used) we have only very limited ability to cure human cancer. The annual meetings of the Princess Takamatsu Symposia are an important contribution to our world-wide attempts to understand, prevent and cure human cancer.

The present symposium deals with the potential relevance of ADP-ribosylation reactions in the genesis, prevention and treatment of cancer. ADP-ribosylation reactions were independently discovered in the laboratories of Paul Mandel, Osamu Hayaishi and Takashi Sugimura at about the same time that phosphorylation and acetylation of chromatin proteins were described. Since then we have acquired collectively a reasonable understanding of the structure, enzymology and some of the functions of ADP-ribosylation reactions. ADP-ribosyl adducts may be monomers and polymers and the latter may be linear or branched. It seems that ADP-ribosylation reactions are quite ubiquitous; they occur in bacteria, bacterial viruses, animals and plants, although DNA-dependent poly ADP-ribosylation seems to be confined to true eukaryotic cells.

Over the last five years some of the physiological functions of ADP-ribosylation reactions have emerged. For example, both diphtheria and cholera toxins are ADP-ribosyl transferase enzymes. The physiological functions of animal mono-ADP-ribosylation are still unknown. Poly ADP-ribosylation reactions however, have yielded up some of their physiological secrets. It is now clear that these reactions are required for efficient cellular responses to some DNA-damaging agents such as alkylating reagents and ionizing radiation. It seems that the nuclear enzyme ADP-ribosyl transferase is sensitive to DNA breaks and may modulate the rate of their rejoining, possibly by regulating the activity of DNA ligase II. In addition to this function in DNA repair, it seems that ADP-ribosylation reactions are also essential steps in certain types of cell differentiation. It has been suggested therefore, that the nuclear ADP-ribosyl transferase may be required for the successful execution of DNA transpositions and rearrangements in animal cells.

The possible involvement of nuclear ADP-ribosyl transferase in cancer may be appreciated better in the context of recent advances in our understanding of human cancer. Carcinogenesis necessarily requires a number of steps, at least some of which are probably genetic mutations. The altered growth behaviour and invasiveness of a malignant tumour requires that the normally mortal cell should become immortal and should also lose its dependence on tissue homeostasis. Most human and murine B-cell lymphomas have specific chromosomal translocations, and a high frequency of chromosomal aberrations is a prominent feature of a progressing malignancy.

It is conceivable that an enzyme like nuclear ADP-ribosyl transferase, which participates in some way in controlling the breaking and rejoining of DNA, might if defective or deficient, lead to chromosomal aberrations. Inhibition of nuclear ADP-ribosyltransferase certainly augments spontaneous sister chromatid exchanges and chromosomal aberrations. DNA transpositions or rearrangements might be important in the pathogenesis of cancer. Consequently, a defective nuclear ADP-ribosyl transferase might be partly responsible for cancer progression and might in certain circumstances be important in the development of the primary tumour.

Poly ADP-ribose metabolism might also provide an opportunity for improved cancer therapy. Inhibitors of ADP-ribosylation potentiate the cell killing effect of some alkylating agents and of radiation. At this symposium we heard of animal experiments in which the efficacy of the anti-cancer antibiotic, bleomycin was potentiated by 3-aminobenzamide. These observations warrant further investigation in animal systems to establish if there is any likelihood of improving the therapy of human cancer.

Finally, ADP-ribose metabolism may be used to identify susceptible individuals in a population. This has been demonstrated with xeroderma pigmentosum heterozygotes. This observation should be extended to all the chromosomal instability diseases. Clearly, one could assess the response of cellular NAD to appropriate DNA- damaging agents and the effect of 3-aminobenzamide on this response. Furthermore, a simple test of DNA repair and of inhibition of DNA synthesis by an appropriate DNA-damaging agent and the effect of 3-aminobenzamide on this test might be a very suitable screening procedure for large populations. This might be appropriate in areas of very high incidence of specific types of cancer such as Burkitt's lymphoma in Africa, or oesophageal or nasopharyngeal carcinoma in China.

In all, it transpires that ADP-ribosylation reactions are potentially important in understanding, preventing and treating human cancer. A substantial effort to achieve a really deep understanding of ADP-ribosylation reactions may contribute to our ultimate control of human cancer as well as enlarging our knowledge of how cells work.

Sydney SHALL
Cell & Molecular Biology Laboratory
University of Sussex
Brighton, England

Contents

ADP-ribosylation and Cancer

## Contributors


| | |
|---|---|
| Nathan A. BERGER | Hematology / Oncology Division of the Department of Medicine, Washington University School of Medicine and The Jewish Hospital of St. Louis, St. Louis, Missouri 63110, U.S.A. |
| Luis O. BURZIO | Instituto de Bioquimica, Facultad de Ciencias, Universidad Austral de Chile, Valdivia, Chile |
| James E. CLEAVER | Laboratory of Radiobiology and Environmental Health, University of California, San Francisco, California 94143, U.S.A. |
| Yoshisada FUJIWARA | Department of Radiation Biophysics, Kobe University, School of Medicine, Kobe 650, Japan |
| D. Michael GILL | Department of Molecular Biology and Microbiology, Tufts University School of Medicine, Boston, Massachusetts 02111, U.S.A. |
| Osamu HAYAISHI | Department of Medical Chemistry, Kyoto University, Faculty of Medicine, Kyoto 606, Japan |
| Helmuth HILZ | Institut für Physiologische Chemie, Universität Hamburg, D-2000 Hamburg 20, West Germany |
| Myron K. JACOBSON | Departments of Chemistry and Biochemistry, North Texas State University, Denton, Texas 76203, U.S.A. |
| Yoshiyuki KANAI | Department of Molecular Oncology, Institute of Medical Science, University of Tokyo, Tokyo 108, Japan |
| Masashi KAWAICHI | Department of Medical Chemistry, Kyoto University, Faculty of Medicine, Kyoto 606, Japan |
| Hisae KAWAMITSU | High-Risk Study Division, National Cancer Center Research Institute, Tokyo 104, Japan |
| William R. KIDWELL | Cell Cycle Regulation Section, Laboratory of Pathophysiology, National Cancer Institute, National Institutes of |

Health, Bethesda, Maryland 20205, U.S.A.

Samuel S. KOIDE — Center for Biomedical Research, The Population Council, The Rockefeller University, New York, New York 10021, U.S.A.

Yoichi KONISHI — Department of Oncological Pathology, Cancer Center, Nara Medical College, Kashihara, Nara 634, Japan

Michael W. LIEBERMAN — Department of Pathology, Washington University School of Medicine, St. Louis, Missouri 63110, U.S.A.

Lawrence A. LOEB — Joseph Gottstein Memorial Cancer Research Laboratory, Department of Pathology, SM-30, School of Medicine, University of Washington, Seattle, Washington 98195, U.S.A.

Paul MANDEL — Centre de Neurochimie du Centre National de Recherche Scientifique, 67084 Strasbourg Cedex, France

Masanao MIWA — Virology Division, National Cancer Center Research Institute, Tokyo 104, Japan

Joel MOSS — Section on Molecular Mechanisms, Laboratory of Cellular Metabolism, National Heart, Lung, and Blood Institute, National Institutes of Health, Bethesda, Maryland 20205, U.S.A.

A. T. NATARAJAN — Department of Radiation Genetics and Chemical Mutagenesis, State University of Leiden, Sylvius Laboratory, 2333 AL Leiden, The Netherlands

Arthur B. PARDEE — Dana-Farber Cancer Institute and Department of Pharmacology, Harvard Medical School, Boston, Massachusetts 02115, U.S.A.

Henry C. PITOT — Department of Oncology, McArdle Laboratory for Cancer Research, University of Wisconsin, Madison, Wisconsin 53706, U.S.A.

Guy G. POIRIER — Centre de Recherche sur les Mécanismes de Sécrétion, Département de Biologie, Faculté des Sciences, Université de Sherbrooke, Québec, Canada JIK 2RI

Sydney SHALL — Cell and Molecular Biology Laboratory, University of Sussex, East Sussex BN1 9QG, England

Mark SMULSON — Department of Biochemistry, Georgetown University Medical Center, Washington, D. C. 20007, U.S.A.

Takashi SUGIMURA — Biochemistry Division, National Cancer Center Research Institute, Tokyo 104, Japan

Masaaki TERADA — Section of Prevention of Carcinogenesis, National Cancer Center Research Institute, Tokyo 104, Japan

Kunihiro UEDA — Department of Medical Chemistry, Kyoto University, Faculty of Medicine, Kyoto 606, Japan

Hiroshi YAMAMOTO — Department of Biochemistry, Toyama Medical and Pharmaceutical University School of Medicine, Toyama, Toyama 930-01, Japan

Koichiro YOSHIHARA — Department of Biochemistry, Nara Medical College, Kashihara, Nara 634, Japan

# Opening Address

H.I.H. Princess KIKUKO TAKAMATSU

It is my great pleasure to welcome all the participants to the 13th International Symposium of the Princess Takamatsu Cancer Research Fund. I would also like to say "Welcome to Japan" to the participants from overseas who have come such a long way to be with us today.

Since cancer is a major disease in many countries, it is my earnest hope that scientists will clarify the mechanisms of carcinogenesis through basic research, thus making possible the prevention and cure of cancer.

The title of the Symposium this year is "ADP-ribosylation, DNA repair and cancer". I have heard that ADP-ribosylation is a unique enzyme reaction that has been found to be related to DNA repair and probably to carcinogenesis. I am also informed that ADP-ribosylation is a new field of basic science that was discovered some 17 years ago, and has now grown into a field of great interest in laboratories all over the world. I understand that the reason why so many distinguished scientists from various countries in the world are assembled here is to discuss their contributions to this important field.

I would like to express my sincere thanks to Professor Mark Smulson of Georgetown University in the United States, Professor Sydney Shall of the University of Sussex in England and the other members of the Organizing Committee for the fine work they did in arranging this Symposium.

It is my wish that the discussions during this Symposium will contribute to a better understanding and early resolution of basic problems in cancer.

Dr. Mark SMULSON

Your Imperial Highness:

You may remember that I had the pleasure of being a speaker last year at the 12th Annual Princess Takamatsu Symposium. Each time I come to Japan, I experience the same fascination with the beauty of its country side, industrious people and its extraordinary mixture of ancient tradition and modern "State of the art," technology.

Many of us here, working on poly(ADP-ribose), visited your wonderful country in 1974 in Tomakomai for a meeting in this same area of science. In fact five international workshops have been held since 1973 on this topic. Some of us wonder whether the *real* biological function of poly(ADP-ribose) is to allow us to come together, every few years, in fascinating parts of the world.

The presence of scientists from many nations at this meeting is a reflection of the international nature of this important scientific area. The excellent opportunity for international cooperation and personal contact offered during this meeting clearly will stimulate even closer relationships than those already underway on poly(ADP-ribose) research with Japanese scientists. My attendance at this meeting last year and again now has clearly been significant in my own scientific career where noted Japanese scientists such as Drs. Sugimura, Hayaishi and Miwa have contributed to my own growth in research.

I am sure that this present, prestigious meeting will provide an unusual opportunity to strengthen such interactions, which should further Cancer Research worldwide.

Poly ADP-ribosylation of nuclear macromolecules has been a fascinating and important topic in the Biochemical field since its discovery, both in Japan and in France, in 1966. The program of this meeting shows work in poly(ADP-ribose) which will form new concepts in regulatory processes in cells, the interaction of carcinogenic agents with DNA and chromatin and thus, ultimately new insights into the *cause of cancer* and its expression.

Continued progress in science relies on hard-work by many of our coworkers in all parts of the world, but it *also* equally relies on the encouragement and promotion of research.

Your Imperial Highness, your leadership in supporting cancer research, and based upon my past experiences your participation in these conferences, in your own elegant manner, has been, and is, an invaluable contribution to science, not only in Japan, but in the world.

On behalf of my fellow scientists, I wish to thank you for it.

# LECTURES

ADP-RIBOSYLATION, DNA REPAIR AND CANCER, M. MIWA ET AL. (EDS.),
JAPAN SCI. SOC. PRESS, TOKYO/VNU SCIENCE PRESS, UTRECHT, PP. 3-25, 1983

KEYNOTE LECTURE

# ADP-ribosylation, DNA Repair, Cell Differentiation and Cancer

Sydney SHALL

*Cell and Molecular Biology Laboratory, University of Sussex, Brighton, East Sussex BN1 9QG, England*

*Abstract:* This review discusses the potential relationships between ADP-ribosylation reactions, DNA repair, cell differentiation, and cancer. ADP-ribosylation of chromatin proteins has been shown to participate in DNA excision repair in all nucleated cells. ADP-ribosylation of chromatin proteins is catalysed by nuclear ADP-ribosyl transferase (ADPRT). This enzyme is entirely dependent on DNA for its activity because it has an absolute requirement for ends or nicks in double-stranded DNA. Exposure of cells to small alkylating agents or to radiation causes a fall in cellular $NAD^+$ levels due to a transient activation of ADPRT and a consequent ADP-ribosylation of chromatin proteins. Inhibitors of ADPRT retard DNA strand-rejoining induced by radiation or by small alkylating agents; such inhibition has at least two biological consequences; a synergistic potentiation of cytotoxicity and an enhancement of sister chromatid exchanges and chromosomal aberrations. No species differences have yet been reported; there are variations between cell types and between different damaging agents.

The enzyme inhibitors do not block early steps in DNA repair, and repair synthesis does not require ADPRT activity. DNA damage increases the activity of both DNA polymerase $\beta$ and DNA ligase II. The activation of DNA ligase II can be blocked by ADPRT inhibitors; presumably ADPRT activity is required for the activation of DNA ligase II. A plausible molecular explanation for the function of ADPRT in DNA repair is that ADPRT regulates the activity of DNA ligase II, the "non-replicative" ligase.

In addition to its function in DNA repair, ADPRT is an obligatory requirement in certain categories of cell differentiation. Inhibitors of ADPRT and nicotinamide starvation both reversibly block cell differentiation. We suggest that a similar mechanism to that of DNA repair may be involved because we observe 100 to 300 single-strand DNA breaks during the cytodifferentiation of primary chick myoblasts. These breaks are not due to a general deficiency in DNA repair.

I suggest that in certain categories of cell differentiation there are rearrangements or transpositions within the mammalian genome, and that ADP-ribosylation reactions have a general function to be sensitive to DNA breaks and to regulate subsequent DNA ligation in DNA repair, in DNA recombination, in sister chromatid exchanges, in chromosome aberrations, in gene rearrangements, in transpositions and in certain categories of cell differentiation. The relevance of these observations and ideas to cancer is discussed.

We are gathered together here in Tokyo to discuss the potential relationships between ADP-ribosylation reactions, DNA repair and cancer. In this lecture I propose to review very briefly the experimental evidence that ADP-ribosylation is required for efficient DNA repair, to describe observations indicating that ADP-ribosylation is also required for certain categories of cell differentiation and then to discuss the possible involvement of these reactions in carcinogenesis and a potential improvement in cancer therapy.

In recent years, a profound change has occurred in the way we view the molecular biology of cancer. There is an increased awareness of the potential importance of DNA repair in carcinogenesis. Recently it has been clearly demonstrated that cancer genes exist not only in tumour viruses but also in tumour cells and even more dramatically, in normal cells. Thirdly there has been considerable and growing interest recently in the potential of DNA transposition, that is movement, of genes or DNA sequences during differentiation and carcinogenesis. A very encouraging development is the convergence of these several, previously independent lines of thought. I shall briefly review some recent evidence that ADP-ribosylation reactions may be related to these trends.

ADP-ribosylation reactions have been exhaustively reviewed in the book edited by Osamu Hayaishi and Kunihiro Ueda (*1*); and I have recently reviewed at length the participation of ADP-ribose in DNA repair (*2*). Please consult those reviews for detailed references. The ADP-ribosylation of chromatin proteins is a recent addition to DNA excision repair in nucleated cells.

ADP-ribosylation of chromatin proteins is catalysed by an enzyme which I shall call nuclear ADP-ribosyl transferase or ADPRT. The enzyme seems to be capable of catalysing three separate related reactions: it can transfer the ADP-ribose portion of $NAD^+$ to a protein acceptor, to form mono-ADP-ribosyl protein (Fig. 1); it can then catalyse the transfer of further ADP-ribose residues to form both linear and branched poly(ADP-ribose) (Fig. 2). Apparently, the unique substrate for this enzyme is $NAD^+$. An essential and central property of this enzyme is its absolute requirement for DNA because the enzyme has an absolute requirement for ends or nicks in DNA. The enzyme is apparently only active when bound to an end or a nick in a double-

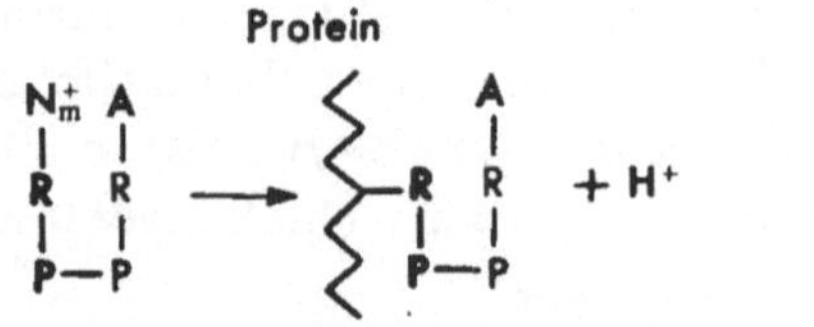


FIG. 1. Synthesis of mono-ADP-ribosyl protein. Nm, nicotinamide; R, ribose; P, phosphate; A, adenine. The corrugated line represents the acceptor protein.

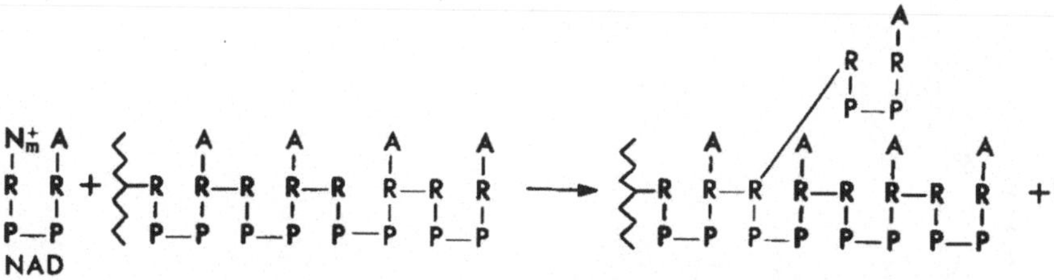

FIG. 2. Synthesis of linear and branching poly(ADP-ribosyl)protein conjugates.

stranded DNA chain. An approximate affinity constant for the binding of the enzyme to double-stranded ends of DNA of about 40 nM can be estimated from the data of Yoshihara and colleagues (*3*). The protein acceptors include ADPRT enzyme molecules themselves, histone H1 and several non-histone chromatin proteins: this topic will doubtless be much discussed at this meeting, so I only make one comment. We should distinguish carefully between those proteins which the enzyme is able to modify in various experimental situations, from those proteins which are in fact covalently modified in intact undisturbed, or damaged or differentiating cells.

A variety of inhibitors of ADPRT are known. Some of these together with their inhibition constants are listed in Table 1.

It has been known for a long time that DNA-damaging agents cause a fall in cellular $NAD^+$ level (*4–6*) (Fig. 3). DNA-damaging agents (both chemicals and radiation) activate ADPRT (*6*) (Figs. 4, 5) probably in proportion to the steady-state level of DNA strand breaks that are associated with these several DNA damaging agents. Some seven years ago we suggested that the loss of $NAD^+$ was due to the biosynthesis of poly(ADP-ribose) (*4*). This accelerated biosynthesis of poly(ADP-ribose) from $NAD^+$ in intact cells exposed to DNA-damaging agents has been directly demonstrated several times. These results imply either a very rapid turnover of poly(ADP-ribose), of the order of 0.1 to 0.2 min, or alternatively we must seek an additional route for the disappearance of the $NAD^+$. The amount of poly(ADP-ribose) detected seems a lot less than the $NAD^+$ which has disappeared.

All the inhibitors of ADPRT prevent the depletion of cellular $NAD^+$ that is caused by DNA-damaging agents (*6*) (Fig. 4). This adds further weight to the notion that DNA damage activates ADPRT resulting in increased synthesis of poly-

TABLE 1. Inhibitors of Nuclear ADPRT

| | $K_i$ (μM) |
|---|---|
| 1. Nicotinamide | 13.0± 1.8 |
| 5-methylnicotinamide | 30.0± 3.5 |
| 2. Thymidine | 13.3± 1.6 |
| 3. Methyl-xanthines | |
| theophylline | 29.8± 3.5 |
| theobromine | 15.2± 1.0 |
| caffeine | 244±16 |
| 4. Benzamide | <1.0 |
| 3-aminobenzamide | 4.4± 1.1 |

The $K_m$ for NAD is 30 μM (but varies with reaction conditions). All these inhibitors are competitive.

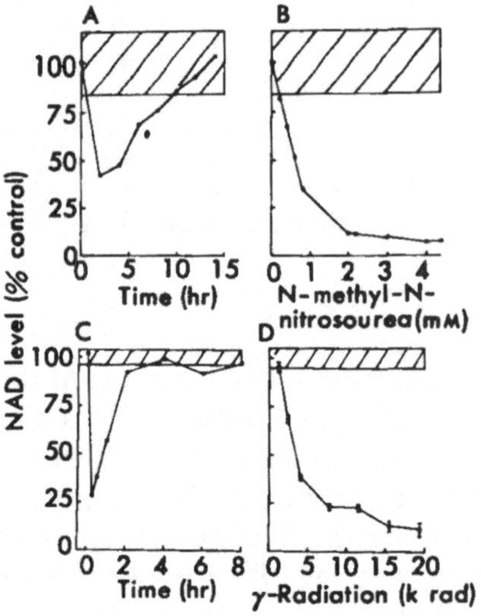


FIG. 3. Effect of N-methyl-N-nitrosourea and of γ-radiation on NAD levels in mouse leukaemia L1210 cells. The hatched area shows the control levels ± 1 S.D. A. The time course of the effect of 800 μM N-methyl-N-nitrosourea. B. The dose response of cell NAD 2 hr after treatment with methylnitrosourea. C. Time course of the effect of 12 krad of γ-radiation. D. The dose response in the drop in cell NAD 15 min after γ-irradiation. (From ref. 6).

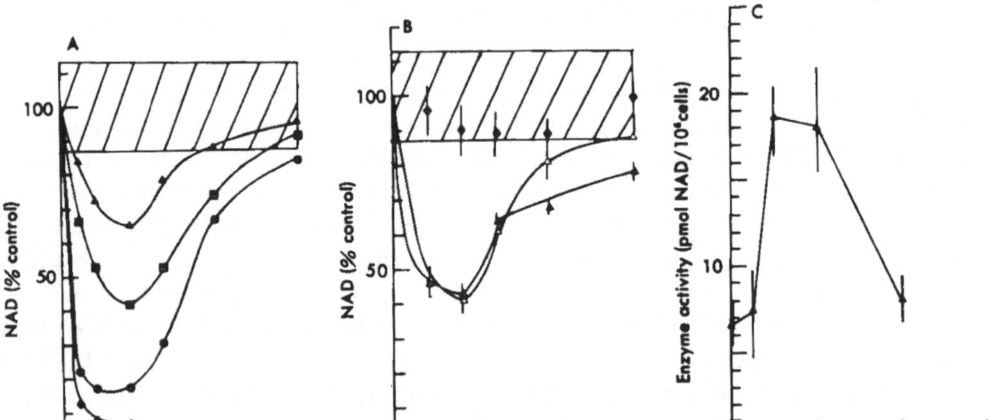


FIG. 4. The effect of dimethyl sulphate on mouse leukaemia (L1210) cells. A. The effect of dimethyl sulphate on the NAD content. ▲ 10.5 μM dimethyl sulphate; ■ 52.6 μM; ● 105.3 μM; ◆ 210.6 μM. B. Inhibition of the lowering of the cellular NAD level by dimethyl sulphate by 3-aminobenzamide but not by 3-aminobenzoic acid. The cells were continuously treated with 53 μM dimethyl sulphate either alone (△) or in the continuous presence of either 2 mM 3-aminobenzamide (◆) or of 2 mM 3-aminobenzoic acid (▲). C. The effect of 1 mM dimethyl sulphate on the ADPRT activity. (From ref. 7).

(ADP-ribose) and a loss of cellular $NAD^+$. In general, the kinetics of $NAD^+$ disappearance and of ADPRT activation correlate rather well (compare Figs. 3 to 5).

The requirement of ADP-ribosylation for efficient DNA excision repair was

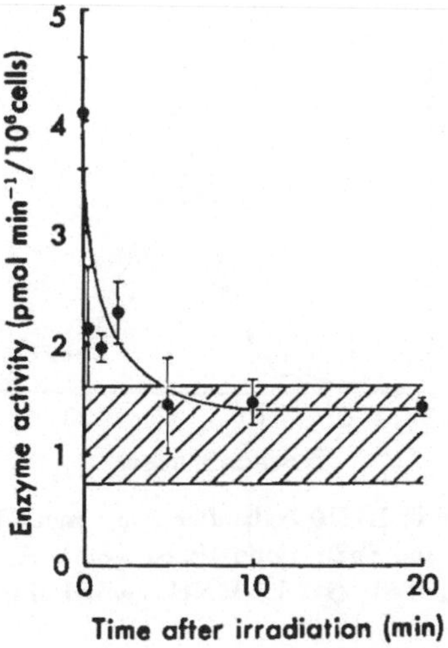


FIG. 5. Time course of ADPRT activity in permeabilised (ref. *19*) L1210 mouse leukaemia cells following 10 krad of $\gamma$-radiation. The hatched area shows the control levels $\pm 1$ S.D. (From ref. *6*).

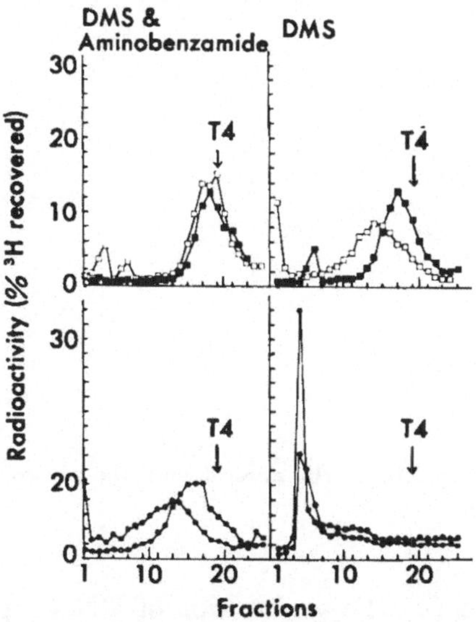


FIG. 6. The inhibition of DNA excision repair after dimethyl sulphate in L1210 cells by 3-aminobenzamide. Strand-rejoining was measured on alkaline sucrose gradients. Centrifugation is from the right to the left. The arrow shows the position of bacteriophage T4 DNA, single-strand molecular weight of $55 \times 10^6$. ■ 0; □ 40; ● 80; ○ 300 min after DMS. DMS, dimethyl sulphate. (From ref. *7*).

directly demonstrated some years ago (*7*); inhibitors of ADPRT inhibit DNA repair as measured on alkaline sucrose gradients. 3 mM 3-aminobenzamide retards DNA strand-rejoining in L1210 cells exposed to 100 $\mu$M dimethyl sulphate (Fig. 6). This enzyme inhibitor will also retard strand-rejoining after exposure to 2 mM methyl nitrosourea (Fig. 7), or to N-methyl-N′-nitro-N-nitrosoguanidine (MNNG). All the

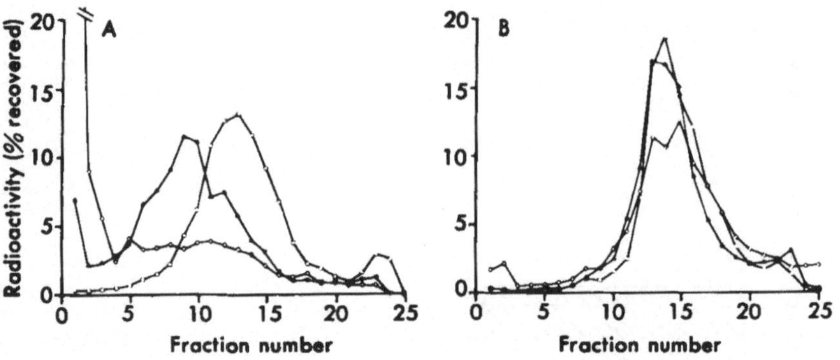


FIG. 7. The inhibition of DNA excision repair in L1210 cells after 2 mM methyl nitrosourea treatment by 3 mM 3-aminobenzamide (Perera and Shall, unpublished work). A. 2 mM MNU control. B. 2 mM MNU+3AB (3 mM). ○ T0; ▲ T40; ◇ T80. MNU, methyl nitrosourea; AB, aminobenzamide.

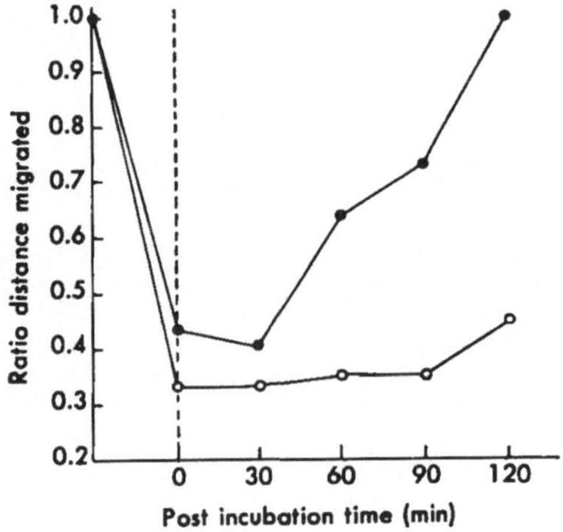


FIG. 8. The inhibition of DNA repair in L1210 cells assessed with the nucleoid technique. ● control; ○ 30 μM DMS+3AB (3 mM).

inhibitors retard DNA repair in L1210 cells (*8*). This inhibition of DNA repair is also evident when tested by following the sedimentation behaviour of nucleoids (*9*) (Fig. 8). This technique is performed at neutral pH and is very sensitive. 30 μM dimethyl sulphate causes very little toxicity by itself. Efficient DNA repair requires ADP-ribosylation not only in rodent cells but in human cells as well: no significant species differences have yet been reported. ADPRT participates in DNA repair in both lymphoid cells and in fibroblast, glial, glioma, carcinoma, and sarcoma cells. It seems to be a general property of nucleated cells.

However, in our limited experimental experience we have observed differences in the responses of different cell types; there are significant variations in the responses of different cell types. The ability of the enzyme inhibitors to retard DNA repair varies in different cell types. In addition, there are clear differences in the responses to the various DNA-damaging agents. Ultraviolet radiation is least efficient at inducing ADPRT activity. I would associate many bulky adducts and bifunctional reagents with ultraviolet radiation in this respect. Ionizing radiation is somewhat more efficient in activating ADPRT; naturally bleomycin and neocarcinostatin are rather similar. By contrast, small monofunctional alkylating- and especially methylating-agents are extremely efficient at activating the nuclear ADPRT. The efficacy of ADPRT inhibitors quite predictably follows the same pattern. I shall describe a plausible explanation for the differences after I have outlined the probable molecular mechanism of ADP-ribosylation in DNA repair.

One particular situation requires comment. Both $\gamma$- and X-radiation lower NAD levels, and the effect of $\gamma$-radiation is inhibited by all the ADPRT inhibitors. Clearly, ionizing radiation stimulates poly(ADP-ribose) biosynthesis. However, it is not easy to demonstrate inhibition of DNA strand-rejoining by the enzyme inhibitors after $\gamma$-radiation. A possible explanation will be presented in a moment. The participation of nuclear ADPRT activity in DNA repair after ionizing radiation may be demonstrated however, by the nucleoid technique. This technique can detect as few as 200 to 300 single-strand breaks per genome equivalent. With the nucleoid technique it was possible to demonstrate that 3-aminobenzamide partially inhibits restoration of supercoiling following $\gamma$-irradiation or dimethyl sulphate treatment (*9*).

These results indicate that we can expect differences between cell types as well as between different DNA-damaging agents. It would be wise not to extrapolate without great care from one cell type to another or from, say, alkylating agents to ionizing or worse still, to ultraviolet radiation.

I want now to consider the possible molecular mechanism by which nuclear ADPRT participates in DNA repair.

The results I have shown you indicate that ADPRT inhibitors do not block

TABLE 2. Stimulation of DNA Ligase Activity by an Alkylating agent, and Prevention of This Stimulation by Inhibitors of ADPRT

| Additions | Number of times each experiment was performed | Relative DNA ligase activity | |
|---|---|---|---|
| | | Controls | Exposed to dimethyl sulphate |
| None | (10) | 1.00 | 2.35 |
| Benzamide | (2) | 1.08 | 0.81 |
| 3-Aminobenzamide | (10) | 0.80 | 1.08 |
| 5-Methylnicotinamide | (2) | 1.10 | 0.85 |
| Theophylline | (2) | 0.94 | 1.06 |
| Thymidine | (2) | 0.94 | 0.97 |
| 3-Aminobenzoic acid | (2) | 0.90 | 1.82 |
| Nicotinic acid | (2) | 1.06 | 1.82 |

Mouse lymphoblastic leukaemia cells (L1210 cells) ($5\times10^5$ cells/ml) were treated with 300 $\mu$M dimethyl sulphate for 1 hr in the presence or absence of the inhibitors (5 mM) or their analogues (5 mM). DNA ligase activity was then extracted from the cells. In control experiments the cells were exposed to the inhibitors or analogues, but were not treated with dimethyl sulphate. DNA ligase activity was estimated with 5'[$^{32}$P] phosphate-DNA as substrate. The DNA ligase assay measures the rate at which [$^{32}$P] 5'-phosphate becomes inaccessible to alkaline phosphatase because of DNA ligation. All enzyme activities are expressed relative to the activity in cells exposed to neither dimethyl sulphate nor to inhibitors nor their analogues. (From ref. *15*).

early steps in DNA repair, because the single-strand breaks appear in the presence of the enzyme inhibitor, as they do in its absence. DNA repair synthesis is not inhibited by the ADPRT inhibitors, as shown by the work of Berger *et al.* (*20*), and Miwa *et al.* (*21*), and ourselves (*10*). We therefore turn our attention to the DNA ligase reaction. All the preceding data led us to propose that ADP-ribosylation regulates the rejoining or the ligation steps of DNA repair.

ADP-ribosylation does not modulate DNA ligation in the course of normal DNA replication, because high concentrations of ADPRT inhibitors or nicotinamide deprivation interfere only very slightly with DNA biosynthesis or with cell growth of undamaged cells.

It has been reported by Söderhäll and Lindahl (*11*) that in mammalian cells (in contrast to bacteria) there are two forms or species of DNA ligase and the two forms have been both chromatographically and serologically separated (*12*). More recent work by Teraoka *et al.*, however, has questioned the presence of the second enzyme which they argue may be the consequence of proteolytic activity (*13, 14*). This point may be taken up later in this meeting.

We have been able to reproduce exactly, the original observations of Söderhäll and Lindahl and observe two ligase activities separable by chromatography on hydroxylapatite. It has been assumed that DNA ligase I is involved mainly in semiconservative replication; its activity fluctuates following physiological or experimental changes in growth behaviour, in a similar way to DNA polymerase α. It has been suggested that the second ligase activity, DNA ligase II, may be a "repair enzyme".

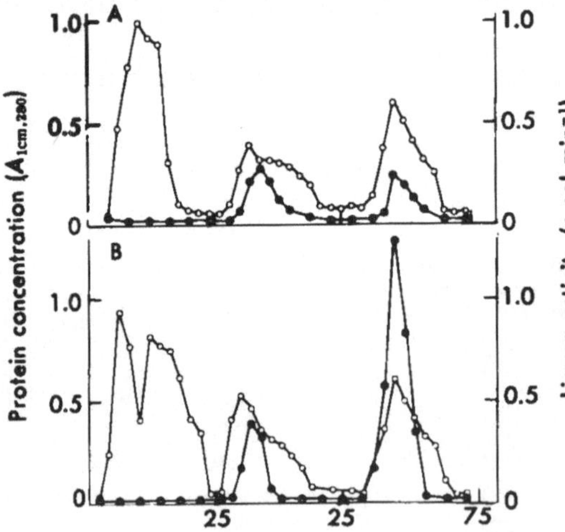


FIG. 9. Increase in DNA ligase II activity after exposure of L1210 cells to 300 μM dimethyl sulphate. The cells were exposed to dimethyl sulphate and then the ligase activity was extracted, partially purified by ammonium sulphate precipitation and phosphocellulose chromatography. The DNA ligase activity from this column was chromatographed on hydroxylapatite; the resulting chromatogram is shown in the figure. The fractions were assayed with purified, previously prepared radioactive nicked DNA. A. Cells not treated with dimethyl sulphate. B. Cells exposed to 300 μM dimethyl sulphate for 1 hr. ○ protein concentration ($A_{280}$ absorbance units); ● DNA ligase activity (pmol/min). (From ref. *15*).

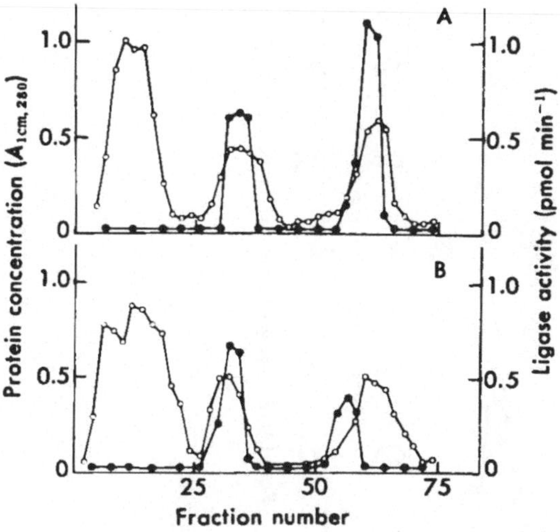


FIG. 10. 3-Aminobenzamide prevents the increase in DNA ligase II activity observed after exposure of L1210 cells to 300 $\mu$M dimethyl sulphate. The experimental protocol is described in the legend to Fig. 9. ○ protein concentration ($A_{280}$ absorbance units); ● DNA ligase activity (pmol/min). (From ref. *15*).

The existence of two ligase activities makes it reasonable to consider whether DNA ligase II, the "repair enzyme activity" is specifically regulated by ADP ribosylation.

Very recently, we have reported direct evidence that ADP-ribosylation regulates DNA ligase II activity in DNA repair (*15*). We assay a partially purified DNA ligase activity from mouse leukaemia cells with purified exogenous DNA. When the leukaemia cells are exposed to the alkylating agent, dimethyl sulphate, the total DNA ligase activity in the cell extracts increases about 2.3-fold (Table 2) (*15*). In addition to this increase in ligase activity there is also an increase in the activity of DNA polymerase $\beta$.

The increase in ligase activity is probably a consequence of ADPRT activity, because all classes of ADPRT inhibitors prevented this increase; non-inhibitory analogues do not do this.

The ADPRT-dependent increase in ligase activity is predominantly due to DNA ligase II. While the ligase I activity is essentially constant, the DNA ligase II activity increased nearly 5-fold (Fig. 9). The ADPRT inhibitors that block the increase in total DNA ligase activity induced by dimethyl sulphate (Table 2), work by specifically preventing the increase in DNA ligase II activity (Fig. 10).

It seems then that alkylation damage increases DNA ligase activity, and predominantly DNA ligase II activity which is presumed to be the repair enzyme. This increase in DNA ligase II activity is totally prevented by inhibiting poly(ADP-ribose) biosynthesis, implying that the increase in DNA ligase activity requires ADPRT activity.

One possible explanation of our results is that DNA ligase II itself or a closely associated protein is ADP-ribosylated. Preliminary experiments indicate that when permeabilised cells are incubated with [$^{3}$H]-NAD$^{+}$ (a specific precursor of poly-

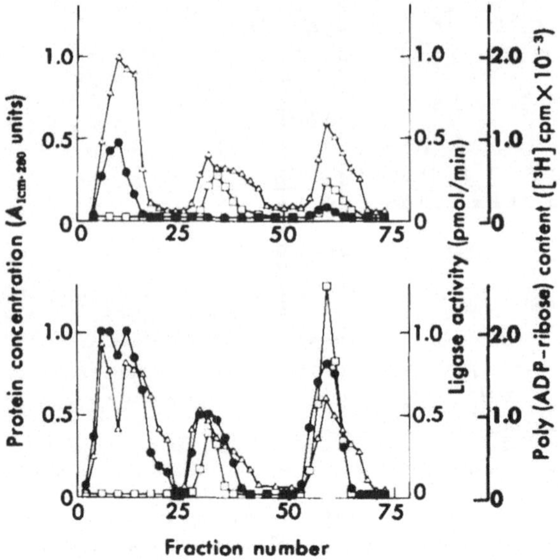


FIG. 11. NAD$^+$ labels the increased DNA ligase activities in mouse leukaemia cells after exposure to 300 μM dimethyl sulphate. The experimental protocol is described in the legend to Fig. 9 with the addition that after exposure to the alkylating agent, the cells were permeabilised and incubated in [$^3$H]-NAD, then the ligase activity was extracted and purified. △ protein concentration ($A_{280}$ absorbance units); □ ligase activity (pmol/min); ● poly(ADP-ribose) ([$^3$H] radioactivity, cpm). The upper panel shows cells not treated with alkylating agent; the lower panel shows cells treated with 300 μM dimethyl sulphate.

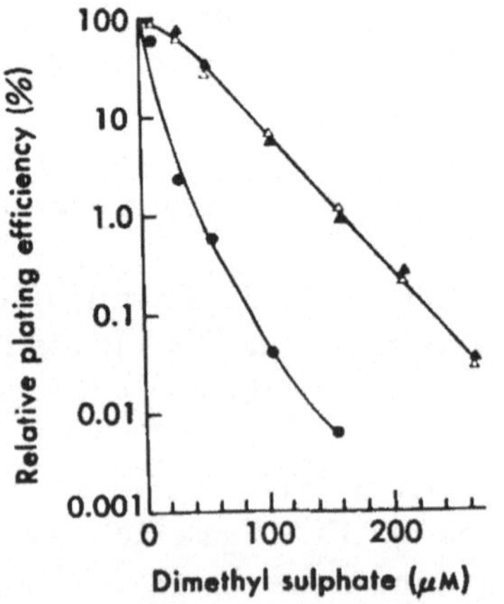


FIG. 12. The synergistic potentiation of cytotoxicity of dimethyl sulphate by 2 mM 3-aminobenzamide. △ exposed to dimethyl sulphate for 1 hr, then washed and plated in soft agar; ● exposed to dimethyl sulphate and 2 mM 3-aminobenzamide for 1 hr, then washed and plated in soft agar containing 2 mM 3-aminobenzamide; ▲ exposed to dimethyl sulphate and 2 mM 3-aminobenzoic acid for 1 hr, then washed and plated in soft-agar containing 2 mM 3-aminobenzoic acid. (From ref. *6*).

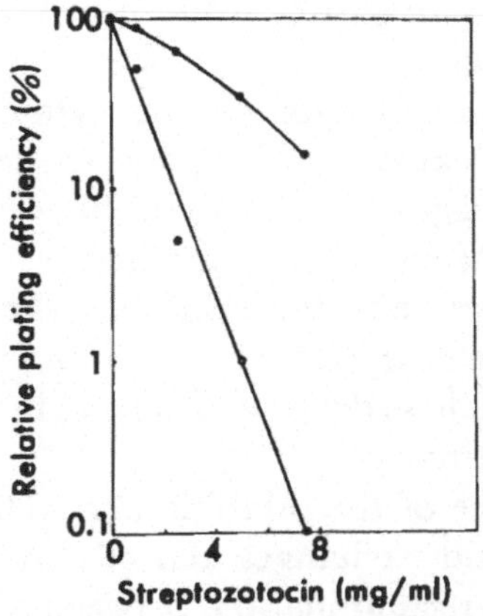


FIG. 13. Synergistic potentiation of the toxicity of streptozotocin by 5-methylnicotinamide. ● streptozotocin alone; ○ streptozotocin plus 2 mM 5-methylnicotinamide.

(ADP-ribose)) a peak of radioactivity co-migrates with the DNA ligase II activity on an hydroxylapatite column (Fig. 11).

An alternative possibility is that a DNA-binding protein is ADP-ribosylated and thus activates the DNA ligase II. A specific example of this hypothesis is of interest. It is known that the ADPRT enzyme can modify ADPRT enzyme molecules; moreover, ADPRT molecules have a very high affinity for nicks in DNA. Thus, ADPRT molecules may be bound to nicks produced during DNA repair making them inaccessible both to nuclease and ligase action. Automodification of enzyme molecules by ADPRT would probably weaken the affinity of the enzyme for the DNA; the enzyme would then dissociate from the break, thus allowing access for the DNA ligase. The increase in DNA ligase activity would then be attributed to the increased availability of substrate.

No doubt, further evidence relating to the mechanism of ADP-ribosylation during repair will be presented and discussed at this meeting.

The data which I have presented does not by itself exclude the possibility that ADP-ribosylation also modifies chromatin structure during repair in addition to its apparent participation in activating DNA ligase. The inhibition of DNA repair which we have observed is totally consistent with the idea that ADPRT activity is required to activate DNA ligase II; but it is not easily reconciled with the idea that ADP-ribosylation is required for efficient DNA repair only because if modifies chromatin structure.

If ADPRT activity is really required for efficient cellular recovery from DNA damage, then inhibition of ADPRT activity would be expected to increase the lethality of DNA-damaging agents. This prediction has been amply confirmed (*7, 16*). 2 mM 3-aminobenzamide enhances the cytotoxicity of dimethyl sulphate for L1210 cells (Fig. 12); there is approximately a 4-fold dose enhancement (*7*). All the inhibitors of ADPRT activity synergistically potentiate the cytoxocity of monofunctional DNA-damaging agents including methyl nitrosourea and MNNG (*16*). There is evidence that the synergistic potentiation by caffeine does not involve ADP-ribosylation, because it and ADPRT inhibitors are synergistic, while the ADPRT inhibitors show only additive effects with one another.

Of very special interest is the synergistic potentiation of the cytotoxicity of streptozotocin by ADPRT inhibitors (Fig. 13), because this drug and its derivatives are used in the treatment of human tumours. I suggest that it would be very worthwhile to investigate the potential of ADPRT inhibitors as chemotherapeutic adjuvants. It would be helpful to know whether these enzyme inhibitors show useful potentiation of the killing of cancer cells in experiments with whole animals. Our cell culture experiments are not a perfect guide to how these drugs will behave in humans. Perhaps we will hear more of this during the meeting.

A further example of a biological consequence of the inhibition of ADPRT activity is the increased frequency of sister chromatid exchanges and of chromosomal aberrations. Sister chromatid exchanges involve recombination between sister duplexes and therefore require breakage and rejoining of DNA. Sugimura and colleagues (*22*) have clearly shown that ADPRT inhibitors increase the frequency of sister chromatid exchanges in proportion to their ability to inhibit ADPRT. A very detailed study of this phenomenon has been made by Natarajan, Csukás, and van Zeeland (*23*); this work will be discussed by Natarajan later in the meeting. The conclusions I draw from their work is that aminobenzamide does not by itself induce sister chromatid exchanges; presumably, it can only induce them when chromatid breaks are engendered by some other process. The increase in sister chromatid exchanges may well be explained by a longer lifetime of gaps in the DNA when ADPRT is inhibited. In the presence of two nearby single-strand breaks, inhibition of ADPRT activity may prolong the life of this double-strand break and therefore increase the probability of a sister chromatid exchange. This notion presupposes that ADPRT is required not only for DNA repair but for other DNA nicking and rejoining phenomena. These results suggest that both $NAD^+$ and ADPRT activity are needed to maintain chromosomal integrity.

I have commented on differences observed between different DNA-damaging agents; simple monofunctional alkylating agents seem to require a large and important involvement of ADP-ribosylation in their repair processes. Ionizing radiation such as X-rays and $\gamma$-irradiation, seem to require much less. For example, it was not possible to observe inhibition of DNA repair after $\gamma$-radiation using alkaline sucrose gradients; it was necessary to resort to the nucleoid technique for an unequivocal demonstration that 3-aminobenzamide inhibits repair after $\gamma$-radiation. Finally, ultraviolet radiation, bifunctional alkylating agents (such as the nitrogen mustards) and very bulky substituents (such as bromomethyl-benzanthracene) do not seem to require ADPRT activity for their repair.

One plausible explanation for these differences is that they reflect varying kinetic behaviour; that is they reflect in each case the steady-state level of single-strand DNA breaks associated with a particular damaging agent. Ultraviolet irradiation does not break DNA directly at physiologically significant doses. DNA strand breaks are produced only during excision repair of ultraviolet damage. But it is known that such repair is slow and that the incision step is rate-limiting. Therefore, only a very small number of breaks are present at any given time. For example, after 10 $J/m^2$ which gives 10 to 20% survival, there are only about 0.3 breaks per $10^9$ base pairs of DNA.

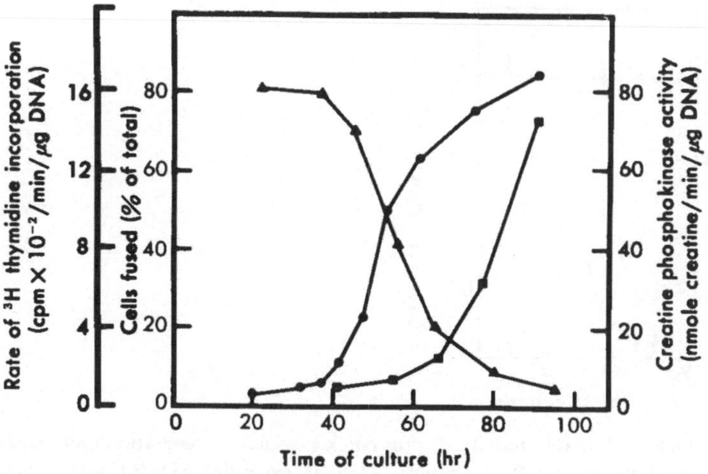


FIG. 14. Muscle cell differentiation in cell culture. Chick myoblasts grown in M199 medium. ● percentage of fused myoblasts; ■ creatine phosphokinase activity; ▲ the rate of DNA synthesis. (From ref. *17*).

At biological doses of ionizing radiation of say, 300 rads which gives again 10 to 20% survival, the number of direct breaks is about 0.6 breaks per $10^9$ base pairs of DNA. After exposure to alkylating agents the enzymatic incision steps are not rate-limiting, and a relatively large number of breaks are seen after exposure to dimethyl sulphate. 100 μM dimethyl sulphate gives about 10 to 20% survival in human fibroblasts; this dose generates about 3 breaks per $10^9$ base pairs of DNA. Of course, this number of breaks is a dynamic equilibrium between the rate of production of the breaks and the rate of their repair by polymerase and ligase activity. Moreover, these breaks will continue to be produced enzymically and by spontaneous hydrolysis for several hours. ADP-ribosylation, if induced by the steady-state level of DNA breaks, will enhance DNA ligase II activity and will therefore shift the equilibrium to a position with fewer DNA breaks. This new dynamic equilibrium will activate ADPRT less and consequently DNA ligase II will be activated to a lesser degree. Gradually, however, the equilibrium will return to that of the undamaged state.

This quantitative discussion shows that at equitoxic doses, the ratio of steady-state DNA breaks is 1: 2: 10 for ultraviolet irradiation, ionising radiation, and monofunctional methylating agent, respectively. The differences in the responses after these different agents may well be found in these different kinetics. In the light of this discussion, quantitative differences in the ADPRT response are to be expected between different cell types and between different damaging agents.

This review has shown that there is clear evidence that ADP-ribosylation is required for efficient DNA repair. Furthermore, current evidence indicates that ADP-ribosylation regulates the ligation step in DNA repair. However, there is circumstantial evidence that ADP-ribosylation is required for functions other than for DNA repair. We may therefore wonder whether some of these other functions may be mediated by similar molecular mechanisms. I wish to describe some observations that indicate that ADP-ribosylation is an obligatory step in certain categories of cell

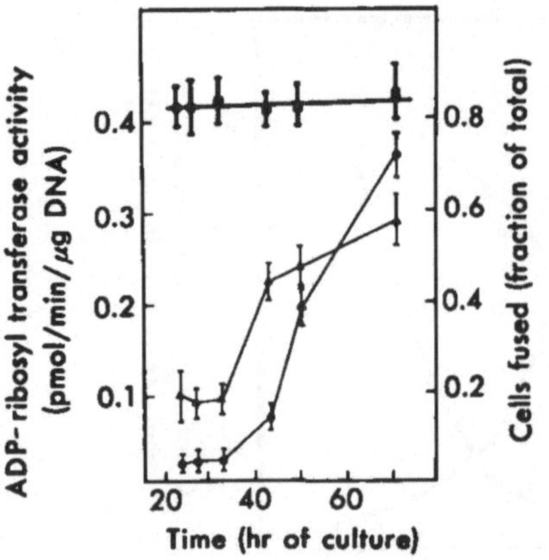


FIG. 15. Nuclear ADPRT activity during chick myoblast proliferation and terminal differentiation. ▲ physiological ADPRT activity; ■ total, potential ADPRT activity; ● fraction of myoblasts fused. (From ref. *17*).

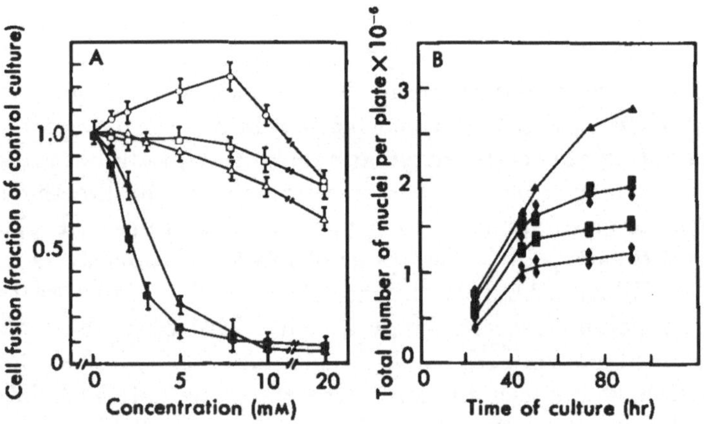


FIG. 16. ADPRT inhibitors block chick myoblast differentiation but not proliferation. A. ▲ 3-aminobenzamide; △ 3-aminobenzoate; ■ 3-methoxybenzamide; □ 3-methoxybenzoate; ○ nicotinamide. B. ▲ 8 mM 3-aminobenzamide; ◆ 10 mM 3-aminobenzamide; ⧫ 20 mM 3-aminobenzamide; ■ 5 mM 3-methoxybenzamide; ▮ 8 mM 3-methoxybenzamide; ● no inhibitor. (From ref. *17*).

differentiation and to suggest that nicking and rejoining of DNA is involved in these differentiation steps (*17*).

Our investigation of the possible participation of ADPRT activity in cell differentiation has used a chick muscle primary cell culture system. When embryonic chick myoblasts are seeded onto collagen-coated petri dishes and grown in a medium containing horse-serum and embryo extract, the majority of the cells pass through an initial phase of proliferation (Fig. 14). The onset of terminal differentiation is marked by the fusion of myoblast cell membranes to form multi-nucleate syncytia of muscle fibres, the cessation of DNA biosynthesis and of cell division. There is also a substantial increase in the activity of creatine phosphokinase, largely due to the *de novo* synthesis of a muscle specific isozyme of this protein. In this work we have made

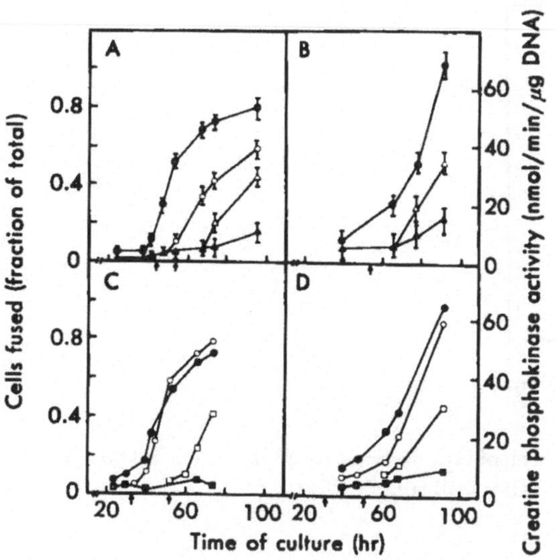


FIG. 17. Reversible inhibition of chick myoblast differentiation by ADPRT inhibitors and by nicotinamide deprivation. Reversible inhibition by 8 mM 3-aminobenzamide (A and B), and by depletion of NAD levels (C and D). A and C show myoblast cell fusion, and B and D show creatine phosphokinase activity. ● control; ▲ continuous presence of 8 mM 3-aminobenzamide; the medium containing inhibitor was removed and replaced with medium from parallel control cultures of the same age at 44 hr (○) or at 54 hr (△). In C and D myoblasts were cultured in nicotinamide-free medium. ■ no nicotinamide; ● 1 mM nicotinamide added at 0 hr; ○ 1 mM nicotinamide added at 32–34 hr of culture; □ 1 mM nicotinamide added at 52 hr of culture. (From ref. *17*).

use of both cell fusion and creatine phosphokinase activity as markers of terminal cell differentiation (Fig. 14).

During the cyto-differentiation of primary chick myoblasts there is a 3-fold increase in the endogenous physiological ADPRT activity, but there is no change in the total, potential enzyme activity (Fig. 15). This suggests that the increase in the physiological activity is due to enzyme activation, and not to an increase in the number of enzyme molecules. The ADPRT activation precedes both cell fusion and the rise in creatine phosphokinase activity.

Inhibitors of ADPRT activity reversibly block both muscle cell fusion and the rise in creatine phosphokinase activity associated with this cell differentiation. Myoblast fusion is inhibited in a concentration-dependent manner in the continuous presence of either 3-aminobenzamide or 3-methoxybenzamide (Fig. 16A). These inhibitors do not inhibit proliferation of myoblasts (Fig. 16B). The non-inhibitory analogues, 3-aminobenzoate and 3-methoxybenzoate caused no significant inhibition of fusion at corresponding concentrations (Fig. 16A). Inhibitors of nuclear ADPRT not only block myoblast fusion (Fig. 17A), but they also prevent the usual increase in creatine phosphokinase activity, (Fig. 17B). Both the inhibition of cell fusion and of the increase in creatine phosphokinase activity by ADPRT inhibitors, is reversible (Fig. 17A, B). Inhibitors of ADPRT activity also inhibit the appearance of acetylcholine receptors on the muscle cell surface.

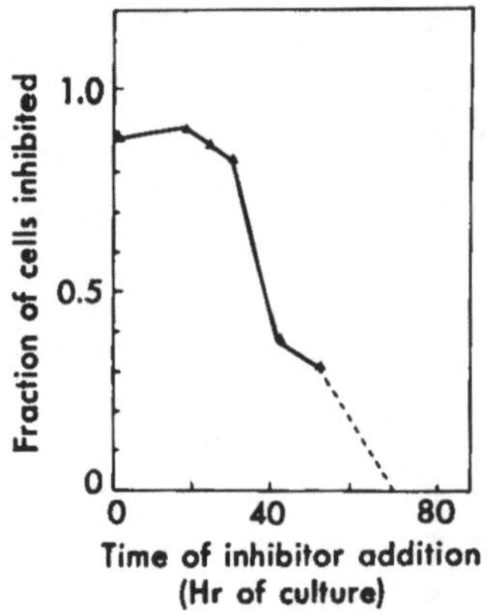


FIG. 18. Time during which myoblasts are sensitive to inhibition by 3-aminobenzamide. (Unpublished work of Farzaneh, Zalin, Brill, and Shall).

We have also experimentally manipulated ADP-ribosylation by partially depleting the cells of their $NAD^+$ content, by nicotinamide starvation. The $NAD^+$-depleted myoblasts do not fuse to form multi-nucleated fibres (Fig. 17C), not do they show the usual rise in creatine phosphokinase activity (Fig. 17D). These blocks to cell differentiation are reversible (Fig. 17C, D). We have determined the time during which the cells are sensitive to the effects of ADPRT inhibitors. The cells become resistant to these enzyme inhibitors between 24 and 40 hr in culture (Fig. 18); cell fusion is only beginning at 40 hr, while the appearance of creatine phosphokinase starts at about 60 hr. Thus, the ADPRT inhibitors are blocking a relatively early event in this example of cell differentiation.

We conclude from these observations that there is an obligatory requirement for ADPRT activity in the terminal differentiation of primary chick myoblast cultures (*17*). These observations have been confirmed and extended by Johnstone and Williams who have found that ADPRT activity is an obligatory early event in the mitogen-induced activation of human peripheral blood lymphocytes (*18*). Althaus *et al.* (personal communication) have demonstrated that ADPRT activity is required for the expression of some foetal proteins in adult rat hepatocytes in culture. Perhaps Dr. Pitot will describe these observations in his talk. It has also been shown by Williams and Johnstone (personal communication) that ADPRT inhibitors will block the differentiation, but not the proliferation of the parasite *Trypanosoma cruzi*, which causes the serious and incurable Chagas' disease in South America. Thus, it is now clear that ADPRT activity is an obligatory requirement for certain categories of cell differentiation. Naturally this has implications both for embryology and for cancer.

Because we had previously shown that the involvement of ADPRT in DNA repair is associated with the breaking and rejoining of DNA strands, we were interested to investigate whether a related mechanism might also be involved in some categories of cell differentiation. There are a number of examples now in which specific gene expression is associated with gene rearrangement or DNA transposition. We have therefore examined the DNA of developing chick muscle.

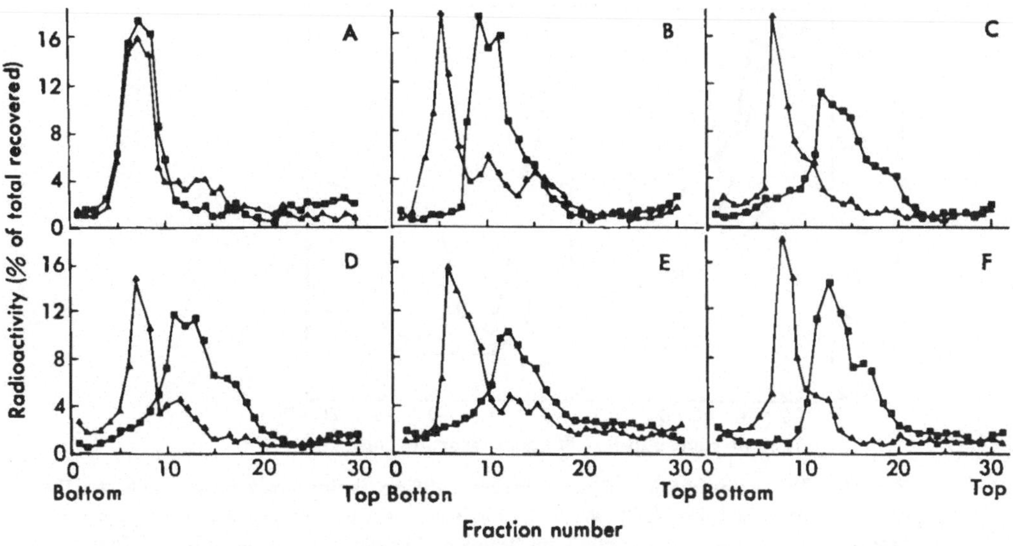


FIG. 19. Molecular weight analysis of muscle cell DNA by alkaline sucrose gradients. ▲ 27 hr cultures labelled with [$^{14}$C] thymidine; ■ 44, 53, 66, 75, and 92 hr cultures labelled with [$^{3}$H] thymidine. (From ref. *17*). ■ $^{3}$H; ▲ $^{14}$C. A, 27 hr; B, 44 hr; C, 53 hr; D, 66 hr; E, 75 hr; F, 92 hr.

We found that between 100 and 300 single-strand DNA breaks appear during the cyto differentiation of primary chick myoblasts. These DNA strand breaks appear in the presence of a proficient DNA repair mechanism, and are therefore not due to a general deficiency in DNA repair.

In alkaline sucrose gradients which estimate the size of single-stranded DNA, the DNA from proliferating cultures has a high molecular weight (Fig. 19A). At 44 hr when only 15% of the cells have fused (see Fig. 14), there is a significant reduction in the molecular weight of the cellular DNA (Fig. 19B). By 53 hr the maximum number of single-strand breaks are present (Fig. 19C), although only 50–60% of the myoblasts have fused. The molecular weight then remains constant up to 92 hr of culture (Fig. 19C–F). Only single-strand breaks appear in the DNA; there is no decrease in the molecular weight of the DNA in neutral sucrose gradients containing ionic detergents between 25 and 92 hr of culture. We have been unable to establish what happens after 92 hr of culture, so we do not yet know whether the DNA breaks persist or are subsequently rejoined. The experiments displayed in Fig. 19 are also 'mixing' experiments, and show that in our experimental conditions, mixing of early (without strand breaks) and late (with strand breaks) cells does not induce strand breaks in the DNA of early cells (see Fig. 19B–F).

Alkaline sucrose gradients detect single-strand breaks that are induced by the alkaline conditions of the gradients (for example, apurinic sites), as well as true single-strand breaks which had been generated enzymatically. We have distinguished between these two phenomena by examining the sedimentation rate of nucleoids. In neutral sucrose gradients containing 1.0 M NaCl, nucleoids from 24 and 38 hr cultures sediment rapidly; in contrast, 6 hr later, at 44 hr the nucleoids sediment more slowly, and by 68 hr a considerable number of metabolically induced single-strand

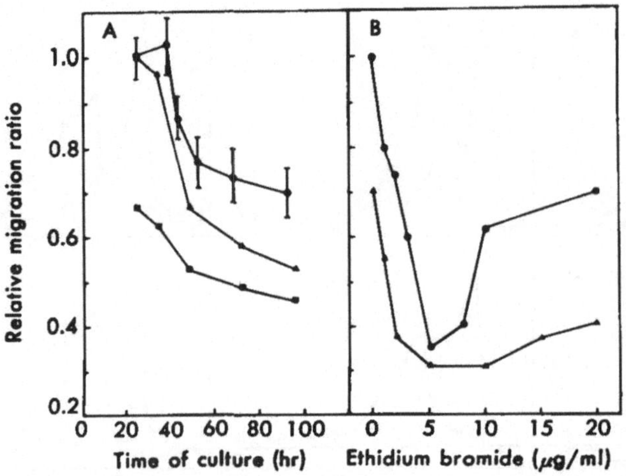


FIG. 20. The sedimentation of nucleoids from differentiating chick muscle cells. A. The lysis mixture contained 1.0 M NaCl (●) or 1.95 M NaCl (▲), or 1.95 M NaCl plus 5 μg/ml ethidium bromide (■). B. The effect of ethidium bromide. The lysis mixture contained 1.95 M NaCl and the indicated concentration of ethidium bromide. ● 24 hr cultures; ▲ 72 hr cultures. (From ref. *17*).

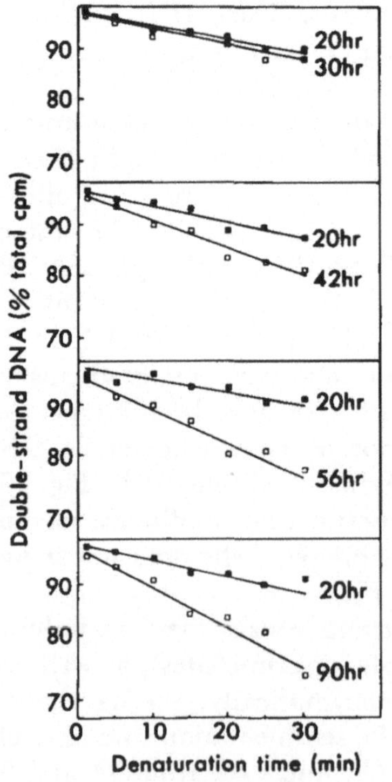


FIG. 21. DNA strand breaks in differentiating muscle cells. ■ 20 hr culture labelled with [$^{14}$C] thymidine; □ 30, 42, 56, or 90 hr cultures labelled with [$^{3}$H] thymidine. □ $^{3}$H; ■ $^{14}$C. (From ref. *17*).

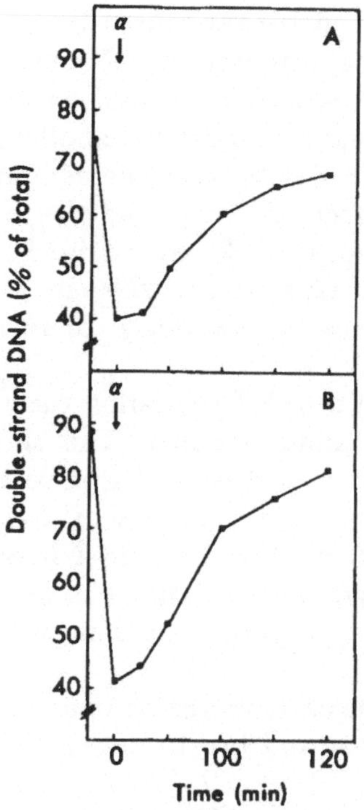


FIG. 22. Rate of DNA repair in developing muscle cells following 300 rads of $\gamma$-irradiation. Repair was analysed by measuring the rate of DNA strand separation in alkali. A, 92 hr; B, 24 hr. (From ref. *17*).

DNA breaks are evident (Fig. 20A). Increasing the salt concentration of the gradients to 1.95 M NaCl, and thus stripping the nuclei of yet more of their protein, does not alter the results significantly (Fig. 20A). Thus the reduction in the sedimentation of the nucleoids is not due to changes in the protein composition of the chromatin.

This inference is also drawn from several experiments with ethidium bromide. In the presence of 5 $\mu$g ml$^{-1}$ ethidium bromide the sedimentation of nucleoids from 24 hr cultures is reduced from 1.0 to 0.67 (Fig. 20A). However, the same concentration of ethidium bromide has a much smaller effect on the sedimentation at 96 hr, which is only reduced from 0.56 to 0.46. In nucleoids from 24 hr cultures, increasing concentrations of ethidium bromide induce the biphasic effect characteristic of DNA molecules under topological constraints (Fig. 20B), whereas nucleoids from 72 hr cultures are affected less (Fig. 20B), probably due to the presence of single-strand breaks. We have estimated the number of single-strand DNA breaks in differentiating muscle cell DNA, from 20–90 hr (Fig. 21A–D); few single-strand breaks were detectable in the DNA at either 20 or 30 hr, but by 42 hr approximately 50–100 single-strand breaks per $10^{12}$ daltons (per diploid genome) had appeared. By 56 hr they had increased to 100–300 per genome (Fig. 21C), and did not increase thereafter (Fig. 21D).

The presence of DNA strand-breaks in differentiating cells may be attributed to a general deficiency of DNA repair. However, these differentiating muscle cells are proficient in repair and can readily rejoin single-strand breaks. There is no detectable decline in the ability to rejoin DNA strand breaks caused by $\gamma$-irradiation (Fig. 22). In the 92 hr cultures (Fig. 22) a significant number of DNA strand breaks exist before irradiation; exposure to 300 rads of $\gamma$-radiation would be expected to introduce a further 600 breaks, about twice as many as were metabolically engendered. The rate at which these breaks were rejoined was the same as that in 24 hr myoblasts—but rejoining stopped when the number of breaks had dropped to the endogenous level. This implies that these endogenous breaks are retained in the face of a proficient DNA strand-rejoining capacity, perhaps because they are of a special character or occur at specific nucleotide sequences; alternatively, there may be a shift in the steady-state level of breaks.

I am suggesting then, in certain categories of cell differentiation there are rearrangements or transpositions within the mammalian genome. The single-strand breaks which we observe are possibly only indicators of these DNA movements. The suggestion that there is lability of the eukaryotic genome is not new; it has been discussed several times before, amongst others by Ruth Sager whom I hope will say something in this connection during the meeting. Such lability has been demonstrated in maize, in yeast and in *Trypanosoma brucei*, so perhaps it is more widespread than we have previously considered.

In mitogen-stimulated human lymphocytes the early increase in nuclear ADPRT activity coincides with a rapid rejoining of single-strand breaks present in the DNA of quiescent lymphocytes (*18*).

Our general conclusion is that perhaps nuclear ADPRT participates commonly in situations where nicking and rejoining of DNA occur, such as in DNA repair, in certain categories of cell differentiation, in transpositions, and in gene rearrangements.

There are several very general notions which relate this tentative conclusion to cancer. Firstly, it is striking that cancer cells frequently show considerable chromosomal lability and sometimes even specific chromosomal translocations. An hypothesis worthy of investigation is that this chromosomal lability is associated with a general defect in the controlled breakage and rejoining of DNA; and we have seen that ADP-ribosylation may be part of this regulation. Secondly, recent evidence from the experiments with both retroviral oncogenes and transforming DNA from human tumour cells has clearly identified human transforming genes. There is some evidence that suggests that perhaps some of these oncogenes are in fact normal cellular differentiation genes. Again, ADP-ribosylation may be required for the proper activity of these "differentiation-transforming" genes and a defect in ADP-ribosylation may contribute both to a generalised chromosomal instability and specifically to erroneous expression of the differentiation genes. I must add that a gross absence of nuclear ADPRT activity is unlikely. Because it seems to be a required function in development, its total absence would probably be lethal.

We may also consider the potential significance of nuclear ADPRT in carcinogenesis because of its involvement in repair. ADPRT inhibitors, and perhaps particularly nicotinamide, are common chemicals and when present in the body may inhibit efficient DNA repair. Whether such interference with DNA repair would increase the risk of carcinogenesis in unclear. There are at the moment, at least two conflicting arguments. Firstly, we may consider the possibility that interference with

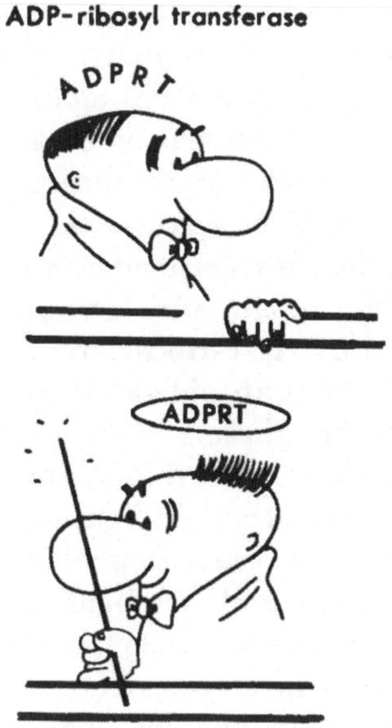


FIG. 23. Cartoon indicating our general view of nuclear ADPRT activity. The enzyme is sensitive to breaks in the DNA and modulates their rejoining.

efficient DNA repair may alter the mutation frequency. I am aware of three independent studies which indicate that ADPRT inhibitors either are not mutagenic or that they lower the frequency of mutagen-induced mutation in the survivors. We shall perhaps hear more of this from Dr. Natarajan. So, on the one hand we observe that ADPRT inhibitors do not increase the mutation frequency in these particular examples. This would lead one to argue that ADPRT inhibitors would not increase the risk of carcinogenesis. Perhaps a wider sample of gene loci would give more confidence to this conclusion.

On the other hand, it is clear that ADPRT inhibitors do increase the frequency of sister chromatid exchanges and of chromosomal aberrations after DNA damage. There has been a recent resurgence of interest in genetic transpositions and in gene amplification as possible mechanisms in carcinogenesis in addition to more conventional mutations. It is possible that ADPRT inhibitors may enhance the risk of carcinogenesis by increasing the frequency or infidelity of genetic transpositions or chromosomal rearrangements.

Finally, I wish to say something about the potential of nuclear ADP-ribosyl transferase inhibitors in the treatment of cancer. I have no doubt that this topic will be raised again during this meeting. I have referred already to the observations that ADPRT inhibitors synergistically increase the cytotoxicity of a variety of DNA-damaging agents, including both nitrosoureas and ionizing-radiation. In particular, the cytotoxicity of streptozotocin is markedly increased by ADPRT inhibitors. This observation suggests that innocuous ADPRT inhibitors may be of clinical value in conjunction with streptozotocin or any other monofunctional alkylating agent. The apparent decrease in mutation frequency may be a distinct advantage, because drug-induced tumours are beginning to appear in patients who have been previously treated with chemotherapy. Furthermore, Sugimura and colleagues (*24*) have shown that both 3-aminobenzamide and benzamide which have no antitumour activity themselves, were able to potentiate the antitumour activity of bleomycin against Ehrlich ascites carcinoma in intact animals. The response of bifunctional or bulky reagents is more difficult to predict; the data available so far would not encourage optimism in these cases, but some experimental investigations would be warranted.

In summary, then, this review has shown that there is clear evidence that ADP-ribosylation reactions are required for efficient DNA repair after certain kinds of damage, probably because DNA ligation is regulated by ADP-ribosylation (Fig. 23). Moreover, there is now evidence that ADPRT activity is an obligatory step in certain categories of cell differentiation. I wonder whether these several functions are mediated by similar molecular mechanisms. The nuclear ADP-ribosyl transferase is stimulated by breaks in the DNA and consequently stimulates their ligation. We have previously suggested (*15*) that ADP-ribosylation reactions have a general function to be sensitive to DNA breaks and to regulate subsequent DNA ligation in DNA recombination, in sister chromatid exchanges, in chromosome aberrations, in gene rearrangements, in transpositions, in DNA repair, and in certain categories of cell differentiation. The importance of ADP-ribosylation reactions lies firstly in its involvement in these reactions, some of which may be relevant in carcinogenesis; secondly, in cancer diagnostics as shown by Hayaishi and collaegues, and thirdly, possibly in cancer treatment as I have indicated. Cancer researchers should be very conscious of the molecular biology of ADP-ribose.

ACKNOWLEDGMENTS

The work in my laboratory has been supported by the Cancer Research Campaign, the Medical Research Council and the Science and Engineering Research Council. I would especially like to thank all my colleagues and students together with whom we have helped to elucidate the mysteries of ADP-ribose.

REFERENCES

1. Hayaishi, O. and Ueda, K. (eds.). ADP-Ribosylation Reactions, Academic Press, New York, 1983.
2. Shall, S. ADP-Ribose in DNA repair: A new component of DNA excision repair. Adv. Rad. Biol., *11*: in press, 1983.
3. Yoshihara, K. and Kamiya, T. Poly(ADP-ribose) synthetase-DNA interaction. *In;* O. Hayaishi and K. Ueda (eds.), ADP-Ribosylation Reactions, pp. 157–171, Academic Press, New York, 1983.
4. Whish, W.J.D, Davies, M. I., and Shall, S. Stimulation of poly(ADP-ribose) polymerase activity by the anti-tumour antibiotic, streptozotocin. Biochem. Biophys. Res. Commun., *65*: 722–730, 1975.
5. Goodwin, P. M., Lewis, P. J., Davies, M. I., Skidmore, C. J., and Shall, S. The effect of gamma radiation and neo-carzinostatin on NAD and ATP levels in mouse leukaemia cells. Biochim. Biophys. Acta, *543*: 576–582, 1978.
6. Skidmore, C. J., Davies, M. I., Goodwin, P. M., Halldorsson, H., Lewis, P. J., Shall, S., and Zai'ee, A-A. The involvement of poly(ADP-ribose) polymerase in the degradation of NAD caused by $\gamma$-radiation and N-methyl-N-nitrosourea. Eur. J. Biochem., *101*: 135–142, 1979.

7. Durkacz, B. W., Omidiji, O., Gray, D. A., and Shall, S. (ADP-ribose)$_n$ participates in DNA excision repair. Nature, *283*: 593–596, 1980.
8. Gray, D. A., Durkacz, B. W., and Shall, S. Inhibitors of nuclear ADP-ribosyl transferase retard DNA repair after N-methyl-N-nitrosourea. FEBS Lett., *131*: 173–177, 1981.
9. Durkacz, B. W., Irwin, J., and Shall, S. The effect of inhibition of (ADP-ribose)$_n$ biosynthesis on DNA repair assayed by the nucleoid technique. Eur. J. Biochem. *121*: 65–69, 1981.
10. Durkacz, B. W., Irwin, J., and Shall, S. Inhibition of (ADP-ribose)$_n$ biosynthesis retards DNA repair but does not inhibit DNA repair synthesis. Biochem. Biophys. Res. Commun., *101*: 1433–1441, 1981.
11. Söderhäll, S. and Lindahl, T. Two DNA ligase activities from calf thymus. Biochem. Biophys. Res. Commun., *53*: 910–916, 1973.
12. Söderhäll, S. and Lindahl, T. Mammalian DNA ligases, serological evidence for two separate enzymes. J. Biol. Chem., *250*: 8438–8444, 1975.
13. Teraoka, J., Tamura, S., and Tsukada, K. Evidence for a single species of DNA ligase localised in nuclei of rat liver. Biochim. Biophys. Acta, *563*: 535–539, 1979.
14. Teraoka, H. and Tsukada, K. Eukaryotic DNA ligase. J. Biol. Chem., *257*: 4758–4763, 1982.
15. Creissen, D. and Shall, S. Regulation of DNA ligase activity by poly(ADP-ribose). Nature, *296*: 271–272, 1982.
16. Nduka, N., Skidmore, C. J., and Shall, S. The enhancement of cytotoxicity of N-methyl-N-nitrosourea and of $\gamma$-radiation by inhibitors of poly(ADP-ribose) polymerase. Eur. J. Biochem., *105*: 525–530, 1980.
17. Farzaneh, F., Zalin, R., Brill, D., and Shall, S. DNA strand breaks and ADP-ribosyl transferase activation during cell differentiation. Nature, *300*: 362–366, 1982.
18. Johnstone, A. P. and Williams, G. T. The involvement of DNA breaks and ADP-ribosyl transferase activity in eukaryotic differentiation, demonstrated in human lymphocytes. Nature, *300*: 368–370, 1982.
19. Halldorsson, H., Gray, D. A., and Shall, S. Poly(ADP-ribose) polymerase activity in nucleotide permeable cells. FEBS Lett., *85*: 349–352, 1978.
20. Berger, N. A. and Sikorski, G. W. Nicotinamide stimulates repair of DNA damage in human lymphocytes. Biochem. Biophys. Res. Commun., *95*: 67–72, 1980.
21. Miwa, M., Kanai, M., Kondo, T., Hoshino, H-O., Ishihara, K., and Sugimura, T. Inhibitors of poly(ADP-Ribose) polymerase enhance unscheduled DNA synthesis in human peripheral lymphocytes. Biochem. Biophys. Res. Commun., *100*: 463–470, 1981.
22. Oikawa, A., Tohda, H., Kanai, M., Miwa, M., and Sugimura, T. Inhibitors of poly-(adenosine diphosphate ribose) polymerase induce sister chromatid exchanges. Biochem. Biophys. Res. Commun., *97*: 1311–1316, 1980.
23. Natarajan, A. T., Csukas, I., and van Zeeland, A. A. Contribution of incorporated 5-bromo deoxyuridine in DNA to the frequencies of sister-chromatid exchanges induced by inhibitors of poly(ADP-ribose) polymerase. Mutat. Res., *84*: 125–132, 1981.
24. Kawamitsu, H., Miwa, M., Tanaka, Y., Sakamoto, H., Terada, M., Hoshi, A., and Sugimura, T. Inhibitors of poly(adenosine diphosphate ribose) polymerase potentiate the antitumour activity of bleomycin against Ehrlich ascites carcinoma. J. Pharm. Dyn., *5*: 900–904, 1982.

ADP-RIBOSYLATION, DNA REPAIR AND CANCER, M. MIWA ET AL. (EDS.),
JAPAN SCI. SOC. PRESS, TOKYO/VNU SCIENCE PRESS, UTRECHT, PP. 27-37, 1983

NAKAHARA MEMORIAL LECTURE

# Poly(ADP-ribose): Structure, Quantification, and Biological Significance

Masanao MIWA, Mieko KATO, Hisae IIJIMA, Yuko TANAKA, Tomoko KONDO, Hisae KAWAMITSU, Masaaki TERADA, and Takashi SUGIMURA

*National Cancer Center Research Institute, Tokyo 104, Japan*

*Abstract:* We have been working on poly(ADP-Rib) since our discovery of it. This lecture will review recent findings on this polymer obtained in our laboratory. Poly (ADP-Rib) has long been thought to be a linear homopolymer attached to nuclear protein. The ribose-ribose bond of poly(ADP-Rib) was shown to be an $\alpha(1''\rightarrow2')$ ribosidic bond. There are two molecular forms of poly(ADP-Rib), namely low molecular weight (L) and high molecular weight (H) poly(ADR-Rib) molecules. The L and H fractions of poly(ADP-Rib) are separable by gel filtration, sucrose density gradient centrifugation and polyacrylamide gel electrophoresis. Physicochemically the L and H forms differ in their molecular size, solubility in high salt, rate of hydrolysis by venom phosphodiesterase and circular dichroism (CD) spectrum. However, the two forms have a similar ultraviolet (UV) absorption spectrum, both show a similar hyperchromicity on heating to 98°C or hydrolysis with snake venom phosphodiesterase and give similar results on ordinary chain length determination. The great discrepancy in size observed by physicochemical methods and by the ordinary chain length method indicates the presence of a branching structure in poly(ADP-Rib). The branching structure was proved by demonstrating the presence of a unique compound, 2'[1''-ribosyl-2''(1'''-ribosyl)]adenosine-5',5'',5'''-tris(phosphate) and by electron microscopy. The amount of poly(ADP-Rib) was determined by our new method, consisting of tritium labeling and high performance liquid chromatography. The merit of this method is that it can determine the recovery of poly(ADP-Rib) exactly and that its sensitivity is high. The amount of poly(ADP-Rib) changed dramatically when human promyelocytic leukemia cells were induced to differentiate by dimethyl sulfoxide or 12-O-tetradecanoylphorbol-13-acetate. The use of inhibitors of poly (ADP-Rib) polymerase in combination with bleomycin, a DNA damaging antitumor drug, potentiated the antitumor activity of belomycin against Ehrlich ascites carcinoma cells *in vivo*.

Since poly(ADP-Rib) was discovered by us and other investigators (*1–3*) as a polymer consisting of ADP-Rib residues, it has been thought to be a nucleic acid-like linear homopolymer like poly(A). As one terminus of poly(ADP-Rib) is attached to nuclear proteins, poly(ADP-Rib) is considered to cause a novel posttranslational modification of protein like phosphorylation, methylation, acetylation, and adenylylation. But modification of the protein by poly(ADP-Rib) is unique in that the modifying moiety is sometimes very big and heterogenous in size and charge compared to other modifications (*1–5*).

As a first step in understanding poly(ADP-ribosyl)ation we have clarified some details of the structure of poly(ADP-Rib) and devised a sensitive new method for measuring poly(ADP-Rib) *in vivo*. A significant change in the poly(ADP-Rib) level was observed during differentiation. This paper describes studies on these topics and discusses the possible clinical use of inhibitors of poly(ADP-Rib) polymerase in combination with a DNA damaging antitumor drug, bleomycin.

### *Structural Analysis*

We devised a method to obtain pure poly(ADP-Rib) in quantity by incubating $NAD^+$ with calf thymus nuclei. This procedure enabled us to obtain sufficient pure polymer for structural analyses. Poly(ADP-Rib) was found to have ribose-ribose

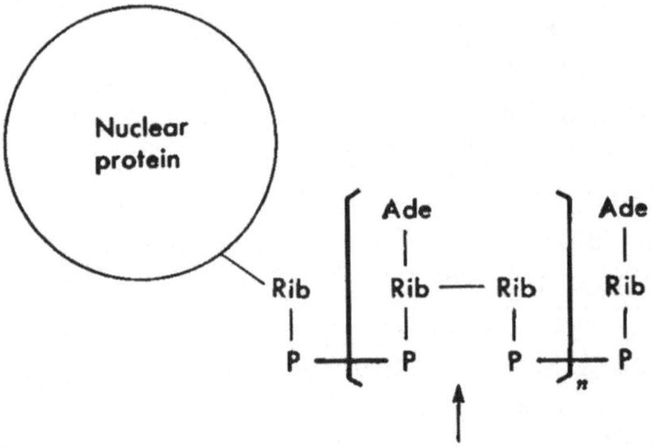


FIG. 1. Linear structure of poly(ADP-Rib) and its monomer unit Ado(P)-Rib-P, namely 2-ribosyladenosine-5′,5″-bis(phosphate) or 5″-phosphoribosyl-5′AMP. Ade, adenine; Rib, ribose; P, phosphate.

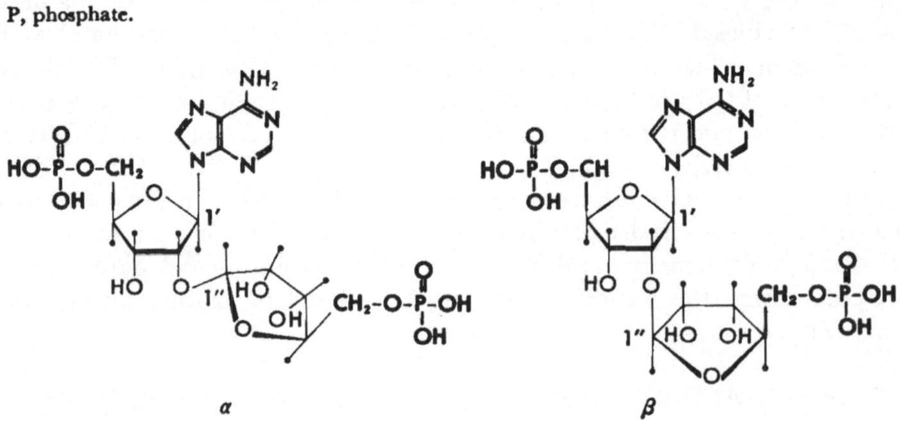


FIG. 2. Structures of α- and β-anomer of Ado(P)-Rib-P. The adenine ring is drawn in the "anti" form.

bonds between ADP-Rib moieties of $NAD^+$. The purified poly(ADP-Rib) was hydrolyzed with snake venom phosphodiesterase yielding 2′-ribosyladenosine 5′,5″-bis (phosphate), or 5″-phosphoribosyl-5′AMP (abbreviated as Ado(P)-Rib-P), the monomer unit of poly(ADP-Rib) (Fig 1). Dr. J. Doly of Dr. P. Mandel's group reported that the ribose-ribose bond of poly(ADP-Rib) synthesized by chick liver nuclei is a 1″-2′ glycosidic bond (*6*). We wished to confirm these results and clarify the anomeric configuration of the ribose-ribose bonds (Fig. 2). The anomeric configuration of the glycosidic bonds between ribose moieties can be determined by optical rotatory dispersion or circular dichroism (CD). However, we found that in addition to the anomeric configuration of the ribose-ribose bond (whether "$\alpha$" or "$\beta$"), the angle of the glycosidic bond between $^7$N of adenine and C-1′ of ribose (whether "anti" or "syn") contributes considerably to the optical rotatory dispersion or CD. Therefore, we used $^{13}$C-NMR spectrometry to analyze the anomeric configuration of C-1″ of ribose which binds to the adenosine residue. With methyl-$\alpha$-D-ribofuranoside and methyl-$\beta$-D-ribofuranoside as reference compounds we found from the chemical shift of C-1″ that the anomeric configuration is $\alpha$ (*7*). Downfield displacement of C-2′ of Ado(P)-Rib-P confirmed that the ribose is linked to the 2′-carbon of the adenosine residue (*7*).

The finding that the linkage is $\alpha$ was confirmed by $^1$H-NMR spectrometry (*8, 9*). Moreover, the angle of the glycosidic bond between $^7$N of adenine and C-1′ of ribose in Ado(P)-Rib-P was shown to be "anti" (*9*). This was the first indication of the existence of an $\alpha$-anomeric configuration in ribose molecules bound to ribose or other sugars.

The structure of poly(ADP-Rib) as a polymer was studied. At least two molecular forms (L and H fractions) were shown to exist (Table 1). The L (low molecular weight) and H (high molecular weight) fractions were separated by gel filtration, sucrose density gradient centrifugation (Fig. 3) and polyacrylamide gel electrophoresis. Since all attempts to convert H to L were unsuccessful, H does not seem to be an aggregated form of L. Analysis of the chain length and size suggested that poly-(ADP-Rib) has a branching structure (*12*). This branching structure was proved by

TABLE 1. Physicochemical Properties of Poly(ADP-Rib)

| | H | L |
|---|---|---|
| SDG (Peak) | 12 S | 5 S |
| Sephadex G-200 (Kd) | 0 | 0.82 |
| Buoyant density (g/cm$^3$) | (precipitate) | 1.62 |
| Solubility in 2 M NaCl | Insoluble | Soluble |
| Rate of hydrolysis by | | |
| venom phosphodiesterase | 0.28 | 1 |
| poly(ADP-Rib) glycohydrolase | 1 | 1 |
| UV spectrum $\lambda_{max}$ pH 2 | 257 nm | 257 nm |
| pH 7, pH 9 | 259 nm | 259 nm |
| Hyperchromicity | | |
| 25°C → 98°C | 13% | 11% |
| enzyme hydrolysis | 15% | 15% |
| "Chain length" | 26 | 20 |

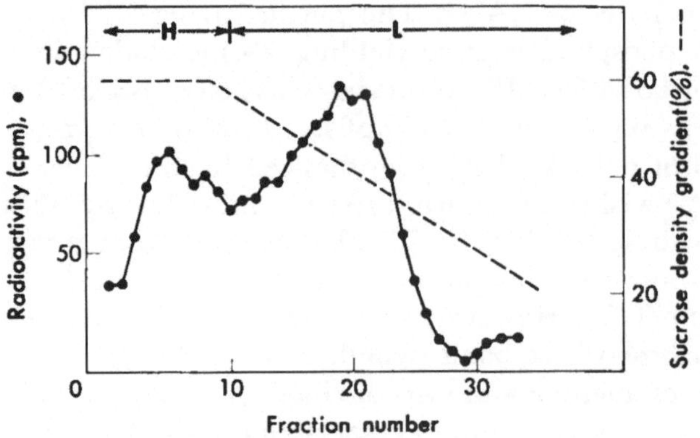


FIG. 3. Separation of poly(ADP-Rib) into H and L fractions. The poly(ADP-Rib) fraction eluted with 0.25–0.40 M sodium phosphate buffer (pH 6.8) from hydroxylapatite column was separated by a 20–60% sucrose gradient centrifugation.

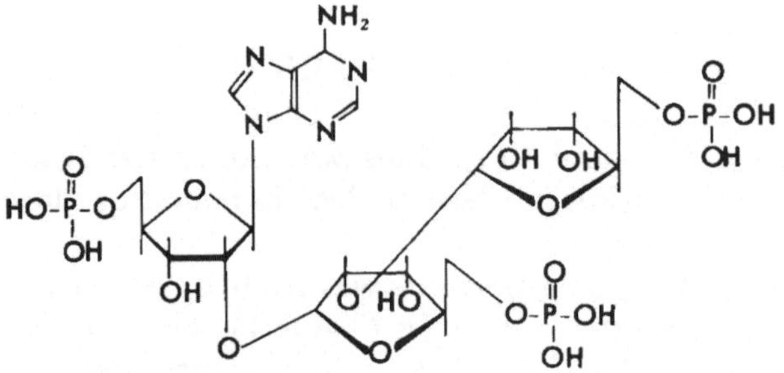

FIG. 4. Structure of a branched portion, Ado(P)-Rib(P)-Rib-P of poly(ADP-Rib), namely 2′[1″-ribosyl-2″(1‴-ribosyl)]adenosine-5′,5″,5‴-tris(phosphate).

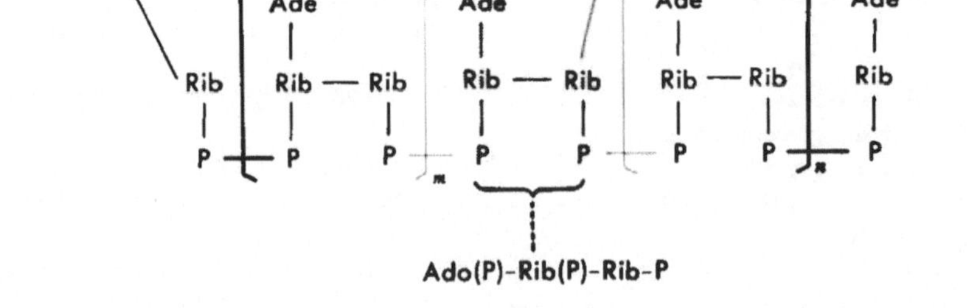


FIG. 5. Branching structure of poly(ADP-Rib). *l*, *m*, *n* denote arbitrary numbers of repeating monomer units.

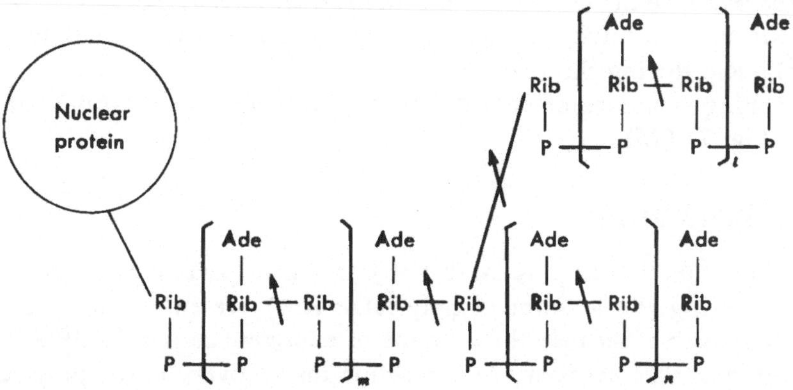


FIG. 6. α(1″→2′)Ribose-ribose bonds which are hydrolyzed by poly(ADP-Rib) glycohydrolase. Arrows denote sites of hydrolysis.

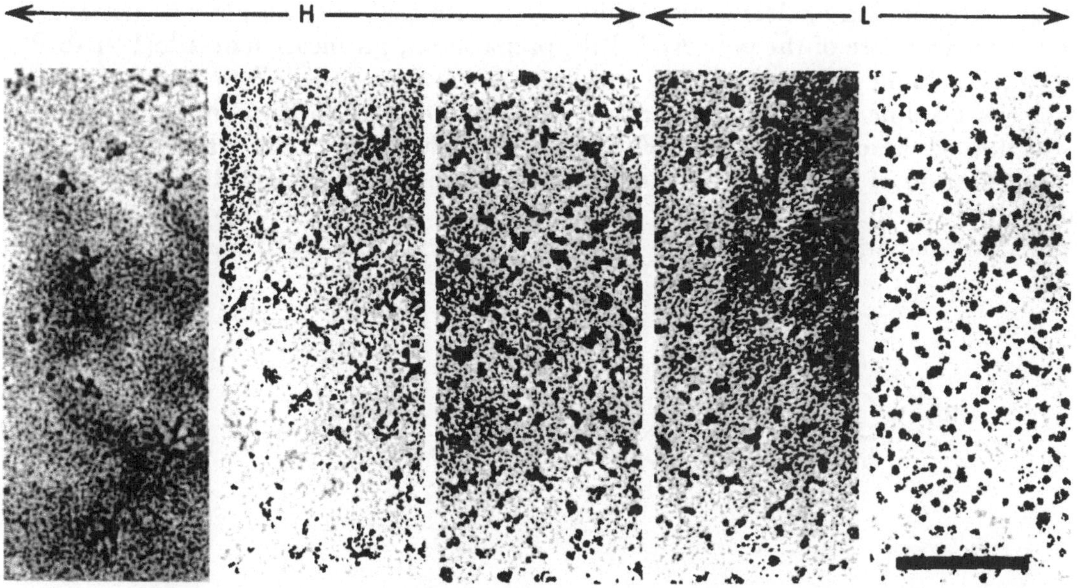


FIG. 7. Electron micrograph of poly(ADP-Rib). Poly(ADP-Rib) eluted from a hydroxylapatite column with 0.25–0.40 M sodium phosphate buffer (pH 6.8) was fractionated on Bio-gel A50m. The fractions eluted faster (H fraction) and slower (L fraction) were subjected to electron micrography. The bar indicates a length of 1 μm length. (From ref. *15*).

demonstrating the presence in the hydrolysates of poly(ADP-Rib) of a new compound, which was found to be 2′[1″-ribosyl-2″(1‴-ribosyl)]adenosine-5′,5″,5‴-tris-(phosphate) (abbreviated as Ado(P)-Rib(P)-Rib-P) (*13*, *14*) (Fig. 4). The branching structure of poly(ADP-Rib) may be drawn as shown in Fig. 5. Anomeric configuration between the three ribose moieties in Ado(P)-Rib(P)-Rib-P was also shown to be α (*14*). The finding that the ribose-ribose linkage had an α-anomeric configuration clarified the stereospecificity of poly(ADP-Rib) glycohydrolase (*10*, *11*) as α(1″-2′) poly(ADP-Rib) glycohydrolase (Fig. 6). Walden inversion of β-$NAD^+$ to form α(1″-

2′)ribose-ribose bonds indicates that the mechanism of the poly(ADP-Rib) polymerase reaction is SNII and may indicate that there is no covalent intermediate of the enzyme and ADP-Rib during the reaction.

The branching structure of poly(ADP-Rib) was also demonstrated by electron micrography (Fig. 7) (*15*).

*Quantification of Poly(ADP-Rib)*

To clarify the function of poly(ADP-Rib), it was important to devise a sensitive method for measuring the amount of poly(ADP-Rib) present *in vivo*. This method had to be very sensitive because the amount of endogenous poly(ADP-Rib) is very small. In addition, it was necessary to determine the recovery of the poly(ADP-Rib) in each sample during extraction.

The method devised (*16*) consists of 1) addition of [$^{14}$C] poly(ADP-Rib) of high specific radioactivity to the sample; 2) digestions with proteinase and nuclease to remove protein and nucleic acids; 3) digestions with snake venom phosphodiesterase and 5′-nucleotidase of the poly(ADP-Rib) preparation, purification of Ado(P)-Rib-P by DE52 column chromatography, and digestion of Ado(P)-Rib-P to Ado-Rib with bacterial alkaline phosphomonoesterase; 4) oxidation of Ado-Rib with $NaIO_4$ and reduction of the compound with KB[$^3$H]$H_4$ of high specific radioactivity (Fig. 8); 5) purification of the tritium labeled Ado-Rib by high performance liquid chromatography (Fig. 9); 6) processing of the sample with a sample oxidizer and counting of the radioactivities of $^3$H and $^{14}$C. The amount of poly(ADP-Rib) is calculated from

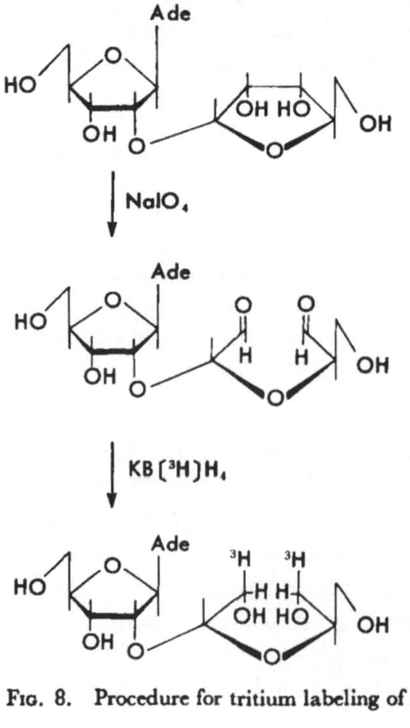


FIG. 8. Procedure for tritium labeling of Ado-Rib for quantification of poly(ADP-Rib).

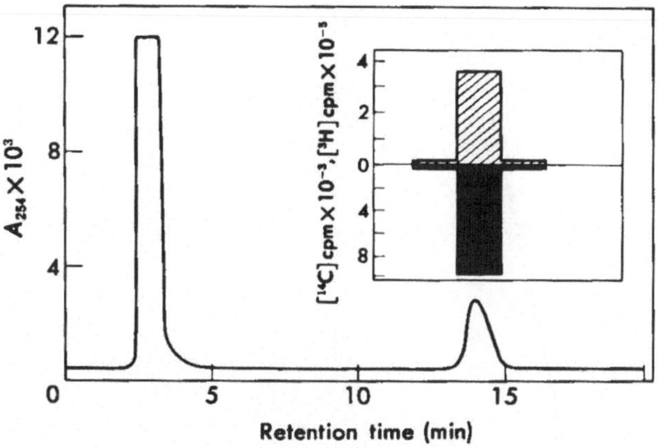


FIG. 9. High performance liquid chromatography of tritium labeled Ado-Rib. [$^{14}$C]Ado-Rib derived from the internal standard, poly([$^{14}$C]ADP-Rib), was also labeled with tritium and served as an internal marker.

the $^3H/^{14}C$ ratio and the recovery is calculated from the recovery of radioactivity. This method is sensitive enough for measurement of 2 pmoles of poly(ADP-Rib).

*Differentiation of Leukemic Cells*

The amount of poly(ADP-Rib) during differentiation of human promyelocytic leukemia cells (HL-60) was determined with the above procedure. The amount before differentiation was around 100 pmoles/mg DNA; 14 hr after addition of dimethylsulfoxide (DMSO), when differentiated phenotypes were hardly observed the poly(ADP-Rib) was decreased to 15% of the uninduced level (Fig. 10) (*17*). In the differentiation of mouse erythroleukemia cells (Friend cells), inhibitors of poly(ADP-Rib) polymerase were reported to induce differentiation. When added at 7.5 mM, the potent inhibitor, benzamide, induced differentiation of 40% of the HL-60 cells after 7 days. At 14 hr after addition of benzamide the amount of poly(ADP-Rib) was decreased to 38% of the uninduced level, and this change in level of poly(ADP-Rib) preceded the appearance of morphological and phagocytic activity. On addition of benzoic acid, which does not inhibit poly(ADP-Rib) polymerase, no change of the poly(ADP-Rib) level or phenotypic change of differentiation was observed (Table 2).

12-O-tetradecanoyl-phorbol-13-acetate (TPA) induces differentiation of HL-60 cells into macrophage-like cells. At 14 hr after its addition, when no significant morphological change or phagocytic activity was yet observed, poly(ADP-Rib) had decreased to 15% of the uninduced level. In contrast phorbol, which does not have tumor promoting activity, did not cause decrease of poly(ADP-Rib) or induction of differentiation. It is interesting that even with DMSO and TPA which does not inhibit poly(ADP-Rib) polymerase, the poly(ADP-Rib) level decreased rapidly before the appearance of differentiated phenotypes. N'-methylnicotinamide, which is a very efficient inducer of Friend cell differentiation, also induced HL-60 differentiation, but to a lesser extent (Table 2). Although this compound does not inhibit poly(ADP-

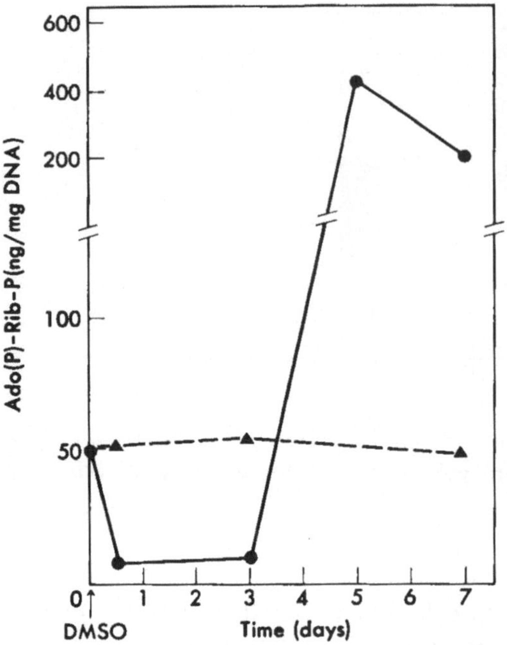


FIG. 10. Change of poly(ADP-Rib) level during differentiation of human promyelocytic leukemia cells, HL-60, induced by DMSO. ● with DMSO; ▲ without DMSO. (From ref. *17*).

TABLE 2. Differentiation of HL-60 Cell and Level of Endogenous Poly(ADP-Rib)

| | Differentiation | Initial decrease of poly(ADP-Rib) |
|---|---|---|
| DMSO | ++ | ++ |
| TPA | ++ | ++ |
| Benzamide | + | + |
| N′-methylnicotinamide | + | ++ |
| Benzoic acid | − | − |
| Phorbol | − | − |

Rib) polymerase, it also reduced the amount of poly(ADP-Rib) in the cell during induction of differentiation.

Poly(ADP-Rib) polymerase activity was measured in permeabilized HL-60 cells during differentiation induced by DMSO. The enzyme activity decreased to half the original level 14 hr after the addition of DMSO. This decrease in enzyme activity may partly explain the great decrease in the poly(ADP-Rib) level observed. However, the polymerase activity gradually decreased during 5–7 days and the low activity does not explain the high level of endogenous poly(ADP-Rib). These findings show that to understand the metabolism of poly(ADP-Rib) it is essential to know its endogenous level (*17*).

The poly(ADP-Rib) hydrolyzing enzyme activity and $NAD^+$ levels 14 hr following addition of these inducers were not changed significantly. Therefore we think that the level of poly(ADP-ribosyl)ation of chromatin proteins is possibly important in terminal differentiation of HL-60 cells induced by such compounds as DMSO, benzamide, N′-methylnicotinamide, and TPA (Table 2).

*Potentiation of the Antitumor Effect of Bleomycin by Inhibitors of Poly(ADP-Rib) Polymerase*

If poly(ADP-Rib) synthesis is indispensable for DNA repair, its inhibition may enhance cell killing by DNA damaging agents as shown by Durkacz *et al.* in L1210 cells *in vitro* (*18*). If DNA damaging anticancer agents are used in combination with inhibitors of poly(ADP-Rib) polymerase *in vivo*, the potentiation of their antitumor effect should be observed. We chose bleomycin as a DNA damaging anticancer agent to test this possibility.

Female ICR mice were inoculated intraperitoneally with $5 \times 10^6$ Ehrlich ascites carcinoma cells and from 24 hr later bleomycin and *m*-aminobenzamide were injected intraperitoneally once a day for 5 days. The ascitic fluid was collected on day 7.

Antitumor effects were calculated as the ratio of the packed cell volume of treated mice to that of untreated mice. *m*-Amionobenzamide alone showed no anticancer effect, but potentiated the anticancer effect of bleomycin as seen in Fig. 11

TABLE 3. Effect of Inhibitors on the Antitumor Activity of Bleomycin to Ehrlich Ascites Carcinoma

| Bleomycin (1 mg/kg/day) | Inhibitors (4.9 mmol/kg/day) | Tumor cell volume (%) |
|---|---|---|
| − | − | 100 |
| + | − | 69 |
| + | Benzamide | 14 |
| + | *m*-Aminobenzamide | 11 |
| + | Nicotinamide | 62 |
| + | N′-methylnicotinamide | 72 |

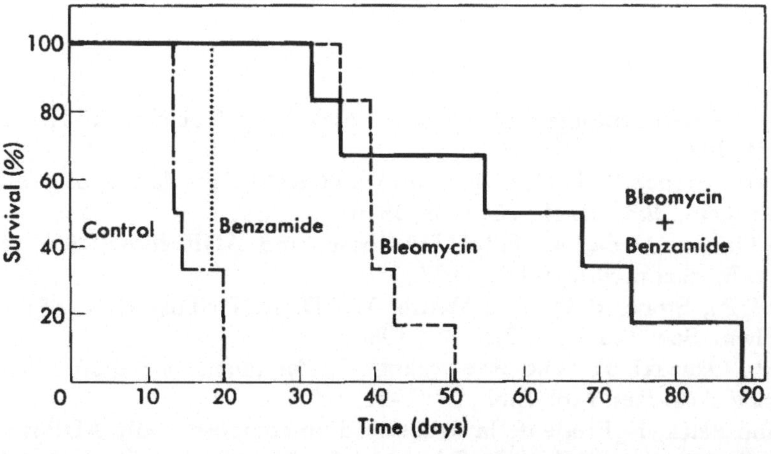


FIG. 11. Potentiation by benzamide of antitumor effect of bleomycin on survival of mice bearing Ehrlich ascites carcinoma. Male $CDF_1$ mice were transplanted with $10^6$ Ehrlich ascites carcinoma cells and were treated with phosphate buffered saline (PBS), bleomycin at 4.7 mg/kg, benzamide at 1.1 mmol/kg, or bleomycin at 4.7 mg/kg plus benzamide at 1.1 mmol/kg. Drugs were injected daily for 10 days and the survival of the groups of 6 mice each was determined. (From ref. *20*).

(*19*). Benzamide also had a potentiating effect, but nicotinamide and N′-methylnicotinamide did not (Table 3). Potentiation of the anticancer effect of bleomycin by benzamide was also demonstrated by its effect on the mean survival of male $CDF_1$ mice (Fig. 11) (*20*).

These potentiating effects of anticancer drugs by inhibitors of poly(ADP-Rib) polymerase seem promising for future chemotherapy, after careful consideration of any possible side effects of the inhibitor.

### *Future Prospects of Poly(ADP-Rib) Study*

The present study on the structure and quantitation of poly(ADP-Rib) revealed various interesting aspects of poly(ADP-Rib), but it is a small step toward clarification of the biological function of poly(ADP-ribosyl)ation. At present, poly(ADP-ribosyl)ation seems to be involved in the regulation of some steps of DNA repair, sister chromatid exchanges, and cell differentiation. However, it is also possible that by changing the chromatin conformation, poly(ADP-ribosyl)ation could be involved in other biological phenomena such as DNA recombination, mutation, and carcinogenesis. In this respect, it is an interesting preliminary observation that application of *m*-aminobenzamide just after treatment with methylazoxymethanol acetate enhanced the induction of hepatomas in small aquarium fish, Medeka (*Oryzias latipes*) (T. Ishikawa *et al.*, unpublished). This might be an example of poly(ADP-ribosyl)ation playing a role in some step of transformation.

Future work should be conducted on isolating and characterizing mutant cells lacking or overproducing poly(ADP-Rib), preparation of monoclonal antibodies to poly(ADP-Rib) polymerase, and also cloning of the gene for poly(ADP-Rib) polymerase and studies of the regulation of its gene expression. These kinds of experiments will assist us in understanding more clearly the function of poly(ADP-Rib) in chromatin.

## REFERENCES

1. Sugimura, T. Poly(adenosine diphosphate ribose). Prog. Nucleic Acid Res. Mol. Biol., *13*: 127–151, 1973.
2. Hilz, H. and Stone, P. Poly(ADP-ribose) and ADP-ribosylation of proteins. Rev. Physiol. Biochem. Pharmacol., *76*: 1–59, 1976.
3. Hayaishi, O. and Ueda, K. Poly(ADP-ribose) and ADP-ribosylation of proteins. Annu. Rev. Biochem., *46*: 95–116, 1977.
4. Purnell, M. R., Stone, P. R., and Whish, W.J.D. ADP-ribosylation of nuclear proteins. Biochem. Soc. Trans., *8*: 215–227, 1980.
5. Mandel, P., Okazaki, H., and Niedergang, C. Poly(adenosine diphosphate ribose). Prog. Nucleic Acid Res. Mol. Biol., *27*: 1–51, 1982.
6. Doly, J. and Petek, F. Étude de la structure d'un composé "poly(ADP-ribose)" synthétisé par des extraits nucléares de foie de poulet. C. R. Hebd. Sénces Acad. Sci. Ser. D. Sci. Nat., *263*: 1341–1344, 1966.
7. Miwa, M., Saitô, H., Sakura, H., Saikawa, N., Watanabe, F., Matsushima, T., and Sugimura, T. A $^{13}C$ NMR study of poly(adenosine diphosphate ribose) and its

monomers: Evidence of $\alpha$-(1″→2′)-ribofuranosyl ribofuranoside residue. Nucleic Acids Res., *4*: 3997–4005, 1977.

8. Ferro, A. M. and Oppenheimer, N. J. Structure of a poly(adenosine diphosphate monomer: 2′-(5″-phosphoribosyl)-5′-adenosine monophosphate. Proc. Natl. Acad. Sci. U.S., *75*: 809–813, 1978.
9. Inagaki, F., Miyazawa, T., Miwa, M., Saitô, H., and Sugimura, T. NMR analysis of conformation of ribosyl adenosine 5′,5″-bis-(phosphate) in aqeuous solution. Biochem. Biophys. Res. Commun., *85*: 415–420, 1978.
10. Miwa, M., Tanaka, M., Matsushima, T., and Sugimura, T. Purification and properties of a glycohydrolase from calf thymus splitting ribose-ribose linkages of poly (adenosine diphosphate ribose). J. Biol. Chem., *249*: 3475–3482, 1974.
11. Ueda, K., Oka, J., Narumiya, S., Miyakawa, N., and Hayaishi, O. Poly ADP-ribose glycohydrolase from rat liver nuclei, a novel enzyme degrading the polymer. Biochem. Biophys. Res. Commun., *46*: 516–523, 1972.
12. Tanaka, M., Hayashi, K., Sakura, H., Miwa, M., Matsushima, T., and Sugimura, T. Demonstration of high molecular weight poly(adenosine diphosphate ribose). Nucleic Acids Res., *5*: 3183–3194, 1978.
13. Miwa, M., Saikawa, N., Yamaizumi, Z., Nishimura, S., and Sugimura, T. Structure of poly(adenosine diphosphate ribose): Identification of 2′-[1″-ribosyl-2″-(or 3″-) (1‴-ribosyl)]adenosine-5′,5″,5‴-tris(phosphate) as a branch linkage. Proc. Natl. Acad. Sci. U.S., *76*: 595–599, 1979.
14. Miwa, M., Ishihara, M., Takishima, S., Takasuka, N., Maeda, M., Yamaizumi, Z., Sugimura, T., Yokoyama, S., and Miyazawa, T. The branching and linear portions of poly(adenosine diphosphate ribose) have the same $\alpha$(1→2)ribose-ribose linkage. J. Biol. Chem., *256*: 2916–2921, 1981.
15. Hayashi, K., Tanaka, M., Shimada, T., Miwa, M., and Sugimura, T. Size and shape of poly(ADP-ribose): Examination by gel filtration, gel electrophoresis and electron microscopy. Biochem. Biophys. Res. Commun., *112*: 102–107, 1983.
16. Kanai, M., Miwa, M., Kuchino, Y., and Sugimura, T. Presence of branched portion in poly(adenosine diphosphate ribose) *in vivo*. J. Biol. Chem., *257*: 6217–6223, 1982.
17. Kanai, M., Miwa, M., Kondo, T., Tanaka, Y., Nakayasu, M., and Sugimura, T. Involvement of poly(ADP-ribose) metabolism in induction of differentiation of HL-60 promyelocytic leukemia cells. Biochem. Biophys. Res. Commun., *105*: 404–411, 1982.
18. Durkacz, B. W., Omidiji, O., Gray, D. A., and Shall, S. (ADP-ribose)$_n$ participates in DNA excision repair. Nature, *283*: 593–596, 1980.
19. Kawamitsu, H., Miwa, M., Tanaka, Y., Sakamoto, H., Terada, M., Hoshi, A., and Sugimura, T. Inhibitors of poly(adenosine diphosphate ribose) polymerase potentiate the antitumor activity of bleomycin against Ehrlich ascites carcinoma. J. Pharm. Dyn., *5*: 900–904, 1982.
20. Sakamoto, H., Kawamitsu, H., Miwa, M., Terada, M., and Sugimura, T. Enhancement of antitumor activity of bleomycin by benzamide *in vitro* and *in vivo*. J. Antibiot., *36*: 296–300, 1983.

# CHROMATIN STRUCTURE AND ADP-RIBOSYLATION

ADP-RIBOSYLATION, DNA REPAIR AND CANCER, M. MIWA ET AL. (EDS.),
JAPAN SCI. SOC. PRESS, TOKYO/VNU SCIENCE PRESS, UTRECHT, PP. 41–47, 1983

# Monoclonal Antibodies against Poly(ADP-ribose) Recognize Different Structures of Poly(ADP-ribose)

Hisae KAWAMITSU, Hiroo HOSHINO, Masanao MIWA, and Takashi SUGIMURA

*National Cancer Center Research Institute, Tokyo 104, Japan*

*Abstract:* The characteristics of two monoclonal antibodies to poly(ADP-Rib) and the various structures of poly(ADP-Rib) recognized by these monoclonal antibodies have been examined.

One antibody, IgG monoclonal antibody 10H, reacted with most parts of poly (ADP-Rib) molecules, and its binding was only slightly inhibited by Ado(P)-Rib-P, a monomer unit of the polymer. On hydrolysis of poly(ADP-Rib) protected by the antibody hydrolytic products, such as Ado(P)-Rib-P and Ado(P)-Rib(P)-Rib-P, were almost the same as those from control poly(ADP-Rib). The other antibody, IgM monoclonal antibody 16B, recognized only some parts of poly(ADP-Rib) molecules of larger size, and its binding was not inhibited by Ado(P)-Rib-P. Branched portions were considerably concentrated in parts protected by this antibody. These two monoclonal antibodies are suggested to recognize different structures of poly(ADP-Rib): possibly IgG antibody 10H recognizes linear portions whereas IgM antibody 16B recognizes branched portions of the polymer. These monoclonal antibodies should be useful in studies on the structures, including unknown structures and the locations of these structures in poly(ADP-Rib).

Monoclonal antibodies are useful in studying the structures and functions of biological substances. We established hybridomas producing monoclonal antibodies to poly(ADP-Rib), a polymer which is known to have linear and branching structures (*1–6*). Various kinds of monoclonal antibodies that recognize different antigenic sites would be helpful for the analysis of known and unknown structures of poly(ADP-Rib).

Inhibitors of poly(ADP-Rib) polymerase have been used in studying the functions of poly(ADP-Rib), such as repair of damaged DNA (*7, 8*), chromosomal aberration (*9*), and cell differentiation (*10*). However, these agents may have various

Abbreviations: Ado(P)-Rib-P, 2′(1″-ribosyl)adenosine-5′,5″-bis(phosphate) or 2′-(5″-phosphoribosyl)-5′-AMP; Ado(P)-Rib(P)-Rib-P, 2′-[1″-ribosyl-2″-(1‴-ribosyl)]-adenosine-5′,5″,5‴-tris(phosphate) or 2′-[2″-(5‴-phosphoribosyl))-5″-phosphoribosyl]-5′-AMP; Ado-Rib-P, 2′-(1″-ribosyl)adenosine-5″-monophosphate or 2′-(5″-phosphoribosyl)adenosine; Ado(P)-Rib, 2′-(1″-ribosyl)adenosine-5′-monophosphate or 2′-(ribosyl)-5′-AMP.

other effects besides inhibiting the polymerase. Thus monoclonal antibodies would also be useful in studying the biological functions of poly(ADP-Rib). Two kinds of monoclonal antibodies are described in the present paper, and it is shown that they react with different antigenic structures of poly(ADP-Rib).

Poly(ADP-Rib) is known to be heterogeneous in size (*11*), and the larger molecules were shown to have potent immunogenicity (*12, 13*). In the present work we separated poly(ADP-Rib) into three main fractions, I, II, and III, by hydroxylapatite column chromatography, and immunized mice with fraction III, which contained poly(ADP-Rib) molecules with more than 20 ADP-Rib residues (*11*).

### *Isolations of Two Kinds of Hybridomas Producing Monoclonal Antibodies to Poly(ADP-Rib)*

Immunization with poly(ADP-Rib) was carried out by the method described previously (*12, 13*), and the hybridization technique was based on the method of Köhler and Milstein (*14*).

Poly(ADP-Rib) coupled to methylated bovine serum albumin and emulsified with complete Freund's adjuvant was injected into the peritoneal cavity of two female BALB/c mice of 6–8 weeks old. Injections were given 3 times at intervals of about 3 weeks. Spleen cells from immunized mice were hybridized with mouse myeloma cells (P3-NSI-1-Ag4-1) with 50% polyethyleneglycol. These hybrid cells were grown in HAT medium in a culture plate with 96 wells from which hybridoma clones were selected. Two to three weeks after hybridization, production of antibodies to poly(ADP-Rib) was detected by passive hemagglutination assay with mouse erythrocytes which had been treated with formalin and coated with poly(ADP-Rib).

Hybridomas grew in 56% (108/192) of the total culture plate wells from one mouse, and in 95% (183/192) of those from the other mouse. Production of antibodies was observed in 15% (16/108) and 3% (6/183), respectively of these wells containing hybridoma clones. Hybridomas producing antibodies to poly(ADP-Rib) were then subcloned three or four times by the limiting dilution method. Most of them lost their antibody producing ability during subclonings or showed a very low antibody titer. Finally, two clones producing monoclonal antibodies to poly(ADP-Rib) were obtained from the two mice. The isotypes of these monoclonal antibodies were determined by the double immunodiffusion method and immunoelectrophoresis. One clone, 10H, produced IgG3, $\kappa$ antibody, and the other clone, 16B, produced IgM, $\lambda$ antibody.

### *Characteristics of Monoclonal Antibodies 10H and 16B*

The binding activities of the monoclonal antibodies to poly(ADP-Rib) were determined using radioactive [$^{14}$C]poly(ADP-Rib) fraction III which bound to the antibodies and was then trapped on a Millipore filter (*12*). On increase in its amount antibody 10H bound 100% of the radioactive poly(ADP-Rib), whereas antibody 16B bound only 50–60% even when present in a large excess. In other words, only a certain population of poly(ADP-Rib) molecules reacted with antibody 16B. The molecules that bound to antibody 16B were found to be of larger sizes by poly-

TABLE 1. Inhibition of Binding of Monoclonal Antibodies

| Competitors[a)] | Monoclonal antibodies | |
|---|---|---|
| | 10H | 16B |
| | % Inhibition | |
| ADP-Rib | 0 | 0 |
| Ado(P)-Rib-P | 21 | 0 |
| Poly(ADP-Rib) | | |
| fraction II | 51 | 14 |
| fraction III | 90 | 99 |
| Poly(A) | 0 | 10 |
| 5′-AMP | 0 | 0 |
| 3′-AMP | 0 | 0 |
| RNA | 2 | 0 |
| DNA | 0 | 0 |

a) The amount of each inhibitor added was 100-fold that of [$^{14}$C]poly(ADP-Rib) fraction III in the assay.

acrylamide gel electrophoresis of the poly(ADP-Rib) immunoprecipitated with the antibody (data not shown).

The competitive inhibition of the binding activity of antibodies by poly(ADP-Rib) itself, or by its analogues, indicated the specificity of the antibodies (Table 1). Poly(ADP-Rib) fractions II and III were strong competitors and the monomer unit Ado(P)-Rib-P was also an competitor, though it only weakly inhibited the binding of antibody 10H to [$^{14}$C]poly(ADP-Rib). With antibody 16B, poly(ADP-Rib) fraction III was a strong competitor, but fraction II was a weak competitor, and Ado(P)-Rib-P was not inhibitory.

These findings suggested that these two monoclonal antibodies react with different antigenic sites in poly(ADP-Rib) molecules. The site recognized by antibody 10H might have a structure that consists of several ADP-Rib residues. This site was found in most poly(ADP-Rib) molecules since the binding activity of the antibody was 100% and was inhibited by Ado(P)-Rib-P. Antibody 16B may recognize only certain molecules of poly(ADP-Rib), since its maximal binding was less than 50–60%. It may bind a structure that is not repeats of ADP-Rib residues, but a rather complicated form since its binding was not inhibited by Ado(P)-Rib-P.

### *Structural Analysis of Poly(ADP-Rib) with Monoclonal Antibodies*

We analyzed the structures of poly(ADP-Rib) recognized by the monoclonal antibodies with different characteristics described above.

Two enzymes, snake venom phosphodiesterase, which hydrolyzes pyrophosphate bonds in poly(ADP-Rib), and bacterial alkaline phosphomonoesterase were used in this analysis, and poly(ADP-Rib) was digested in three steps:

1) Poly(ADP-Rib) fraction III was incubated with or without monoclonal antibodies and hydrolyzed with snake venom phosphodiesterase (Table 2). The poly(ADP-Rib) that was not protected by antibodies was hydrolyzed to Ado(P)-Rib-P, Ado(P)-Rib(P)-Rib-P, AMP or ribosephosphate while the protected parts were re-

TABLE 2. Hydrolysis of Poly(ADP-Rib) Preincubated with Monoclonal Antibodies

| Time of hydrolysis (hr) | % hydrolysis[a] of poly(ADP-Rib) preincubated with: | | | |
|---|---|---|---|---|
| | 10H | 16B | Normal mouse γ-globulin | No antibody |
| 1 | 0 | 54 | 100 | 100 |
| 2 | 0 | 63 | 100 | 100 |
| 3 | 0 | 75 | 100 | 100 |
| 4 | 0 | 75 | | |

[a] % hydrolysis shows the ratio of the amounts of Ado(P)-Rib-P and AMP produced by hydrolysis, to the amounts of total ADP-Rib residues.

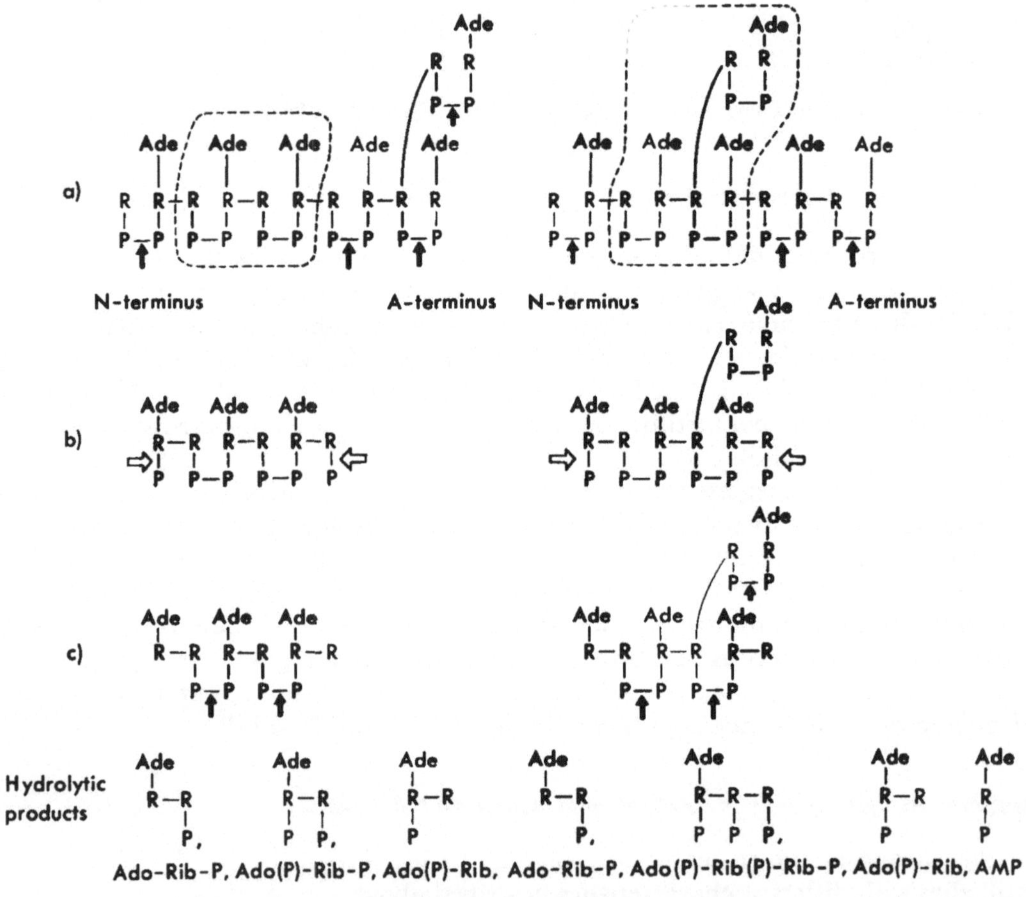


FIG. 1. Strategy for structural analysis of poly(ADP-Rib) protected by monoclonal antibodies. Two kinds of cases are shown in the figure. The analytical processes followed the order of a to c. Areas enclosed by broken lines are the parts of poly(ADP-Rib) protected by monoclonal antibodies. ↑ and ⇧ show hydrolysis with snake venom phosphodiesterase and bacterial alkaline phosphomonoesterase, respectively.

sistant to hydrolysis (Fig. 1a). The main hydrolytic products, Ado(P)-Rib-P and AMP, were analyzed by thin layer chromatography.

2) The unhydrolyzed parts were extracted with water-saturated phenol and sub-

TABLE 3. Hydrolytic Products of Poly(ADP-Rib) Protected by Monoclonal Antibodies

| Poly(ADP-Rib) protected with: | Hydrolytic products | | | |
|---|---|---|---|---|
| | Ado-Rib-P | Ado(P)-Rib(P)-Rib-P | Ado(P)-Rib-P | AMP, Ado(P)-Rib |
| Antibody 10H | 1.1 | 1[a] | 45 | 2.9 |
| Antibody 16B | 0.2 | 1[a] | 26 | 1.4 |
| Control[b] | 0.2 | 1[a] | 45 | 1.7 |

[a] The amounts of Ado(P)-Rib(P)-Rib-P are taken as 1. [b] Untreated poly(ADP-Rib) was hydrolyzed as described in the text.

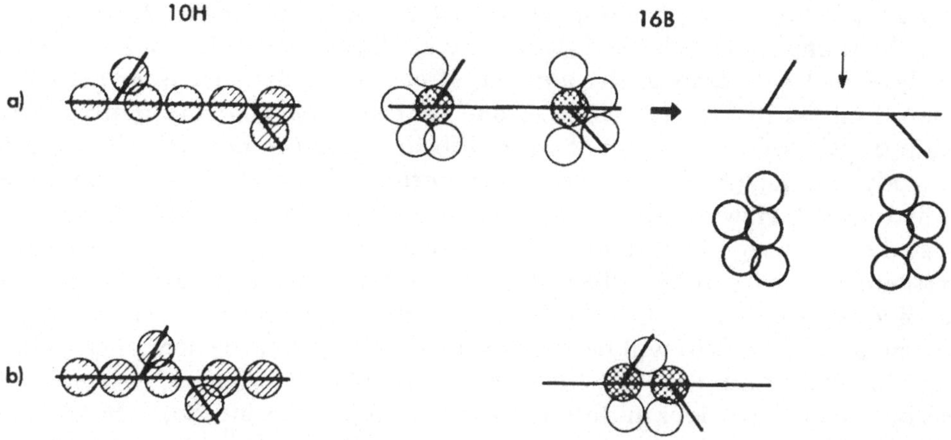


FIG. 2. The possible binding patterns of monoclonal antibodies to poly(ADP-Rib) molecules with two branched portions. One circle indicates one monomer of an immunoglobulin molecule.

jected to gel filtration. Phosphates should be released from terminal sites of protected parts by alkaline phosphomonoesterase treatment (Fig. 1b).

3) The alkaline phosphomonoesterase was removed with water-saturated phenol and the unhydrolyzed parts were then digested with snake venom phosphodiesterase. After treatment with the enzyme, the following products were produced: Ado-Rib-P from the N-terminus, Ado(P)-Rib(P)-Rib-P from branched portions, Ado(P)-Rib-P from linear portions, and Ado(P)-Rib and AMP from the A-terminus (Fig. 1c). These hydrolytic products were analyzed by paper chromatography. The ratios of the products to Ado(P)-Rib(P)-Rib-P are shown in Table 3.

When poly(ADP-Rib) was hydrolyzed with snake venom phosphodiesterase after incubation with antibody 10H(Step 1), no production of Ado(P)-Rib-P or AMP was observed, whereas 100% of poly(ADP-Rib) was hydrolyzed after incubation with normal mouse $\gamma$-globulin or with buffered saline (pH 7.4) (Table 2). This means that antibody 10H protected most poly(ADP-Rib) molecules from hydrolysis. Table 3 shows the hydrolytic products of parts protected by antibody 10H after Steps 2 and 3; these are almost the same as those of control poly(ADP-Rib). This finding seems compatible with the results of the hydrolysis in Step 1, indicating that antibody 10H may recognize linear portions of poly(ADP-Rib) (Fig. 2). If antibody 10H bound to branched portions, it would not protect most of the polymer, since the branched portions of poly(ADP-Rib) constitute only 2% of the total ADP-Rib residues.

In the experiment with antibody 16B, 25% of the portions of poly(ADP-Rib) molecules were protected from hydrolysis by venom phosphodiesterase after incubation with the antibody (Table 2), and branched portions in the parts protected by antibody 16B were twice as concentrated as those protected by antibody 10H (Table 3). If this antibody binds to branched portions, it might cover parts of about 26 residues around a branched portion, since the ratio of Ado(P)-Rib-P to Ado(P)-Rib(P)-Rib-P is 26: 1 in the parts protected by the antibody. In this case, about 50% with the portions (2%×26) of poly(ADP-Rib) should be recovered after digestion with snake venom phosphodiesterase (Step 1). However, in fact the recovery was only 25%. This finding could be explained by the model in Fig. 2. As antibody 16B is IgM, its affinity to poly(ADP-Rib) may be dependent on how many monomers of the IgM molecule bind to the antigen. If only one monomer (shown by the crosshatched circles in Fig. 2a) binds to one branched portion, the antibody molecule might be released as the result of an equilibrium of association and dissociation of the antibody and poly(ADP-Rib), and some portions of poly(ADP-Rib) might gradually be hydrolyzed. However, in a condition whereby two or more branched portions are located close enough for two or more monomers of the IgM molecule (shown by the crosshatched circles in Fig. 2b) to bind to each branched portion at the same time, the binding between the IgM molecule and the poly(ADP-Rib) will be very strong, and the poly(ADP-Rib) will be resistant to attack by snake venom phosphodiesterase (Fig. 2b). Poly(ADP-Rib) consists of molecules of different sizes, with various numbers of branched portions in different locations in a molecule. So, IgM antibody, like 16B, should have different affinities to each molecule. It is also possible that poly-(ADP-Rib) has unknown structures besides the known linear and branching structures and that antibody 16B recognizes one of such unknown structure that exists in a small population of the molecules.

From this study on the characteristics of binding of two kinds of monoclonal antibodies to poly(ADP-Rib) we suggested that antibody 10H binds to linear structures of poly(ADP-Rib), whereas antibody 16B binds to structures containing branched portions. Further investigations using these two antibodies are planned on locations and functions of these structures *in vivo*.

ACKNOWLEDGMENT

This work was supported in part by Grants-in-Aid from the Ministry of Education, Science and Culture, and from the Ministry of Health and Welfare, Japan.

REFERENCES

1. Doly, J. and Petek, F. Étude de la structure d'un composé "poly(ADP-ribose)" synthétisé par des extraits nucléaires de foie de poulet. C. R. Hebd. Séances Acad. Sci. Ser. D. Sci. Nat., *263*: 1341–1344, 1966.
2. Miwa, M., Saitô, H., Sakura, H., Saikawa, N., Watanabe, F., Matsushima, T., and Sugimura, T. A $^{13}$C NMR study of poly(adenosine diphosphate ribose) and its monomers: Evidence of $\alpha$-(1″→2′) ribofuranosyl ribofuranoside residue. Nucleic Acids Res., *4*: 3997–4005, 1977.

3. Inagaki, F., Miyazawa, T., Miwa, M., Saitô, H., and Sugimura, T. NMR analysis of conformation of ribosyl adenosine 5′,5″-bis(phosphate) in aqueous solution. Biochem. Biophys. Res. Commun., *85*: 415–420, 1978.
4. Ferro, A. M. and Oppenheimer, N. J. Structure of poly(adenosine diphosphate ribose) monomer: 2′-(5″-phosphoribosyl)-5′-adenosine monophosphate. Proc. Natl. Acad. Sci. U.S., *75*: 809–813, 1978.
5. Miwa, M., Saikawa, N., Yamaizumi, Z., Nishimura, S., and Sugimura, T. Structure of poly(adenosine diphosphate ribose): Identification of 2′-[1″-ribosyl-2″-(or 3″-)(1‴-ribosyl)]adenosine-5′,5″,5‴-tris(phosphate) as a branch linkage. Proc. Natl. Acad. Sci. U.S., *76*: 595–599, 1979.
6. Miwa, M., Ishihara, M., Takishima, S., Takasuka, N., Maeda, M., Yamaizumi, Z., Sugimura, T., Yokoyama, S., and Miyazawa, T. The branching and linear portions of poly(adenosine diphosphate ribose) have the same $\alpha(1\rightarrow2)$ribose-ribose linkage. J. Biol. Chem., *256*: 2916–2921, 1981.
7. Durkacz, B. W., Omidiji, O., Gray, D. A., and Shall, S. (ADP-ribose)$_n$ participates in DNA excision repair. Nature, *283*: 593–596, 1980.
8. Miwa, M., Kanai, M., Kondo, T., Hoshino, H., Ishihara, K., and Sugimura, T. Inhibitors of poly(ADP-ribose) polymerase enhance unscheduled DNA synthesis in human peripheral lymphocytes. Biochem. Biophys. Res. Commun., *100*: 463–470, 1981.
9. Oikawa, A., Tohda, H., Kanai, M., Miwa, M., and Sugimura, T. Inhibitors of poly(adenosine diphosphate ribose) polymerase induce sister chromatid exchanges. Biochem. Biophys. Res. Commun., *97*: 1311–1316, 1980.
10. Farzaneh, F., Zalin, R., Brill, D., and Shall, S. DNA strand breaks and ADP-ribosyl transferase activation during cell differentiation. Nature, *300*: 362–366, 1982.
11. Tanaka, M., Hayashi, K., Sakura, H., Miwa, M., Matsushima, T., and Sugimura, T. Demonstration of high molecular weight poly(adenosine diphosphate ribose). Nucleic Acids Res., *5*: 3183–3194, 1978.
12. Kanai, Y., Miwa, M., Matsushima, T., and Sugimura, T. Studies on anti-poly-(adenosine diphosphate ribose) antibody. Biochem. Biophys. Res. Commun., *59*: 300–306, 1974.
13. Kanai, Y., Miwa, M., Matsushima, T., and Sugimura, T. Comparative studies on antibody and antibody production to poly(ADP-ribose) in mice. Immunology, *34*: 501–508, 1978.
14. Köhler, G., and Milstein, C. Derivation of specific antibody producing tissue culture and tumor lines by cell fusion. Eur. J. Immunol., *6*: 511–519, 1976.

ADP-RIBOSYLATION, DNA REPAIR AND CANCER, M. MIWA ET AL. (EDS.),
JAPAN SCI. SOC. PRESS, TOKYO/VNU SCIENCE PRESS, UTRECHT. PP. 49-70, 1983

# Fine Resolution of the Poly ADP-ribosylated Domains of Polynucleosomal Chromatin: DNA Gene and Integrity Analysis; Mechanism of Histone Hl Modification

Mark SMULSON, N. MALIK, M. WONG, N. POMATO, and P. THRAVES

*Department of Biochemistry and Division of Molecular Biology, Lombardi Cancer Center, Georgetown University Medical Center, Washington, D.C. 20007, U.S.A.*

*Abstract:* The focus of our laboratory has been to ascertain how the poly(ADP-ribosyl)ation reaction influences the structure and biological function of nucleosomal chromatin.

Antibody to poly(ADP-ribose (ADP-Rib)) was coupled to Sepharose to prepare an immunoaffinity column. The following new information concerning poly(ADP-Rib) and chromatin was obtained with this column: 1) those limiting domains of nucleosomal chromatin undergoing the modification (*circa* 10%) could be isolated in the "bound" fraction, 2) antibody-bound nucleosomes contained all the poly ADP-ribosylated acceptors and polymerase activity of the bulk chromatin, 3) bound nucleosomes contain significant numbers of internal DNA strand breaks, and [$^3$H]-thymidine (TdR) repair incorporation from *in vivo* "DNA repair label," compared to the unbound, bulk of the chromatin, 4) the presence of actively transcribed genes in "bound" nucleosomes is being investigated, 5) the acetylation modification of histones occurs in the same domains of chromatin as does poly(ADP-ribosyl)ation, and 6) poly ADP-ribosylated histone H1 can be selectively purified by the immunoaffinity method. These same histone H1 molecules appear to be equally accessible to the histone kinase, phosphorylation modification.

In addition, new information has been obtained concerning histone H1 cross-linking by poly(ADP-ribosyl)ation, and on polymerase binding sites to chromatin. We have reconstituted histone H1-depleted chromatin with intact H1 and peptide domains of H1, and subsequently studied H1-poly(ADP-Rib)-complex synthesis. The data indicate that elongation of poly(ADP-Rib) proceeds on the amino terminal region of this histone.

By utilizing the new techniques of DNA technology, huge advances in our understanding of the programmed structure of the eukaryotic genome have been accomplished in a relatively short time. To complement this explosion of information, it is important to ascertain how these recently appreciated properties of eukaryotic DNA are packaged within extended and condensed domains of chromatin. The research to be discussed is directed, in part, at this latter topic. We have initiated a program aimed at furthering our comprehension of chromatin structure by studying one specific, enzymatically active chrosomal protein.

The chromatin-associated enzyme, poly(ADP-Rib) polymerase catalyzes the successive transfer of the ADP-ribose moiety of NAD to various nuclear acceptors, including core nucleosomal histones, histone H1, and a 112,000 dalton non-histone protein, shown to be poly(ADP-Rib) polymerase. With increasing concentrations of NAD, the polymer reaches chain lengths exceeding 100 units long. This reaction thus represents a considerable physical alteration of the structure of those nucleosomes undergoing this modification, since recent data suggests that histones are cross-linked upon extensive poly(ADP-ribosyl)ation. There is increasing evidence that this enzyme is activated *in vivo* by DNA strand breaks, and functions in some step of DNA repair. A direct correlation between NAD concentration, the level of chromatin aggregation, and the length of poly(ADP-Rib) chains in the complex was detected.

Experimentation, aimed at relating biological function to structural features of nucleosomal chromatin, has suffered from the lack of methods to isolate, selectively, large quantities of highly enriched nucleoprotein domains undergoing such reactions. In the present study, the use of anti-poly(ADP-Rib) antibody covalently attached to Sepharose has permitted the selective enrichment of chains of oligonucleosomes proximal to poly(ADP-ribosyl)ation sites in chromatin. Using this technique we have elucidated new information concerning the regions around this modification, site, related to gene activation, DNA repair/replication and alternative nuclear protein modification reactions.

### *Poly(ADP-Rib) Polymerase Is Bound to Specific Chromatin Domains*

One interest in our laboratory has been the placement of polymerase within chromatin. Previous *in vitro* data indicated that the specific activity of poly(ADP-Rib) polymerase increases with the size of nucleosomal chains. Utilizing the forementioned incorporation of ADP-Rib to the 112K protein as a marker, it was tentatively concluded that the maximal enzymatic "activity" on oligonucleosomes (8–10N) represents preferred binding of this enzyme to specialized domains of chromatin at a periodicity of approximately 8–10 nucleosomes. It is of interest that one turn of the 300 Å chromatin fiber is reported to consist of approximately 7–8 nucleosomes. Accordingly, information on the interaction of poly(ADP-Rib) polymerase with specialized regions of polynucleosomes could provide a useful and sensitive probe for elucidation of general relationships between non-histone proteins and chromatin structural domains.

To facilitate investigations on the organization of poly(ADP-Rib) polymerase in chromatin, and to elucidate its biological function, polymerase purified from HeLa nuclei was used to elicit antibodies in mice. The association of polymerase with oligonucleosomes of differing chain size was determined by the specific binding of polymerase antibody (and as control, anti-histone H3) to nitrocellulose transfers of native electrophoretic gels of these particles. By this technique, polymerase appears to be associated with a select subclass of nucleosomes. Further resolutions of the polymerase binding sites within nucleosome classes was achieved with the antibody by

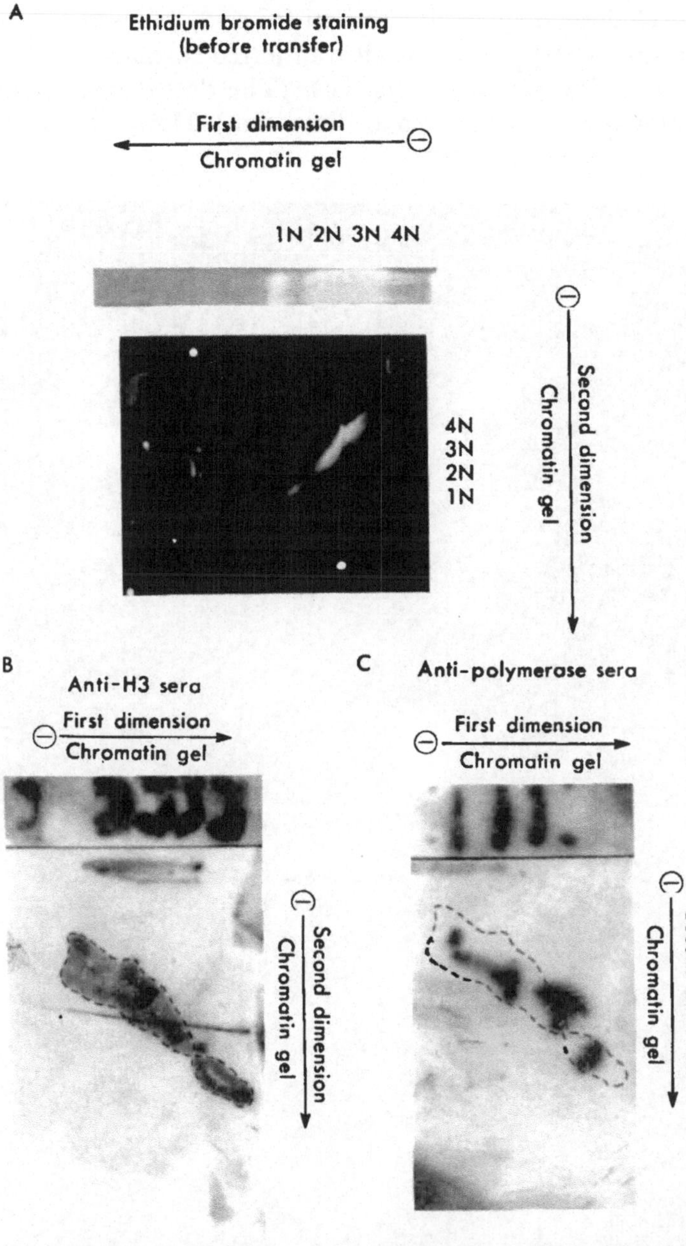


FIG. 1. Reactivity of anti-H3 or anti-polymerase sera with nitrocellulose blots of oligonucleosomes resolved on polyacrylamide gels by two dimensional electrophoresis. Mono-tetra nucleosomes were separated on a polyacrylamide gel. The slab gel lane was cut and fused to 3% acrylamide, 0.5% agarose gel and electrophoresed in second dimension. The two-dimensionally resolved nucleosomes, as well as nucleosomes, resolved in one direction were transferred to nitrocellulose strips. A. Ethidium bromide staining pattern. B. Autoradiogram of transfer treated with anti-H3. C. Autoradiogram of transfer after treatment with anti-polymerase sera. (From ref. 4).

two-dimensional native polyacrylamide gel electrophoresis and nitrocellulose transfer.

The resolution of mono- through tetra-nucleosomes in two dimensions is shown in the data of Fig. 1A. When one- and two-dimensionally resolved nucleosomes were transferred onto nitrocellulose, incubated with anti-H3 sera, followed by detection of the immune complex using [$^{125}$I] protein A, all transferred subclasses of nucleosomes were noted to react with H3 antibody (Fig. 1B). (The dotted lines represent the ethidium bromide staining area of the second dimension). However, when an iden-

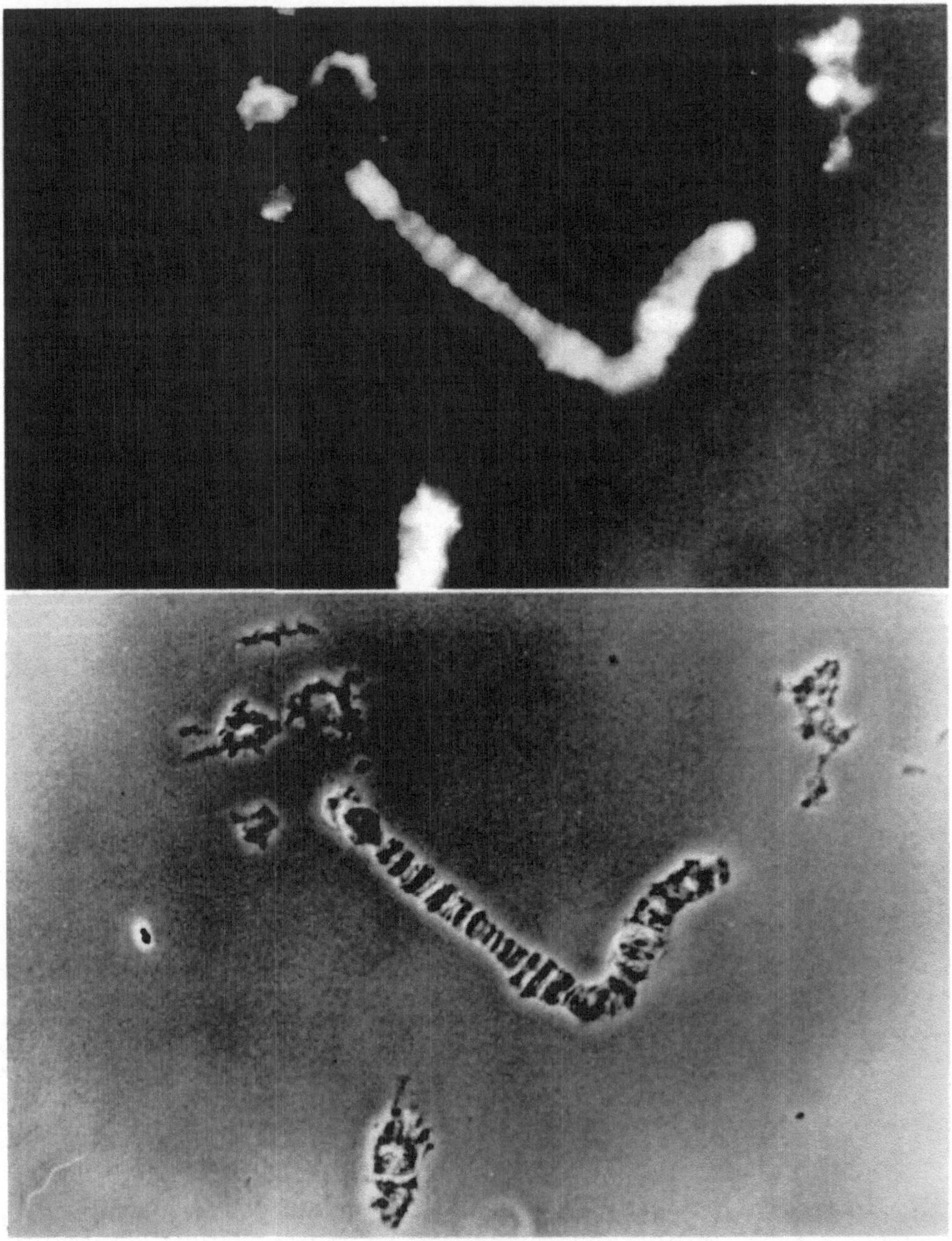

FIG. 2. Reactivity of poly(ADP-Rib) polymerase antibody with the chironomus polytene chromosome. Reaction was detected using a secondary fluorescent antibody.

tical nitrocellulose strip was treated with anti-polymerase sera (Fig. 1C), only specific subsets of nucleosome reacted with polymerase antibody.

In another series of experiments (in collaboration with Dr. Bustin of NIH) antibody was utilized to visualize enzyme binding sites in the chironomus polytene chromosome (Fig. 2). Using immunofluorescence, the association of polymerases, mainly with the puffed regions of the chromosome were observed.

At this point, we turned our attention to experiments with chromatin using antibody to poly(ADP-Rib), very graciously supplied by Drs. Miwa and Sugimura. We believe the new approaches we have taken have allowed a closer resolution of the interaction of poly(ADP-Rib) with the genome.

### *Separation of Poly ADP-ribosylated Nucleosomes by Immunoaffinity Chromatography*

Based upon a high turnover rate and resultant low level of poly(ADP-ribosyl) ation, a very sensitive method is required to purify those domains of chromatin undergoing this reaction. In the past, we have partially solved this problem *via* selective micrococcal nuclease digestion procedures (*1*), and by the use of gel electrophoresis of crosslinked poly ADP-ribosylated nucleosomal particles (*2*). In the latter case, it was observed that, upon long chain poly(ADP-ribosyl)ation, those limited oligonucleosomes undergoing the modification, became condensed, as analyzed by either sedimentation velocity centrifugation or gel electrophoresis of native chromatin particles and could be partially separated from bulk chromatin (*2*).

The development of highly active antisera, directed specifically against the modified moiety of this reaction, poly(ADP-Rib) polymer (*3*), led us to test whether an immunological affinity method could be employed to bind selectively, those domains of chromatin undergoing this reaction.

It was of importance to ascertain whether the purified anti-poly(ADP-Rib) IgG fraction was capable of interaction with endogenous poly ADP-ribosylated nuclear proteins bounded within the nucleosomal structure. Accordingly, oligonucleosomes were incubated in the presence and absence of 10 $\mu$M NAD, subsequently separated by native polyacrylamide electrophoresis, and transferred electrophoretically to nitrocellulose sheets by a method recently described (*4*). The nitrocellulose sheet was incubated with the antibody to the modified moiety, poly(ADP-Rib) polymer and subsequently with [$^{125}$I] protein A (data not shown). Significant binding of the IgG to each respective chain size of nucleosomes was noted in the chromatin sample preincubated with NAD. These data suggested that *in vitro* elongation of poly(ADP-Rib) on nucleosomal acceptor histones would be desirable for the anticipated immunoaffinity experiments. since limited endogenous poly(ADP-Rib) appeared to be present on the isolated HeLa cell nucleosomes. Having shown that the anti-poly (ADP-Rib) IgG fraction was capable of reactivity with nucleosome-associated polymer, the antibody was covalently coupled to cyanogen bromide activated Sepharose 4B. As a control, a preimmune fraction of IgG was bound to Sepharose in a similar manner. The antibody-bound Sepharose (6 ml, bed-volume) was washed and applied to a column. To test the binding of ADP-ribosylated chromatin to the column, a preparation of oligonucleosomes (1.5 $A_{260}$) was incubated with [$^{32}$P]NAD (10 $\mu$M). Under these conditions, chains of (ADP-Rib) units from 2 to greater than 15 are

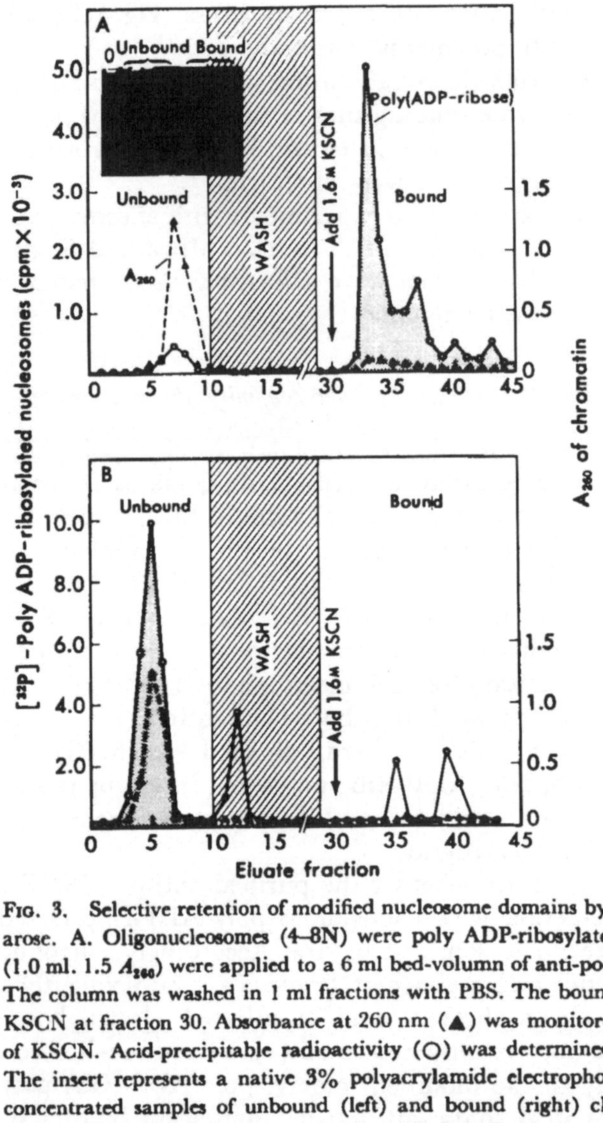


FIG. 3. Selective retention of modified nucleosome domains by anti-poly (ADP-Rib) IgG-Sepharose. A. Oligonucleosomes (4–8N) were poly ADP-ribosylated with [$^{32}$P]NAD. Nucleosomes (1.0 ml. 1.5 $A_{260}$) were applied to a 6 ml bed-volumn of anti-poly (ADP-Rib) IgG-Sepharose 4B. The column was washed in 1 ml fractions with PBS. The bound material was eluted with 1.6 M KSCN at fraction 30. Absorbance at 260 nm (▲) was monitored on fractions prior to additions of KSCN. Acid-precipitable radioactivity (O) was determined as previously described (*2, 5*). The insert represents a native 3% polyacrylamide electrophoretic separation of dialyzed and concentrated samples of unbound (left) and bound (right) chromatin, stained with ethidium bromide. B. A similarly prepared poly ADP-ribosylated oligonucleosomal sample was analyzed on a preimmune IgG coupled Sepharose column. (From ref *13*).

enzymatically coupled to nucleosomal histone acceptors by the chromatin-associated polymerase (*2, 5*). However, only a limited domain of chromatin is accessible for poly (ADP-ribosyl)ation at any one time, accounting for approximately 0.5% of the total histone acceptors (*6, 7*).

The *in vitro* incubated oligomers, (4–8 N; $4.95\times10^{6}$ cpm) in 1 ml, were partitioned through Sephadex G-25, to remove unincorporated [$^{32}$P]NAD, and subsequently slowly applied to the column over a 45 min period (Fig. 3A). Washing was initiated with PBS; a peak of $A_{260}$ material eluted between fractions 5–10. This represented greater than 90% of the original polynucleosome sample. In addition to absorbance, acid precipitable radioactivity, representing poly ADP-ribosylated nucleosomal acceptors (see below), was monitored in all fractions. Negligible [$^{32}$P] incorporation was noted in the unbound fractions, where the majority of nucleosomes eluted. When no further [$^{32}$P](ADP-Rib)$_n$, nor $A_{260}$ eluted, the column was washed with 20 ml of PBS. The bound material was released by the addition of 1.6 M KSCN (fraction 30). An almost immediate elution of acid-precipitable [$^{32}$P]poly ADP-ribosylated material was noted. Approximately 95% of the total nucleosomal radioactive incorporation was represented in five 1.0 ml fractions (fractions 32–37). The presence of KSCN prevented reliable quantitation of $A_{260}$ absorbance; however, when bound fractions were dialyzed, nucleoprotein structures and detectable $A_{260}$ were observed (insert, and Fig. 3).

*1. Specificity of immuno-binding*

When a similarly prepared column containing preimmune IgG coupled to Sepharose was utilized (Fig. 3B), the majority of the acid-insoluble [$^{32}$P]poly ADP-ribosylated material remained unbound to the column and eluted coincidental with the bulk of the nucleosomes, which passed directly through the column (fractions 3–7). Only negligible radioactivity was eluted with KSCN.

The above data demonstrated that the immunoaffinity column possesses the desired specificity for binding of poly ADP-ribosylated components derived from *in vitro* labeled nucleosomes. The procedure appeared to offer the potential to isolate nucleosomal domains proximal to this one specific modification region of chromatin. Accordingly, subsequent experiments were designed to characterize the poly ADP-ribosylated chromatin remaining bound to the column in comparison to unbound chromatin.

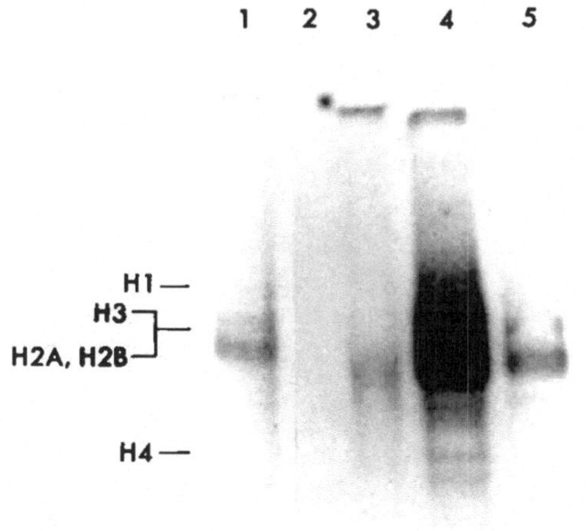


FIG. 4. Identification of poly ADP-ribosylated histones in immunoaffinity-bound chromatin. A separation of [$^{32}$P]NAD-labeled nucleosomes on the anti-poly(ADP-Rib) column was performed, as described above. The acid-soluble material was separated by acetic acid/urea polyacrylamide gels, and exposed for autoradiography. Lane 1, nucleosomes applied to column; lane 2, unbound nucleosomes; lanes 3–5, bound nucleosomes, fractions 34–36 (Fig. 3), respectively. (From ref. *13*).

*2. Analysis of immuno-complexed radioactive histones*

Fractions from unbound and bound material of an experiment similar to Fig. 3A were acid extracted and subjected to electrophoresis and autoradiography to examine histone acceptors (Fig. 4). Utilizing nucleosomes, we had previously determined that all five classes of histones act as acceptors for poly(ADP-Rib) polymerase, although the inner core histone, H4, is only weakly ADP-ribosylated (*2*). Additionally, utilizing high NAD substrate concentration, poly(ADP-ribosyl)ation of histone H1 has been shown to lead to a H1 dimer complex (*7*). In the experiment in Fig. 4, histones isolated from the labeled nucleosomes, prior to immuno-fractionation, showed the typical profile of modification, as noted above (lane 1). However, only a limited percentage of total nucleosomes in this population are shown (Fig. 3A) to undergo poly(ADP-ribosyl)ation. Accordingly, no poly ADP-ribosylated acceptors were detected in the histone fraction isolated from the unbound fraction (lane 2), although this fraction constituted the vast majority of the chromatin applied to the column. In contrast, well defined poly ADP-ribosylated histones were noted in the region near the peak radioactive fraction of bound material (lane 4). The best resolved modified histone were noted to elute slightly after the peak tube; however, this varied to some extent amongst experiments. These results would imply a high level of enrichment of the limited domains of modified chromatin, and this was directly quantitated in experiments described subsequently.

*3. Non-histone protein association with enriched nucleosomes*

Data from this laboratory and others have shown that the major non-histone protein (112,000 dalton) acceptor for poly(ADP-Rib) polymerase constitutes the

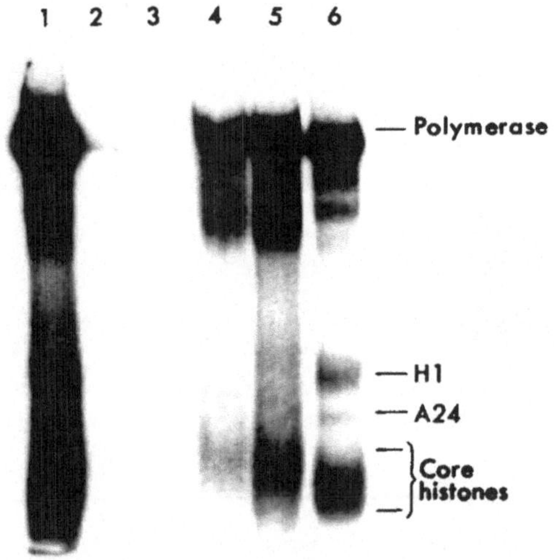


FIG. 5. Total nucleosomal poly (ADP-Rib) acceptors of chromatin purified by anti-poly (ADP-Rib). The experiment was performed essentially as described in Fig. 3 except that carrier-containing samples were ethanol precipitated and subjected to Na DodSO$_4$ polyacrylamide gel electrophoresis, and autoradiography. Lane 1, nucleosomes applied to column; lanes 2, 3, unbound nucleosomes; lanes 4–6, bound nucleosomes, fractions 34–36 (Fig. 3), respectively. (From ref *13*).

enzyme itself, undergoing automodification (*8–10*). Antibody to the polymerase has been shown to react with blots of electrophoretically separated nucleosomes (Fig. 1); catalytic activity of the protein can be detected by *in situ* [$^{32}$P]NAD incorporation on gels of separated, enzymatically active nucleosome species (*4*). Accordingly, it might be anticipated, especially in view of the above mentioned automodification that the polymerase would be found with those nucleosomes showing affinity for the anti-poly(ADP-Rib) column.

Based upon the automodification reaction, the enzyme may be conveniently identified by the NAD-labeling of a 112,000 dalton non-histone protein (*8, 10, 11*). Total poly ADP-ribosylated acceptors were examined in both bound and unbound chromatin by ethanol precipitation of the respective labeled fractions and $NaDodSO_4$ polyacrylamide gel electrophoresis (Fig. 5). As before (Fig. 4), all poly ADP-ribosylated acceptors were detected only in the bound fractions (lanes 4–6). Labeled histones and possibly protein A-24 (*12*) were noted in these fractions, although the resolution of the core histones was more accurate in the data provided in Fig. 4. These results suggest that the immunoaffinity column not only partitions poly ADP-ribosylated domains of chromatin, but in addition, enzymatically active regions of chromatin, containing the 112,000 dalton poly(ADP-Rib) polymerase. This possibility was explored in greater detail below.

### 4. *Degree of enrichment of poly ADP-ribosylated nucleosomes*

Not all nucleosomes within an oligonucleosome chain would be expected to contain poly ADP-ribosylated histones. To precisely quantitate the level of enrichment by the antibody of nucleosomal chains in which some, or all histones contain poly(ADP-Rib), the specific activity of poly(ADP-Rib), with respect to histone mass (*i.e.*, from cells grown in [$^{3}$H]lysine and [$^{3}$H]arginine) was determined.

To precisely analyze the degree of enrichment of unbound and bound fraction for poly ADP-ribosylated histone acceptors, samples were acid-extracted and histones purified (data not shown; see ref. *13*). The [$^{32}$P]/[$^{3}$H] ratio of histones extracted from the nucleosomes, prior to the antibody column was 2.4. After fractionation, the specific activity of the bound fraction was 14.1, representing a 5.8-fold enrichment. These results were expected, since not all nucleosomes, within a domain of chromatin (Table 1), might be anticipated to undergo poly(ADP-ribosyl)ation at any one time.

It was not clear whether nucleosomal organization of chromatin was maintained after dissociation of bound material from the antibody with KSCN, although ethidium bromide staining of bound material (Fig. 3, insert) had indicated this to be the case.

Dialyzed and concentrated bound, unbound and unfractionated samples were labeled, in a poly(ADP-Rib) polymerase assay, with high specific activity [$^{32}$P]-NAD. After termination of the reactions with the polymerase inhibitor nicotinamide, the nucleosomes were subjected to native 3.5% polyacrylamide gel electrophoresis, ethidium staining (Fig. 6A), and autoradiography (Fig. 6B). As indicated by the strong ethidium bromide staining pattern (Fig. 6A, lane 2), the unbound nucleosomes contained the majority of the nucleosomal chromatin applied to the column, yet *no polymerase activity remained* with these classes of chromatin (Fig. 6B, lane 2). Accord-

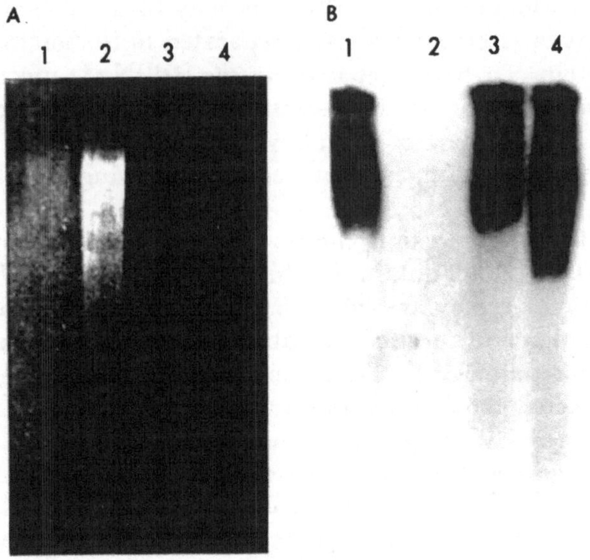


FIG. 6. Poly ADP-ribosylation of enzymatically active nucleoprotein complexes before and after immuno-fractionation. Lanes 1–4 represent samples of bound, unbound, bound (duplicate), and unfractionated (*i.e.*, sample before column) chromatin fractions, respectively. After fractionation, samples (*i.e.*, lanes 1–3) were dialyzed overnight against PBS and concentrated. The chromatin preparations were incubated with 25 nM [$^{32}$P]NAD (2 $\mu$Ci) under conditions optimal for poly (ADP-Rib) polymerase activity. The assays were terminated with nicotinamide and subjected to electrophoresis in 3.5% polyacrylamide gels as previously described (*2*). The $A_{260}$ of the chromatin samples applied to the gels were: unfractionated, 0.03; unbound, 0.065. A. Ethidium bromide. B. Autoradiography. (From ref. *13*).

ingly, the immunoaffinity fractionation procedure provides us with nucleosomal chromatin for future studies, which is totally depleted of poly(ADP-ribosyl)ation potential.

The bound fraction was noted to possess essentially similar nucleoprotein electrophoretic porperties as the original sample, as assessed by ethidium bromide-staining (Fig. 6A, lanes 1, 3). Although the quantities of material are much less, than the other samples, the bound fraction showed considerable poly(ADP-Rib) polymerase activity (Fig. 6B, lanes, 1, 3), which co-migrated with the ethidium bromide-staining chromatin bands. The specific activity (*i.e.*, [$^{32}$P]/ethidium bromide intensity) was noted to be significantly enhanced in the bound material. These data suggest that nucleoprotein complexes, not only total poly(ADP-Rib) nucleosomal acceptors, but also enzymatically active polymerase, are selectively purified by this technique, and that upon elution, the active fractions retain electrophoretic properties similar to that of intact chromatin.

*5. Poly ADP-ribosylated regions of chromatin contain significant DNA strand breaks*

Total cellular levels of poly(ADP-ribosyl)ation have been shown to become greatly elevated by a variety of cellular treatment which lead to the production of DNA strand breaks (*14–17*). Accordingly, the status of internal DNA single-strand breaks was compared in chromatin totally lacking poly ADP-ribosylated species

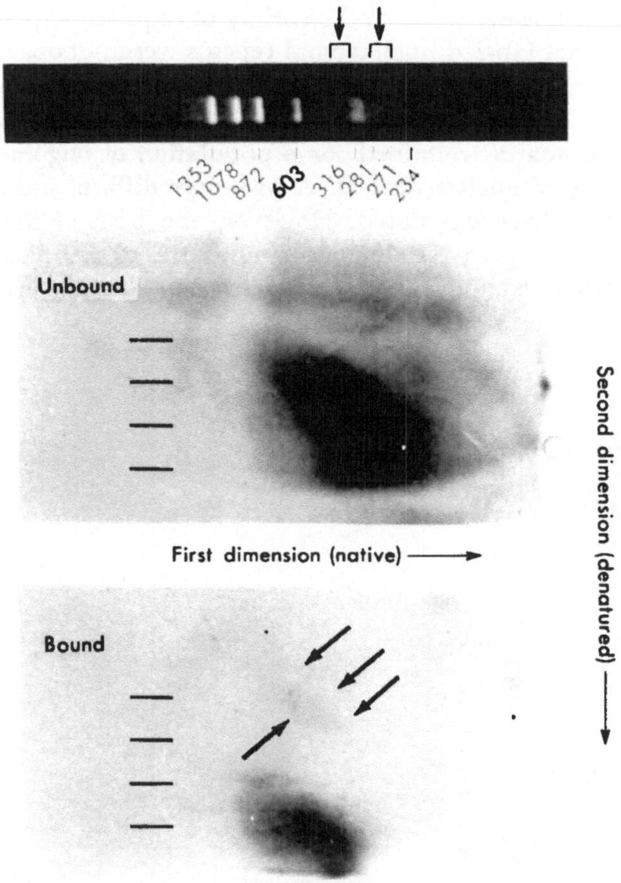


FIG. 7. Poly(ADP-ribosylated nucleosomes contain significant internal DNA strand breaks. Unbound nucleosomes (4.4 $A_{260}$) and bound nucleosomes (0.47 $A_{260}$) were isoamyl alcohol, chloroform, phenol-extracted for DNA. The DNA was treated with alkaline phosphatase, phenol extracted, and incubated with polynucleotide kinase (100 units) and 100 $\mu$Ci[$^{32}$P]ATP (4,500 Ci/mmol). The labeled DNA samples were subjected to native 1.6% agarose gel electrophoresis as described by Modak and Beard (*18*). A duplicate lane of unbound DNA and a lane of Hae III restricted $\phi$X-174 DNA were stained with ethidium bromide (top). The arrows indicate staining at mono- and dinucleosome regions of the gel. The first dimension lanes were subsequently electrophoresed, under denaturating conditions, at pH 11, in the second dimension (*18*). The gels were dried and exposed for autoradiography. The arrows in the bound (lower) autoradiograph outline a low level of nucleosomal DNA "diagonal," which is in the same position as the major diagonal of the top, unbound DNA sample. (From ref. *13*).

(*i.e.*, unbound chromatin) with bound chromatin. DNA from the respective fraction was phenol extracted, treated with alkaline phosphatase (to cleave external and internal, free phosphates), and [$^{32}$P] labeled with polynucleotide kinase. The DNA was analyzed on a native 1.6% agarose gel, in the first dimension, and an alkaline agarose gel (*18*), to detect strand breaks, in the second dimension (Fig. 7). An ethidium bromide stain of the first dimension of the unbound DNA (Fig. 7, top) showed distinct stained nucleosomal DNA species at approximately 270 and 400 base pairs. The autoradiograph (Fig. 7, unbound, first dimension direction) indicated a range of base pairs from 200–800 base pairs; however, due to the variety of required experimental treatments of the sample, distinct labeled nucleosomal repeats were not observed. In the second dimension (denaturating), the unbound sample migrated as a diagonal. Distinct, faster moving, bands off the diagonal, and lower, diffuse labeled DNA indicated that the unbound chromatin contained some population of oligomers with internal strand breaks. The bound nucleosomal DNA was more diffuse, and a minor diagonal band (denoted by arrows) was noted. However, the bulk of the bound DNA migrated considerably faster than the diagonal and appeared to be composed almost entirely of nucleoprotein complex (possible nucleosome) with considerable internal strand breaks.

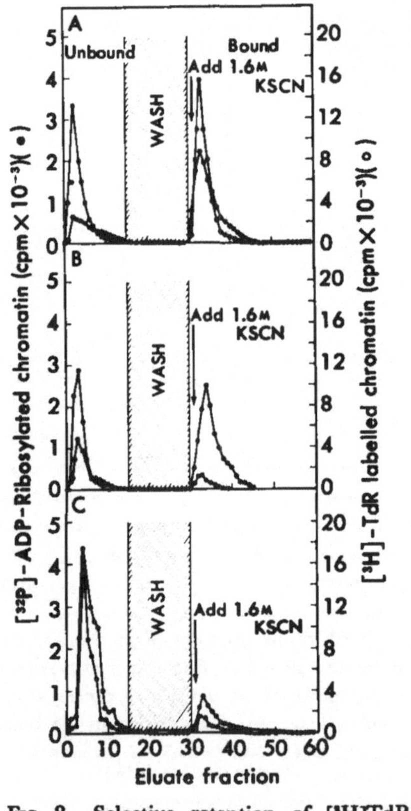


FIG. 8. Selective retention of [$^3$H]TdR labeled chromatin by anti-poly(ADP-Rib) IgG-Sepharose. Cells were incubated in the presence or absence of MNU, labeled with [$^3$H]TdR during unscheduled DNA synthesis, and nucleosomes isolated. [$^3$H]TdR-labeled chromatin samples (1.5 $A_{260}$ units) from either control (B) in MNU-treated samples (A) were suspended in 10 mM Tris-HCl, pH 8.0, containing 2 mM DTT and 5 mM $Mg^{2+}$ and ADP-ribosylated by the addition of 10 $\mu$M [$^{32}$P]NAD (0.2 $\mu$Ci) and incubated at room temperature for 5 min. Antibody column chromatography was performed by slowly applying appropriate amounts of *in vitro* [$^{32}$P]ADP-ribosylated oligonucleosomes to antibody columns over a 30 min period. A similarly prepared poly ADP-ribosylated oligonucleosomal sample (C) from MNUtreated cells was analyzed on a preimmune IgG coupled Sepharose column. ○ [$^{32}$P]poly ADP-ribosylated; ● [$^3$H]TdR unscheduled DNA synthesis.

TABLE 1. Fold Enrichment of Poly ADP-ribosylated Histones in Bound Oligonucleosomes

| Nucleosome fraction | Incorporation into purified histone fraction (dpm × $10^{-3}$) | | | | |
|---|---|---|---|---|---|
| | [$^3$H]amino acid | [$^{32}$P]ADPR | [$^{32}$P]/[$^3$H] | Fold enrichment | % protein recovered |
| Unfractionated (*i.e.*, applied) | 29.2 | 70.0 | 2.4 | 1 | — |
| Unbound | 25.5 | 0.1 | 0.004 | 0.002 | 85.1 |
| Bound | 3.5 | 49.4 | 14.1 | 5.8 | 11.7 |

The oligonucleosomes fractionated on the anti-poly(ADP-Rib) column were derived from HeLa cells which had been labeled during growth with [$^3$H]arginine and [$^3$H]lysine. The nucleosomes were incubated under conditions optimal for poly(ADP-ribosyl)ation with [$^{32}$P]NAD prior to immuno-fractionation. After fractionation, the original sample bound and unbound samples were extracted for histones with 0.4N $H_2SO_4$. Gel electrophoretic analysis indicated that >90% of the [$^3$H] and [$^{32}$P] incorporation in these samples resided in histones.

*6. Poly ADP-ribosylated chromatin domains are enriched in DNA repair labeling*

HeLa cells were incubated in the presence or absence of MNU and subsequently "repair labeled" with [$^3$H]thymidine (TdR) (with hydroxyurea). Nucleosomes were prepared and immuno-fractionated as described above. The data (Fig. 8) indicate that there was considerable enrichment of repair label in the bound nucleosomes. Negligible repair label or chromatin was bound to a preimmune column.

*7. Do Poly ADP-ribosylated domains of chromatin selectively contain actively transcribed DNA sequences?*

We have begun to answer this important question by two approaches: directly investigating for the presence of active genes, and alternatively relating the poly (ADP-ribosyl)ation modification to acetylation and phosphorylation of histones. We are currently fractionating, by use of the poly(ADP-Rib) antibody column, polynucleosomal preparations from two sources: fibroblasts, where the collagen gene is transcriptionally active, and K562 human leukemia cells, induced for the globin gene transcription by cell incubation with hemin. We plan to use dot-blot hybrization with cDNA probes of globin and collagen genes. The data (not shown) indicate initial dot-blots showing that our nuclease digested preparations of starting material chromatin are enriched in both genes of interest. These experiments should ultimately ascertain whether or not active genes are proximal to poly ADP-ribosylated signals within cells.

The acetylation modification of histones has been intimately related to domains of chromatin undergoing transcription. We asked the question whether these same regions of chromatin were enriched in poly(ADP-ribosyl)ation. HeLa cells were labeled *in vivo* with [$^3$H]acetate for 30 min and subsequently chased with sodium butyrate for 18 hr to induce hyperacetylation. [$^3$H]acetylated nucleosomes (labeled *in vitro* with [$^{32}$P]NAD) were fractionated by the poly(ADP-Rib) antibody column.

It was of great interest that a significant fraction of acetylated chromatin appeared also to be poly ADP-ribosylated. The acetylated domains were not retained on a preimmune column, and their binding to polymer IgG column was decreased when poly(ADP-Rib) was removed from the chromatin with phosphodiesterase (Fig. 9).

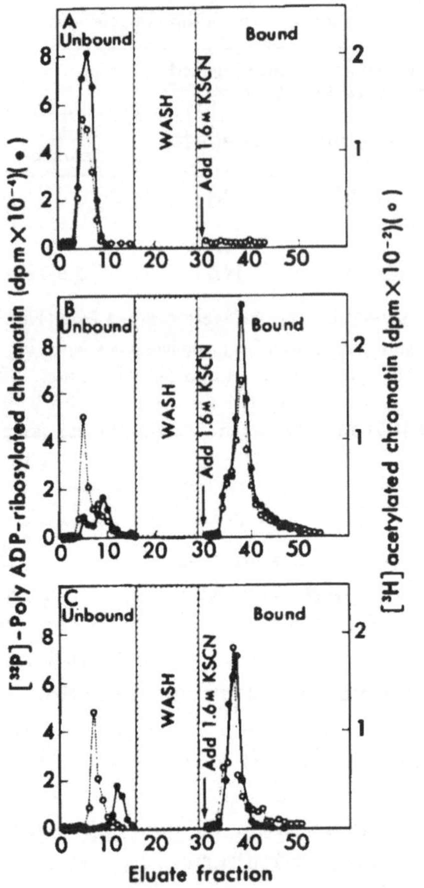


FIG. 9. Association of acetylation, poly(ADP-ribosyl)ation and HMG-17 presence in oligonucleosomes as determined by immuno-fractionation on anti-poly(ADP-Rib) IgG-Sepharose and anti-HMG IgG-Sepharose columns. HeLa cells ($4 \times 10^7$) were acetylated *in vivo* using 200 $\mu$Ci [$^3$H] acetate according to the method of Perry and Chalkley (*24*). The oligonucleosomes were isolated. [$^3$H] acetylated nucleosomes 0.75 $A_{260}$ were poly ADP-ribosylated using [$^{32}$P]NAD according to the method previously described (*13*). Subsequently, [$^3$H] acetylated and [$^{32}$P]-poly ADP-ribosylated nucleosomes were fractionated on (A) preimmune IgG-Sepharose, (B) anti-poly(ADP-Rib) IgG-Sepharose and (C) anti-HMG Sepharose columns.

## *Poly ADP-ribosylation Histone of H1: Relationship to H1 Phosphorylation*

Whitlock *et al.* (*19*) have found that acetylated forms of histone H3 are particularly susceptible to phosphorylation. With our recent development of affinity methods for the antibody directed against the modified moiety, poly(ADP-Rib), we thought it timely to test whether an immunological affinity method could be employed to bind selectively, those very limited subpopulations of histone H1 within chromatin which undergo ADP-ribosylation. Secondly, it was of importance to ascertain whether or not phosphorylation and poly(ADP-ribosyl)ation of histone H1 are coordinated within the cell.

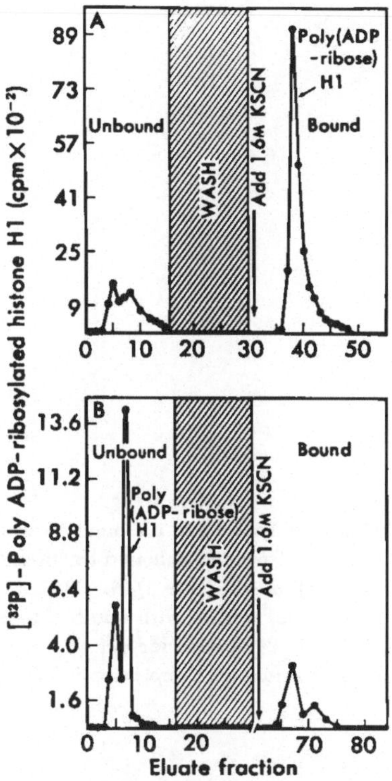


FIG. 10. Selective retention of poly ADP-ribosylated histone H1 by anti-poly(ADP-Rib) IgG-Sepharose. Nuclei ($1 \times 10^8$) were incubated with 50 $\mu$M [$^{32}$P]NAD (20 $\mu$Ci) and poly ADP-ribosylated H1 selectively extracted. The sample was subsequently applied to either anti-poly (ADP-Rib) IgG-Sepharose 4B (A) or preimmune IgG-Sepharose 4B (B). The column was washed with fourteen 1 ml of saline, the bound material was eluted with 1.6 M KSCN. The amount of radioactivity incorporated into acid-insoluble material was determined in selected fractions. (From ref. *25*).

*1. Selective affinity of poly ADP-ribosylated H1 by anti-poly(ADP-Rib)-Sepharose*

In the experiment performed in Fig. 10, nuclei were incubated with 50 $\mu$M [$^{32}$P]-NAD and selectively extracted for histone H1. Electrophoretic analysis of this sample (Fig. 11, lane 1) indicated the presence of the typical family of sequentially poly ADP-ribosylated H1 molecules, including H1 complex. This sample was applied to an anti-poly(ADP-Rib)-Sepharose column. The column was eluted with 1 ml fractions of phosphate buffered saline; a small flow-through peak of acid-insoluble radioactivity was noted in fractions 5–10. In addition, the bulk of the unmodified H1 sample was eluted in these early unbound fractions. An acid-insoluble radioactive peak was noted in fraction 36–42 which was subsequently (Fig. 11, below) identified as poly ADP-ribosylated histone H1 by electrophoretic analysis. As a control, the [$^{32}$P]poly ADP-ribosylated H1 fraction was chromatographed on a preimmune IgG-Sepharose column (Fig. 10B). In contrast to the selective retention noted in Fig. 10A, essentially the entire sample passed directly through the column, with only minimal non-specific binding.

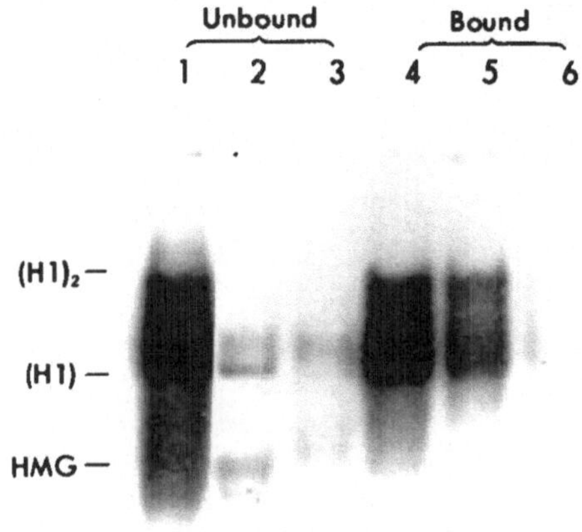


FIG. 11. Analysis of poly ADP-ribosylated H1 species bound and unbound by anti-poly(ADP-Rib) IgG-Sepharose 4B immunofractionation. Samples from the unbound fractions 4–6 (lane 2), 7–8 (lane 3) of unbound and fractions 37–38 (lane 4), 39–40 (lane 5), 41–42 (lane 6) of bound material in Fig. 10 were precipitated with 20% TCA and washed with acidified acetone/acetone. Electrophoresis of the sample on an acetic acid/urea polyacrylamide gel and autoradiography were performed. Lane 1 represents the original, unfractionated sample applied to the column in Fig. 10. (From ref. *25*).

*2. Characterization of immuno-purified H1*

The specifically bound [$^{32}$P]labeled samples (lanes 4–6) contained a series of poly ADP-ribosylated H1 species containing progressively longer polymer chains and extending to the H1 poly(ADP-Rib)-complex position in the gel (Fig. 11). It was of interest that the minimal quantity of radioactivity observed not to bind to the antibody column (fraction 4–8, Fig. 10A) appeared to be predominately H1 species with very short poly(ADP-Rib) units (lanes 2, 3). In addition, two other 5% perchloric acid soluble bands, possibly mono ADP-ribosylated high mobility group (HMG) proteins, present in small quantities in the unfractionated sample (lane 1), were not adsorbed by the antibody column.

The above experiments indicated that poly ADP-ribosylated species of histone H1 could be selectively examined by the immunoaffinity technique and further suggested that the method might be expanded toward studying the relationship between histone H1 phosphorylation and poly(ADP-ribosyl)ation.

*3. Binding of phosphorylated H1 to anti-poly(ADP-Rib)-Sepharose*

Nuclei were incubated with [$\gamma$-$^{32}$P] ATP in the presence (Fig. 12A) or absence (Fig. 12B) of cAMP and selectively extracted with 5% perchloric acid for histone H1. The samples were subsequently subjected to chromatography on anti-poly(ADP-Rib)-Sepharose as described earlier. Unexpectedly, greater than half of the phosphorylated H1 sample was observed to bind to the poly(ADP-Rib) antibody column (Fig. 12A, B).

The phosphorylated histone H1 was treated with snake venom phosphodiesterase. The sample contained 3,150 acid-insoluble cpm prior to treatment and 3,100

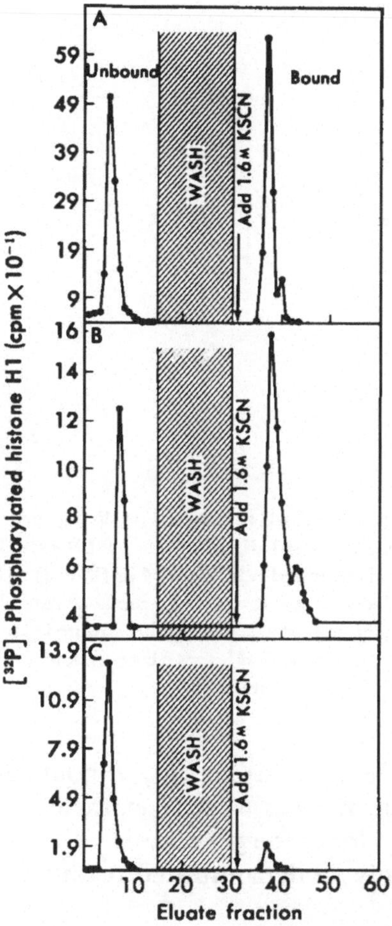


FIG. 12. Fractionation of phosphorylated H1 on anti-poly(ADP-Rib) IgG-Sepharose. A. $2\times10^8$ nuclei were incubated with 25 $\mu$Ci [$\gamma$-$^{32}$P]ATP and 10 $\mu$M cAMP for 10 min at 37°C. Histone H1 was selectively extracted with 5% perchloric acid and analyzed on the poly(ADP-Rib) antibody column. B. $2\times10^8$ nuclei were incubated with 25 $\mu$Ci [$\gamma$-$^{32}$P]ATP for 10 min at 37°C and processed as described in A. C. $2\times10^8$ nuclei were incubated with 25 $\mu$Ci [$\gamma$-$^{32}$P]ATP and extracted for histone H1 as described in A. Phosphorylated H1 was treated with 0.2 units of phosphatase free snake venom phosphodiesterase (Worthington) for 10 min at 25°C prior to loading onto the column. (From ref. *25*).

cpm after the treatment. Accordingly, little if any $^{32}$P-labeled poly(ADP-Rib) was present in the sample. Subsequently, the phosphodiesterase treated [$^{32}$P]ATP-labeled sample was re-chromatographed on the antibody column (Fig. 12C). In contrast to the earlier data, essentially all of the phosphorylated H1 sample eluted directly through the column. These experiments suggested that a large percentage of HeLa cell nuclear histone H1 occupy domains in chromatin accessible to both phosphorylation and poly(ADP-ribosyl)ation.

### 4. *The natural occurrence of H1-dimer*

We have recently utilized the poly(ADP-Rib) antibody to demonstrate the *in*

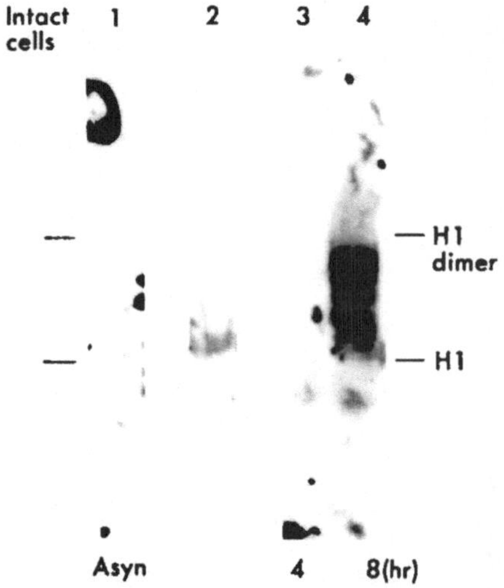


FIG. 13. The natural occurrence of poly(ADP-Rib)-H1 complex in living cells-relationship to the cell cycle. The histone H1 fraction was extracted directly from $1 \times 10^8$ asynchronous cells in duplicate (lanes 1 and 2) or from $1 \times 10^8$ synchronized cells, either 4 hr (lane 3) or 8 hr (lane 4) after the entry of cells into S phase. The cells were synchronized by methotrexate. The H1 fractions were electrophoresed and subsequently incubated with either diluted anti-poly(ADP-Rib) IgG fraction (lanes 1, 3, and 4), or diluted anti-H1 sera (lane 2) and [$^{125}$I]protein A. The transfers were exposed for autoradiography. (From ref. *20*).

*vivo* existence of poly ADP-ribosylated complexes of histone H1, without cell labeling or permeabilization techniques (*20*). Histone H1 was directly extracted from growing cells, separated by electrophoresis and a blot of the gel reacted with the antibody (Fig. 13). The data show that this could only be demonstrated readily at the S/G2 boundary of the cell cycle.

*Mechanism of H1 Dimer Synthesis: Use of Reconstitution and H1 Peptide Domains*

A major series of questions, still unanswered concerns the enzymatic mechanism by which the H1-dimer complex is constructed within chromatin. In the following work (*21*), we have attempted to gain an insight into this process by two related approaches: reconstitution of H1-depleted chromatin with defined fragments of H1 and subsequent *in vitro* poly(ADP-ribosyl)ation of the H1 peptides, and secondly, analysis of the polarity of poly(ADP-Rib) elongation on H1 (*i.e.*, amino or carboxyl growth) by selective NBS cleavage. With respect to the first approach, we have previously shown that H1-derived peptide which contained an intact folded (*i.e.*, globular) domain, subsequent to reconstitution, are correctly located on polynucleosomes (*22*). This was demonstrated by their ability to protect the H1-dependent 168 base pair DNA fragment during micrococcal nuclease digestion. In contrast, a C-terminal fragment, which lacks the globular domains, was found not to be correctly positioned subsequent to reconstitution (*22*).

When H1-depleted chromatin, reconstituted with peptide NG-H1 was incubated with 0.1 $\mu$M [$^{32}$P]NAD (Fig. 14, lane 4) incorporation was noted in a band which

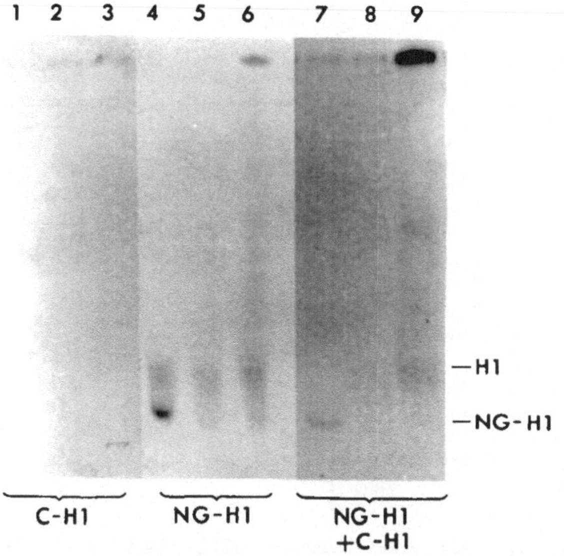


FIG. 14. Reconstitution of peptides C-H1, NG-H1, and a mixture of C-H1/NG-H1 to H1-depleted chromatin. Lane 1, oligonucleosomes reconstituted with C-H1 fragment were incubated with 0.1 $\mu$M [$^{32}$P]NAD for 30 sec; lane 2, with 10 $\mu$M [$^{32}$P]NAD for 5 min; lane 3, with 100 $\mu$M [$^{32}$P]NAD for 10 min; lane 4, oligonucleosomes reconstituted with peptide NG-H1 were incubated with 0.1 $\mu$M [$^{32}$P]NAD for 30 sec; lane 5, with 10 $\mu$M [$^{32}$P]NAD for 5 min; lane 6, with 100 $\mu$M [$^{32}$P]NAD for 10 min; lane 7, reconstituted oligonucleosomes with peptides NG-H1 and C-H1 were incubated with 0.1 $\mu$M [$^{32}$P]NAD for 30 sec; lane 8, with 10 $\mu$M [$^{32}$P]NAD for 5 min; lane 9, with 100 $\mu$M [$^{32}$P]NAD for 10 min. (From ref. *21*).

migrated in the same position as the unmodified NG-H1. This band presumably represents mono-ADP-ribosylated NG-H1. This is the first demonstration of *in vitro* poly (ADP-ribosyl)ation of a histone fragment. The data are consistent with previous demonstrations of primary modification sites on the N-terminal domain of this histone (*23*). With increasing concentration of NAD (Fig. 14, lanes 5, 6), a discrete series of more slowly migrating modified bands was observed. There was, however, no build up of a limit complex, as was observed with intact H1. NG-H1 associated poly(ADP-Rib) chains extending to 11–12 residues in length, (the complex has been shown to contain a chain of 15–16 units) were clearly resolved on the gel. These data suggest that the intact H1 molecule is not required for recognition by the polymerase, and secondly that initiation may occur at the N-terminal domain of H1 with elongation of chains extending to close to full length chains.

When H1-depleted polynucleosomes were reconstituted with C-H1 (Fig. 14, lanes 1–3) no incorporation of (ADP-Rib) was observed. The use of $^3$H-labeled C-H1 binding procedures, and the extraction of fragment from purified reconstituted samples both indicated that C-H1 had associated with depleted nucleosomes under these conditions (data not shown). The presence (Fig. 14, lanes 7–9) of NG-H1 and C-H1 (*i.e.*, the complementary total components of H1) did not alter the pattern noted with NG-H1 alone. The use of these reconstitution techniques have established that the presence of the C-H1 domain maintained intact in H1, is necessary for the formation of a limit complex. Experiments below, using complementary approach, utilizing

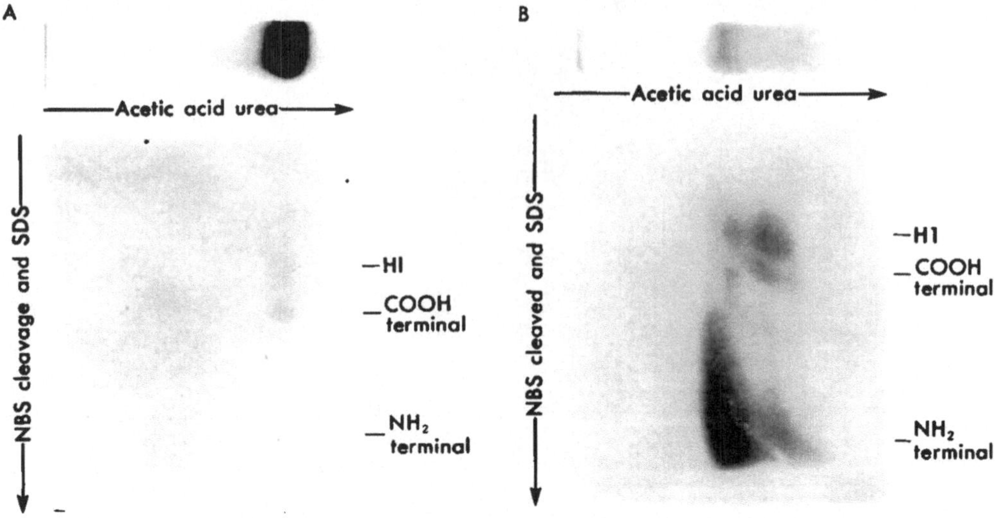


FIG. 15. Mapping segments of poly ADP-ribosylated $(H1)_2$ which participates in the cross-linking. Poly ADP-ribosylated H1 were prepared and electrophoresed in the first dimension on an acetic acid/urea polyacrylamide gel. The gel strip was subsequently dehydrated, incubated 0.7 mg/ml NBS for 1 hr and loaded onto a second dimension $NaDodSO_4$ polyacrylamide. Shown are autoradiograms of the gel. A. Pulse; histone H1 extracted from nuclei incubated with 0.1 $\mu$M [$^{32}$P]NAD. B. Chase; with 100 $\mu$M [$^{32}$P]NAD. (From ref. *21*).

intact H1, indicated that some minimal ADP-ribosylation of the C-terminal end occurs early in the initiation of this modification, however, as before, major elongation occur on the N-terminal domains of H1. In an attempt to clarify further which termini of the H1 is utilized for initiation and elongation of complex synthesis, a two-dimensional gel analysis involving a first dimension gel analysis in acetic acid/urea, followed analysis of NBS cleavage products in a second dimension SDS polyacrylamide gel was performed (Fig. 15).

In the experiment in Fig. 15, a labeled H1 sample, composed predominately of $^{32}$P histone H1 dimer as indicated by the first-dimensional analysis, was cleaved with NBS in the gel, and then subjected to a second dimension analysis by SDS gel electrophoresis. Two major labeled NBS cleavage products were noted. From their mobility in the SDS gel, it is assumed, that the fragments with slower mobility corresponded to C-terminal fragment produced by the cleavage. Label was noted in both cleavage products, suggesting initial ADP-ribosylation in both terminal domains of the H1 molecule at early stages in H1 complex synthesis. As the poly(ADP-Rib) chain became progressively longer, ultimately approaching the dimer, the amount of label existing in the C-terminal fragment remained relatively constant, whereas there was a marked increase in the labeled incorporation into the N-terminal cleavage product. In addition, the mobility of the N-terminal fragment became slower, suggesting that longer poly(ADP-Rib) chains were associated with these molecules. The data are suggestive, but not conclusive at this time, that the limit linkage occurs between the extended N-terminal chain and a singly modified C-terminal residue.

A knowledge of poly(ADP-ribosyl)ation of histone H1 should help complement the variety of other approaches taken by our laboratory, as well as many other colleagues in the field, towards understanding the biological function of this important nuclear enzyme system.

ACKNOWLEDGMENTS

This work was supported by NIH grants CA 25344 and CA 13195. The author wishes to gratefully thank Dr. Michael Bustin for histone antibodies and Drs. Sugimura, Kanai, and Miwa for poly(ADP-Rib) antibody. The excellent typing of Ms. Faith Keenan is gratefully appreciated.

REFERENCES

1. Butt, T. R., Jump, D. B., and Smulson, M. E. Nucleosome periodicity in HeLa cell chromatin as probed by micrococcal nuclease. Proc. Natl. Acad. Sci. U.S., *76*: 1628–1632, 1979.
2. Butt, T. R. and Smulson, M. E. Relationship between nicotinamide adenine dinucleotide concentration and *in vitro* synthesis of poly(adenosine diphosphate ribose) on purified nucleosomes. Biochemistry, *19*: 5239–5242, 1980.
3. Kanai, Y., Miwa, M., Matsushima, T., and Sugimura, T. Studies on anti-poly (adenosine diphosphate ribose) antibody. Biochem. Biophys. Res. Commun., *59*: 300–306, 1974.
4. Malik, N., Bustin, N., and Smulson, M. Antibody to poly(ADP-Rib) polymerase and its use in chromatin analysis. Nucleic Acids Res., *10*: 2939–2950, 1982.
5. Butt, T. R., DeCoste, B., Jump, D. B., Nolan, N., and Smulson, M. E. Characterization of a putative poly(adenosine diphosphate ribose) chromatin complex. Biochemistry, *19*: 5243–5249, 1980.
6. Stone, P. R., Lorimer, W. S., III, and Kidwell, W. R. Properties of the complex between histone H1 and poly(ADP-ribose) synthesized in HeLa cell nuclei. Eur. J. Biochem., *81*: 9–18 1977.
7. Nolan, N. L., Butt, T. R., Wong, M., Lambrianidou, A., and Smulson, M. E. Characterization of poly(ADP-ribose) histone H1 complex formation in purified polynucleosomes and chromatin. Eur. J. Biochem., *113*: 15–25, 1980.
8. Jump, D. and Smulson, M. E., Purification and characterization of the major non-histone protein acceptor for poly(adenosine diphosphate ribose) in HeLa cell nuclei. Biochemistry, *19*: 1024–1031, 1980.
9. Jump, D., Butt, T. R., and Smulson, M. E. Reconstitution of HeLa cell poly (adenosine diphosphate ribose) polymerase with purified oligonucleosomal chromatin. Biochemistry, *19*: 1031–1037, 1980.
10. Ogata, N., Ueda, K., Kawaichi, M., and Hayaishi, O. Poly(ADP-ribose) synthetase, a main acceptor of poly(ADP-Ribose) in isolated nuclei. J. Biol. Chem., *256*: 4135–4137, 1981.
11. Jump, D. B., Butt, T. R., and Smulson, M. E. Nuclear protein modification and chromatin substructure. 3. Relationship between poly(adenosine diphosphate) ibosylation and different functional forms of chromatin. Biochemistry, *18*: 983–990, 1979.
12. Goldknopf, I. L., Taylor, C. W., Baum, R. M., Yeoman, L. C., Olson, M.O.J., Prestayko, A. W., and Busch, H. Isolation and characterization of protein A24, a "histone-like" non-histone chromosomal protein. J. Biol. Chem., *250*: 7182–7187, 1975.
13. Malik, N., Miwa, M., Sugimura, T., Thraves, P., and Smulson, M. E. Immuno-

affinity fractionation of the poly ADP-ribosylated domains of chromatin. Proc. Natl. Acad. Sci. U.S., *80*: 2554–2558, 1983.

14. Smulson, M. E., Schein, P., Mullins, D. W., Jr., and Sudhaker, S. A putative role for nicotinamide adenine dinucleotide-promoted nuclear protein modification in the antitumor activity of N-methyl-N-nitrosourea. Cancer Res., *37*: 3006–3012, 1977.
15. Sudhakar, S., Tew, K. D., and Smulson, M. E. Effect of 1-methyl-1-nitrosourea on poly(adenosine diphosphate ribose) polymerase activity at the nucleosomal level. Cancer Res., *39*: 1405–1410, 1979.
16. Durkacz, B. W., Omidiji, O., Gray, D. A., and Shall, S. (ADP-Ribose)$_n$ participates in DNA exision repair. Nature, *283*: 593–596, 1980.
17. Benjamin, R. C. and Gill D. M. ADP-ribosylation in mammalian cell ghosts. Dependence of poly(ADP-ribose) synthesis on strand breakage in DNA. J. Biol. Chem., *255*: 10493–10501, 1980.
18. Modak, S. P. and Beard, P. Analysis of DNA double- and single-strand breaks by two dimensional electrophoresis: action of micrococcal nuclease on chromatin and DNA, and degradation *in vivo* of lens fiber chromatin. Nucleic Acids Res., *8*: 2665–2678, 1980.
19. Whitlock, J. P., Jr., Augustine, R., and Schulman, H. Calcium-dependent phosphorylation of histone H3 in butyrate-treated HeLa cells. Nature, *287*: 74–76, 1980.
20. Wong, M., Kanai, Y., Miwa, M., Bustin, M., and Smulson, M. Immunological evidence for the *in vivo* occurrence of a crosslinked complex of poly ADP-ribosylated histone H1. Proc. Natl. Acad. Sci. U.S., *80*: 205–209, 1983.
21. Wong, M., Allan, J., and Smulson, M. The mechanism of histone H1 crosslinking by poly ADP-ribosylation: reconstitution with peptide domains. In preparation.
22. Allan, J., Hartman, P. G., Crane-Robinson, C., and Aviles, F. X. The structure of histone H1 and its location in chromatin. Nature, *288*: 675–679, 1980.
23. Ogata, N., Ueda, K., Kagamiyama, H., and Hayaishi, O. ADP-ribosylation of histone H1. J. Biol. Chem., *255*: 7616–7620, 1980.
24. Perry, M. and Chalkley, R. Histone acetylation increases the solubility of chromatin and occurs sequentially over most of the chromatin. J. Biol. Chem., *257*: 7336–7343, 1982.
25. Wong, M., Miwa, M., Sugimura, T., and Smulson, M. The Relationship between histone H1 poly ADP-ribosylation and phosphorylation using anti-poly(ADP-Rib) antibody. Biochemistry, *22*: 2384–2389 (1983).

ADP-RIBOSYLATION, DNA REPAIR AND CANCER, M. MIWA ET AL. (EDS.), JAPAN SCI. SOC. PRESS, TOKYO/VNU SCIENCE PRESS, UTRECHT, PP. 71-81, 1983

# Some Electron Microscopic Aspects of Poly(ADPR) Polymerase-DNA Interactions and of Auto-poly(ADP-ribosyl)ation Reaction

P. MANDEL,*[1] J. JONGSTRA-BILEN,*[1] M. E. ITTEL,*[1] G. de MURCIA,*[2] E. DELAIN,*[3] C. NIEDERGANG,*[1] and H. P. VOSBERG*[4]

*Centre de Neurochimie du CNRS, 67084 Strasbourg Cédex,*[1] Laboratoire de Biophysique de l'Institut de Biologie Moléculaire et Cellulaire, 67084 Strasbourg Cedex,*[2] Institut Gustave-Roussy, 94800 Villejuif (Val de Marne), France,*[3] and Max Planck Institut für Medizinische Forschung, Abteilung für Molecular Biologie, D-6900 Heidelberg, West Germany*[4]*

*Abstract:* Interaction of calf thymus poly(ADP-ribose(ADPR)) polymerase with a copurified DNA fraction (sDNA) was investigated. Electron microscopic studies of sDNA which appeared to be a powerful poly(ADPR) polymerase activator have shown that the purified poly(ADPR) polymerase-DNA complexes possess a "nucleosome like structure", with DNA wrapping around the enzyme molecule.

Examination of the DNA linked poly(ADPR) polymerase preparations revealed the presence of Y-structures in sDNA. The enrichment in the sDNA fraction of the Y shape DNA suggests the existence of replication fork structures in the poly(ADPR) polymerase linked DNA and or in the vicinity of the enzyme.

With increasing auto-poly(ADP-ribosyl)ation the enzyme molecule becomes much denser, increases in size and detaches from the DNA. When poly(ADPR) formed was purified and examined by electron microscopy, branched polymers of different sizes were observed. The formation of these polymers may explain the size gained by poly ADP-ribosylated enzyme molecules.

When the interaction of poly(ADPR) polymerase with the plasmid pBR 322 was tested, a slight contamination of our enzyme preparation with topoisomerase I was detected. The contaminant topoisomerase I activity, however, was completely abolished by ADP-ribosylation. Further experiments with purified calf thymus topoisomerase I confirmed that this enzyme loses its activity following ADP-ribosylation with poly(ADPR) polymerase. These results may suggest that ADP-ribosylation of topoisomerase I can be one of the regulatory mechanisms of its activity. Furthermore, these results confirm that a topoisomerase I contaminant does not interfere with the ADP-ribosylation experiments of purified poly(ADPR) polymerase preparation.

Since the discovery of poly(ADP-ribose) polymerase (poly ADPRP) (*1*) its biological role remains unclear (see for review *2–6*). Attempts to correlate the activity of the enzyme with a particular cellular state could hardly establish a causative relationship. Enzyme activity measurements *in vitro* express a potentiality rather than an effective metabolic level *in vivo*. The studies employing inhibitors have proved to be of limited significance since the inhibitor's specificity is often questioned. The results obtained with isolated nuclei, apparently an interesting model to investigate the role of poly(ADPR) polymerase, may support that ADP-ribosylation has either no effect or decreases, or increases dTTP incorporation (see for review *2–6*). It was suggested by Berger *et al.* (*7*) that permeabilized cells may be a good model for studying ADP-ribosylation. These authors conclude that inhibition of poly(ADPR) polymerase has effect on DNA synthesis and that inhibitors of DNA synthesis have no effect on poly-(ADPR) synthesis in permeabilized L cells. It also appeared that the rate of synthesis of poly(ADPR) was greater in stationary cells than in exponential phase cells (see for review *5*). However, phytohemoagglutinine stimulation of normal and CLL lymphocytes causes an increase of ADP-ribosylation in parallel to an increase of thymidine incorporation (*8*, *9*). According to Bredehorst *et al.* (*10*) transition from exponential to stationary phase of Ehrlich ascitic tumor cells is associated with an increase in both intrinsic poly(ADPR) polymerase activity and hydroxylamine resistant mono-ADP-ribosylated proteins. It was also reported that NAD depleted cells were able to proliferate but unable to repair DNA damage (*11*, *12*). Nevertheless, evidence has accumulated suggesting that ADP-ribosylation of nuclear proteins seems to play a role in repair of DNA although the precise mechanism remains to be known (*13*). Moreover, conflicting reports have emerged to assign a role played by ADP-ribosylation in DNA transcription (*14–16*) (see also for review *5*, *6*).

In view of the conflicting data and the lack of direct causative and mechanistic evidence concerning the involvement of poly(ADPR) polymerase in multiple nuclear functions it appeared useful to visualize some morphological aspects of poly (ADPR) polymerase activity on the chromatin level.

Since we have succeeded in copurifying thymus poly(ADPR) polymerase together with a DNA fraction (sDNA) which stimulates poly(ADPR) synthesis 100 times as much as bulk thymus DNA (*17*, *18*), we have examined morphological aspects of the activity of the poly(ADPR) polymerase DNA complex. Thus, the electron microscopic aspects 1) of the localisation of sDNA *versus* poly(ADPR) polymerase, 2) the automodification of poly(ADPR) polymerase during poly(ADPR) synthesis, and 3) the interaction of poly(ADPR) polymerase with sDNA were investigated (*19*). The effect of ADP-ribosylation on nucleosomal structure (the basic unit of chromatin) was also examined (*20*). The influence of ADP-ribosylation on one of the topoisomerases controlling the topological state of DNA: topoisomerase I was studied.

### *Interaction between Poly(ADPR) polymerase and DNA*

Figure 1 shows the morphological features of the poly(ADPR) polymerase preparation. The enzyme (Fig. 1A) appears as globular dark dots irregularly spaced on the specific DNA (sDNA) which is heterogenous in size. From the comparison between the DNA deprived enzyme (Fig. 1B) and DNA linked enzyme (Fig. 1A) as observed in bright field, it appears that the electron density of the enzyme deprived of its sDNA is less than that of the DNA bound enzyme. The external diameter of poly(ADPR) polymerase is significantly smaller in the former (11.5±1.5 nm) (Fig. 1B) than in the latter (14.5±1.5 nm) (Fig. 1A). This difference may account for the

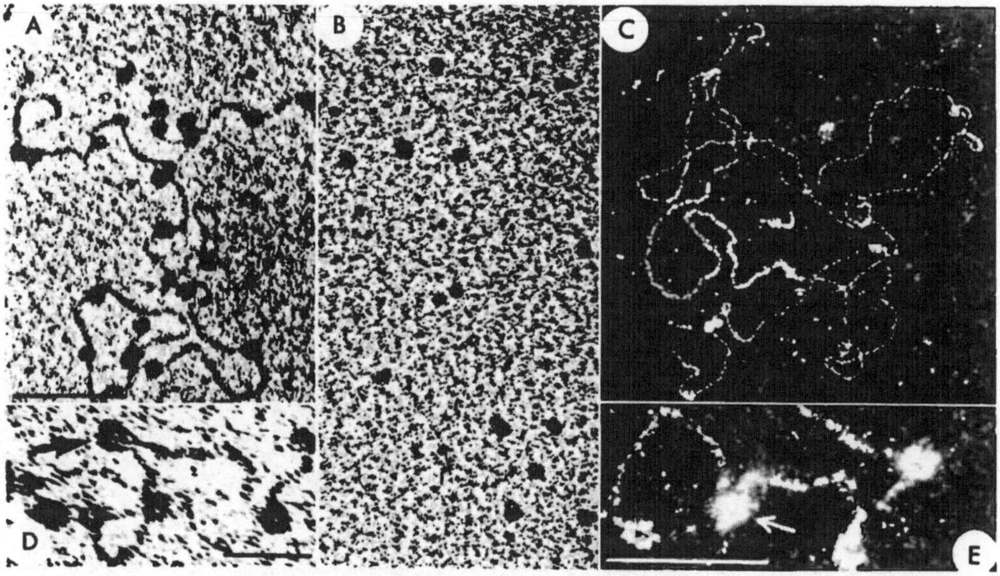


FIG. 1. Poly(ADPR) polymerase (DNA associated) visualized: A) in bright field (*22*); C) in dark field (*21*); D, E inserts of A and C (respectively). Arrows indicate the wrapping of DNA around the enzyme molecule. The bars indicate 50 nm. B. DNA-deprived enzyme (enzyme without DNA) visualized in bright field (*22*). The same bar indicates 100 nm in A, B, and C. Poly(ADPR) polymerase was prepared according Mandel *et al.* (*17*).

presence of one or more turn(s) of DNA wrapped around the poly(ADPR) polymerase molecule.

Actually the DNA linked poly(ADPR) polymerase when observed by the dark field mode (Fig. 1C, E) shows that most of the uranyl acetate, which stains DNA, is located at the periphery of the enzyme (Fig. 1E, arrow), as in the case of nucleosomes observed by the same technique (*23, 24*). These results suggest an external binding of the sDNA to the enzyme.

Moreover, when the DNA linked poly(ADPR) polymerase was digested with micrococcal nuclease and submitted to gel electrophoresis, a part of sDNA appeared to be protected by the enzyme like nucleosomal DNA (data not shown).

### *Some Ultrastructural Aspects of sDNA*

Examination of the DNA linked poly(ADPR) polymerase preparations (*18*) (Fig. 2) revealed the presence of Y structures in sDNA as demonstrated in Fig. 2A (arrows). To rule out possible spreading artefacts due to occasional binding of a protein to two DNA molecules, fully deproteinized sDNA was spread using the cytochrome *c* spreading technique (*25*) (Fig. 2B, C). The existence of Y shaped DNA molecules is supported by the results shown in Fig. 2B and 2C. The proportion of Y structure was roughly evaluated by counting the number of these structures per electron micrographs field (12 $\mu m^2$) at the magnification of 20,000. On the average, one Y structure per $\mu m^2$ negative was found. Total calf thymus DNA was spread under the same conditions. In this case, no such structures were detected. The en-

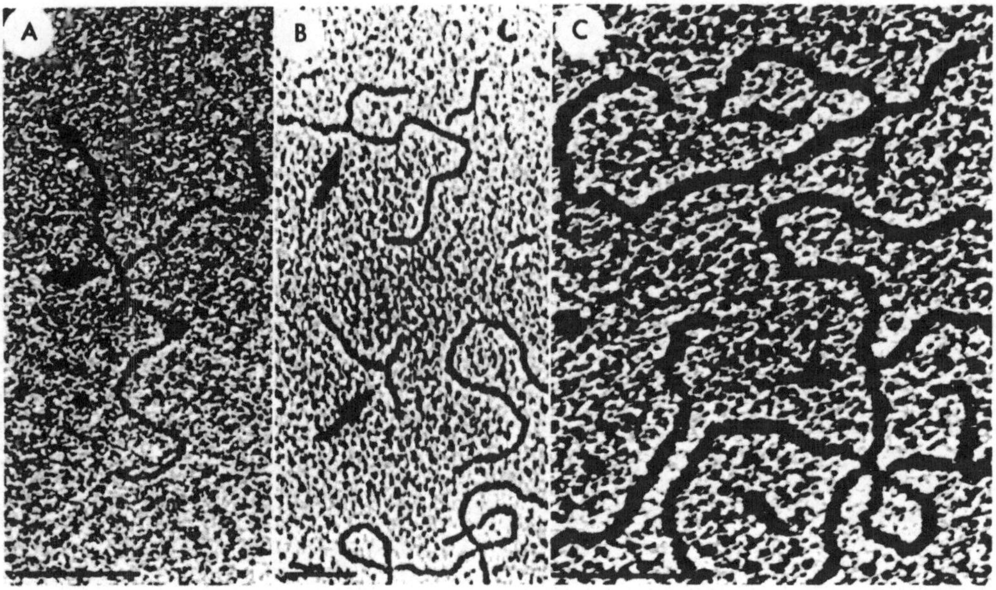


FIG. 2. Visualization of replication forks (indicated by arrows). A) in sDNA of DNA-linked enzyme in bright field (*22*). B, C) in deproteinized sDNA by cytochrome *c* method (*25*). The bars indicate 100 nm.

richment of the sDNA fraction in the Y shape DNA suggests the existence of replication fork structures in the poly(ADPR) polymerase linked DNA or in the vicinity of the enzyme molecules.

In order to understand the physical properties of sDNA as well as the high rate of stimulation of poly(ADPR) polymerase by sDNA (*18*, *26*) the UV spectroscopic properties of sDNA have been compared to those of calf thymus DNA. It appeared that both DNAs have practically the same base composition (when Tm values are compared). The lower hyperchromicity value (37%) as compared to the control (41%) suggests that sDNA contains denatured or single stranded DNA regions.

The effect of digestion by $S_1$ endonuclease is under investigation in order to confirm the presence of single stranded DNA fragments in correlation with structures presumed to be replication forks.

### *Automodification of Poly(ADPR) polymerase*

When purified DNA linked poly(ADPR) polymerase was incubated with 100 $\mu$M NAD under varying time of incubation at 37°C a plateau value was reached at 45 min. This shows an extensive auto ADP-ribosylation of the enzyme as reported earlier (*18*, *27*). Cold NAD (100 $\mu$M) was used in parallel experiments in order to visualize the automodification reaction by electron microscopy.

As observed in the dark field (Fig. 3) DNA bound enzyme with the linked poly-(ADPR) chains becomes more dense (in terms of staining) after 15 min of ADP-ribosylation (Fig. 3B) when compared to the control (Fig. 3A). Poly(ADPR) polymerase molecules become surrounded by numerous points of uranyl acetate which

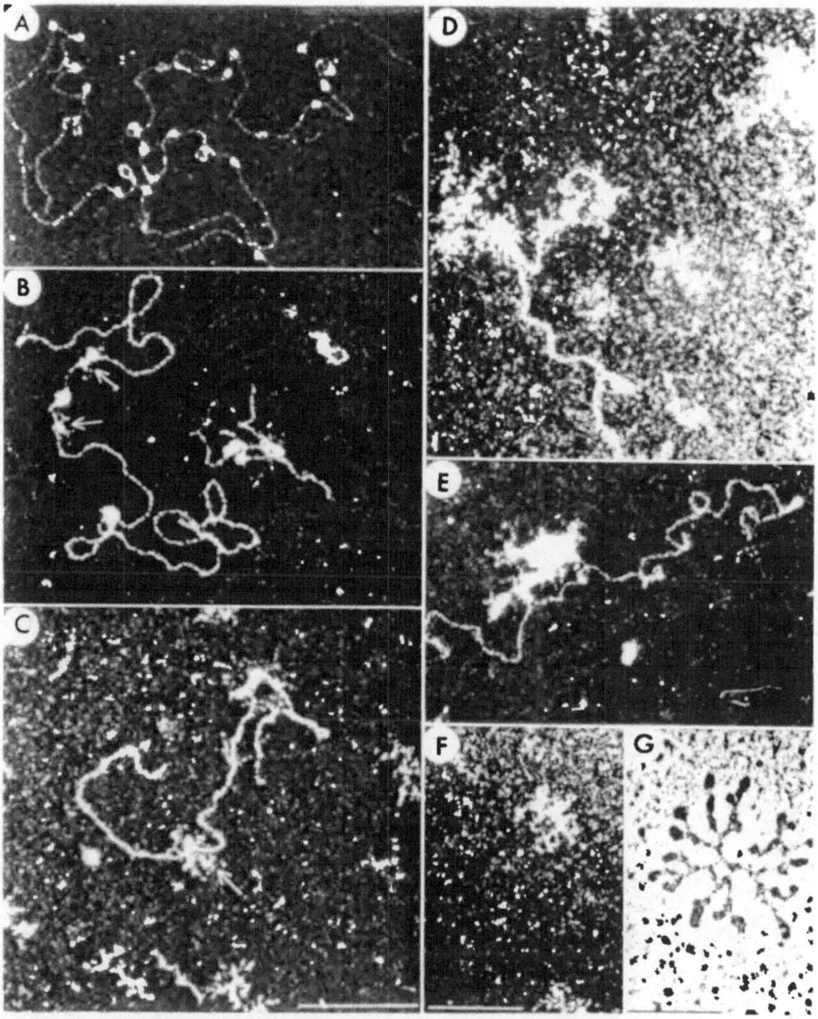


FIG. 3. Visualization of the kinetics of auto ADP-ribosylation at 37°C in presence of 100 $\mu$M NAD (in dark field) and of the purified product; poly(ADPR) A) $t=0$ min; B) $t=15$ min; C) $t=30$ min; D) $t=45$ min; E) $t=60$ min. F, G) purified poly(ADPR) after 60 min of auto ADP ribosylation. F) using Dubochet's method in dark field (*21*), G) using cytochrom *c* spreading (*25*). The bars indicate 100 nm.

also stains the products of the reaction (see arrow in Fig. 3B, C). After 45 min of ADP-ribosylation the electron density of the enzyme molecules increases (Fig. 3D), some material is detached from DNA and adsorbed onto the carbon film. With increasing time (Fig. 3E) the quantity of this material increases. In order to visualize only the product of the automodification reaction: poly(ADPR), the DNase I and proteinase K resistant material was observed by the dark field technique (Fig. 3F) and by the cytochrome *c* spread technique (Fig. 3G). The star like structures observed by both techniques were not seen when the control enzyme fraction (Fig. 3A) was treated in the same manner either directly or after incubation in the absence of NAD. Due to the difference in resolution of the cytochrome *c* and the dark field methods (protein free spreading) the branching of the polymer appears only in Fig. 3G. The number of branches increases with the size of the synthesized polymer. These branches are up to 100 nm long for the largest molecules observed.

### *Poly(ADP-ribosyl)ation of Polynucleosomes*

When native pancreatic chromatin was incubated with NAD and poly(ADPR) polymerase in the requisite medium a relaxation of the native structure started after 1 min reaching a plateau within 7 min. The polynucleosomes occupy a greater area as compared to the control, *i.e.*, incubation in the presence of poly(ADPR) polymerase without NAD. The native zigzag structure of chromatin became converted to a-beads-on-a-string structure which is characteristic of $H_1$ depleted chromatin (Fig. 4) (*20*). Under conditions of high ionic strength no condensation of polyADP-ribosylated chromatin was observed (*20*) as compared to native chromatin which folds towards the solenoid conformation as described by Thoma *et al.* (*28*). It was controlled that the relaxation of the chromatin is not due to spreading artefact. Sedimentation velocity analysis also confirmed the relaxed state of poly ADP-ribosylated nucleosomes as compared to condensed forms at the same ionic strength ($s^0_{20,w}$ of 59.4±1.55 control, 50.6±0.3 S poly ADP-ribosylated nucleosomes). When poly ADP-ribosylated material was analysed by sucrose gradient centrifugation, gel electro-

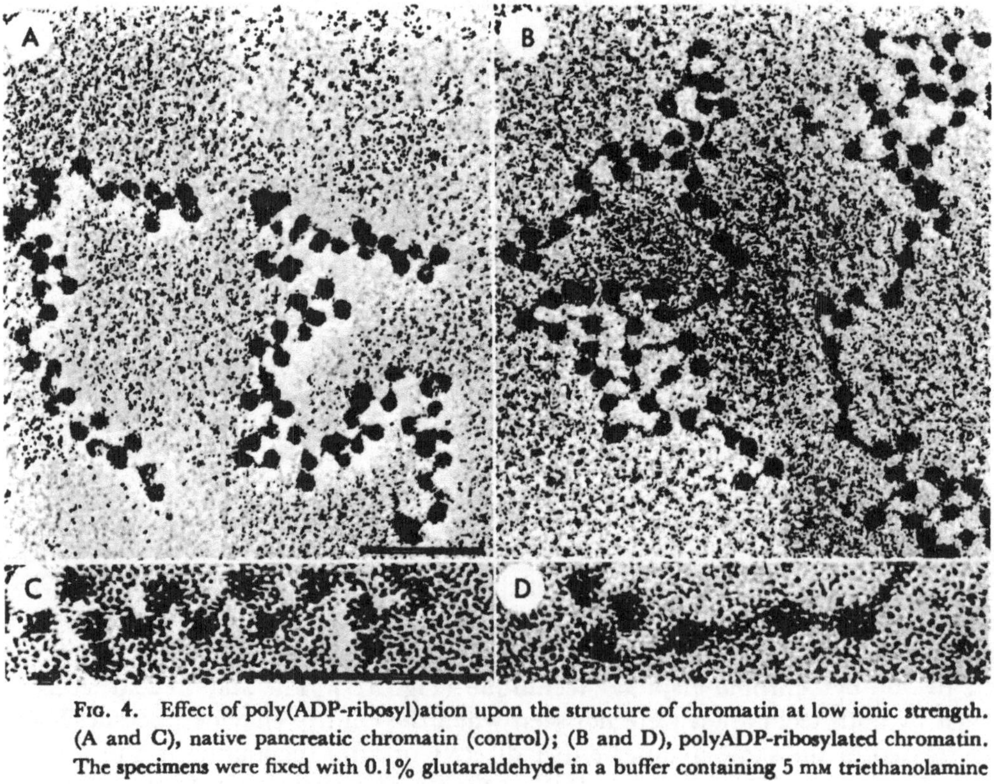


FIG. 4. Effect of poly(ADP-ribosyl)ation upon the structure of chromatin at low ionic strength. (A and C), native pancreatic chromatin (control); (B and D), polyADP-ribosylated chromatin. The specimens were fixed with 0.1% glutaraldehyde in a buffer containing 5 mM triethanolamine at pH 7.4 and 0.2 mM EDTA. The bars indicate 100 nm.

phoresis, and autoradiography after polynucleosome ADP-ribosylation using [$^{32}$P] NAD, the main ADP-ribosylated histones were found to be histone $H_1$ with several $H_1$ complexes (*20*). Detachment of histone $H_1$ from the chromatin is unlikely because no histone $H_1$ was found on the top of the sucrose gradient loaded with polynucleosomes submitted to ADP-ribosylation (*20*).

### *ADP-ribosylation of Purified Topoisomerase I*

Figure 5 illustrates the increase of [$^{32}$P]ADP-ribosylation of topoisomerase I with increasing amounts of [$^{32}$P] NAD. The two labelled bands (T) below the position of poly(ADPR) polymerase were previously shown to contain topoisomerase activity (Vosberg, personal communication). Auto-ADP-ribosylation of poly(ADPR) polymerase (with 5 and 10 $\mu$M NAD), results in a slower migration of the enzyme molecule both in the presence and absence of topoisomerase I (Fig. 5, slots 2, 3 and 5, 6, respectively). ADP-ribosylation of topoisomerase I does not seem to change its relative migration on the gel, under the same conditions.

The activity of topoisomerase I decreases upon ADP-ribosylation (Fig. 6). The degree of inhibition is proportional to the amount of NAD used (1–100 $\mu$M, slots 1–3). There is no inhibition of activity when topoisomerase I was incubated under the same conditions without poly(ADPR) polymerase (slots 4–6). This indicates that the observed inhibition is not due to a thermal or other kind of inactivation during preincubation of topoisomerase I or due to the presence of NAD but can be attributed to the ADP-ribosylation of topoisomerase I molecule itself.

Purified polymer poly(ADP-ribose) could inhibit topoisomerase I activity only at a very high concentration (minimum 15 $\mu$g/$\mu$l). This indicates that the observed

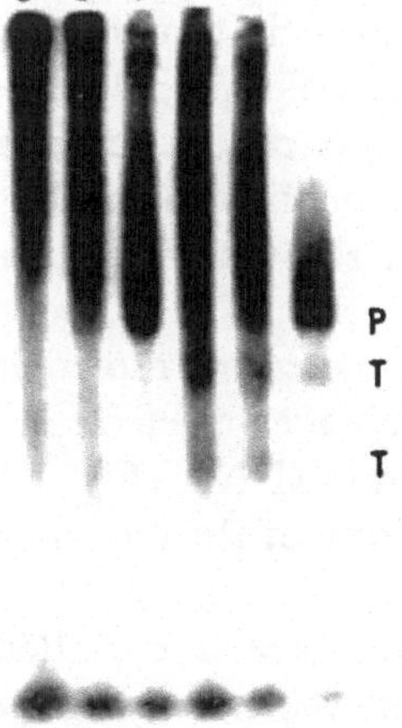


FIG. 5. Autoradiogram of SDS 7% polyacrylamide gel (*29*): slots 1–3, topoisomerase I after ADP-ribosylation; slots 4–6, poly(ADPR) polymerase DNA linked. ADP-ribosylation was performed under standard conditions with 1 $\mu$M NAD (slots 1, 4), 5 $\mu$M NAD (slots 2, 5), and 10 $\mu$M NAD (slots 3, 6). The positions of topoisomerase I (T) and of poly(ADPR) polymerase (P) are indicated. Topoisomerase I was prepared according to Prell and Vosberg (*30*).

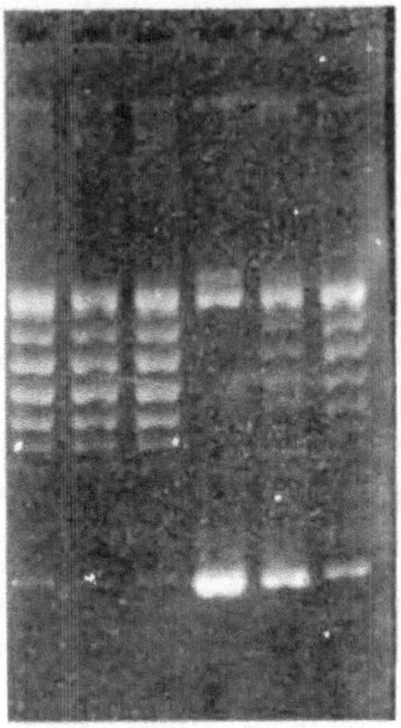


FIG. 6. Activity of topoisomerase I following its ADP-ribosylation with increasing amounts of NAD. Topoisomerase I was incubated with 1 $\mu$M NAD (slots 1, 4), 10 $\mu$M NAD (slots 2, 5), and 100 $\mu$M NAD (slots 3, 6) with (slots 1–3) and without (slots 4–6) DNA-linked poly(ADPR) polymerase. After 15 min of ADP-ribosylation aliquots were taken and topoisomerase activity was measured. The positions of supercoiled (s) and relaxed (r) plasmids with their respective dimer molecules (Ds and Dr) are indicated.

inhibition in the presence of poly(ADPR) polymerase is not due to the synthesis of poly(ADPR) in the incubation mixture (data not shown).

## DISCUSSION

It appears from our investigations that characteristic structural and functional interactions may exist between poly(ADPR) polymerase molecules and a DNA fraction copurified with the enzyme. The enzyme is apparently irregularly spaced on the DNA strands. This may suggest a differential localization related either to an activation phenomenon produced by specific DNA sequences or to involvement in the repair mechanism in vicinity of DNA nicks.

The existence of DNA turns wrapped around the enzyme molecule as in the case of nucleosomes is strongly suggested: a) by the differences in electron density and in the diameter of the enzyme either bound or deprived of sDNA (Fig. 1A, B), b) by the location of uranyl acetate stains mainly at the periphery of the enzyme molecule (Fig. 1C, E).

The sDNA copurified with poly(ADPR) polymerase is enriched in Y shaped structures (Fig. 2) which are not detected in total calf thymus DNA spread under the same conditions. This observation may suggest a selection of operating sites of DNA rare in whole calf thymus DNA, during the purification of poly(ADPR) polymerase-DNA complexes. Moreover, the enrichment of Y structures in sDNA further suggests a preferential distribution of poly(ADPR) polymerase in chromatin in vicinity of

DNA replication forks providing an additional evidence for the role of poly(ADPR) polymerase in DNA replication.

A simple addition of NAD is sufficient to start poly(ADPR) synthesis by the sDNA bound enzyme. As the reaction proceeds the electron density of the enzyme molecules and its environment increases. Finally the poly ADP-ribosylated enzyme is detached from the DNA strand suggesting a mechanism of arrest of the enzymatic activity. It is also noteworthy that the branching of the polymers previously described by Miwa *et al.* (*31*) appears clearly visible on electron microscopic preparations. The high charge of such structures may explain the repulsion from the DNA strands and detachment of the enzyme.

Although the precise cellular function of eukaryotic type I topoisomerase is unknown, topoisomerases have been suggested like poly(ADPR) polymerase to be involved in DNA replication, transcription, recombination, and repair (*32*).

Our observation that poly(ADP-ribosyl)ation is able to produce an inhibition of topoisomerase I activity suggests a possible involvement of ADP-ribosylation in regulation of the activity of this enzyme. In view of the fact that topoisomerase I can be found as a minor contaminant in preparations of DNA linked poly(ADPR) polymerase and that poly(ADPR) polymerase can be found as a contaminant of topoisomerase I preparations (*33*) may suggest a close localization of the two enzymes.

Studies concerning chromosomal structure and functions suggest a cascade regulation of the enzymatic activities attached to chromatin: involved in DNA replication, transcription and repair. It seems likely that ADP-ribosylation plays a basic role in this sequential regulation.

## REFERENCES


1. Chambon, P., Weill, J. D., Doly, J., Strosser, M. T., and Mandel, P. On the formation of a novel adenylic compound by enzymatic extracts of liver nuclei. Biochem. Biophys. Res. Commun., *25*: 638–643, 1966.
2. Hilz, H. and Stone, P. Poly(ADP-ribose) and ADP-ribosylation of proteins. Rev. Physiol. Biochem. Pharmacol., *76*: 1–58, 1976.
3. Hayaishi, O. and Ueda, K. Poly(ADP-ribose) and ADP-ribosylation of proteins. Annu. Rev. Biochem., *46*: 95–116, 1977.
4. Sugimura, T., Miwa, M., Saito, H., Kanai, Y., Ikejima, M., Terada, M., Yamada, M., and Utakoji, T. Studies of nuclear ADP-ribosylation. *In;* G. Weber (ed.), Advances in Enzyme Regulation 18, pp. 195–220, Pergamon Press, New York, 1980.
5. Purnell, M. R., Stone, P. R., and Whish, W. J. D. ADP-ribosylation of nuclear proteins. Biochem. Soc. Trans., *8*: 215–227, 1980.
6. Mandel, P., Okazaki, H., and Niedergang, C. Poly(adenosine diphosphate ribose). Prog. Nucleic Acid Res. Mol. Biol., *27*: 1–51, 1982.
7. Berger, N. A., Weber, G., and Kaichi, A. S. Characterization and comparison of poly-(adenosine diphosphoribose) synthesis and DNA synthesis in nucleotide-permeable cells. Biochim. Biophys. Acta, *519*: 87–104, 1978.
8. Rochette-Egly, C., Ittel, M. E., Bilen, J., and Mandel, P. Effect of nicotinamide on RNA and DNA synthesis and on poly(ADP-ribose) polymerase activity in normal and phytohemagglutinin stimulated human lymphocytes. FEBS Lett., *120*: 7–11, 1980.

9. Berger, N. A., Adams, J. W., Sikorski, G. W., Petzold, S. I., and Shearer, W. Synthesis of DNA and poly(adenosine diphosphate ribose) in normal and chronic lymphocytic leukemia lymphocytes. J. Clin. Invest., *62*: 111–118, 1978.
10. Bredehorst, R., Goebel, M., Renzi, F., Kittler, M., Klapproth, K., and Hilz, H. Intrinsic ADP-ribose transferase activity *versus* levels of mono(ADP-ribose) protein conjugates in proliferating Ehrlich ascites tumor cells. Hoppe-Seyler's Z. Physiol. Chem., *360*: 1737–1743, 1979.
11. Durkacz, B. W., Nduka, N., Omidiji, O., Shall, S., and Zia'ee, A. A. ADP-ribose and DNA repair. *In;* M. E. Smulson and T. Sugimura (eds.), Novel ADP-ribosylations of Regulatory Enzymes and Proteins. (Developments in Cell Biology, Vol. 6), pp. 207–216, Elsevier/North-Holland Publisher Co., Amsterdam, 1980.
12. Berger, N. A. and Sikorski, G. W. Poly(adenosine diphosphoribose) synthesis in ultraviolet-irradiated xeroderma pigmentosum cells reconstituted with micrococcus luteus UV endonuclease. Biochemistry, *20*: 3610–3614, 1981.
13. Creissen, D. and Shall, S. Regulation of DNA ligase activity by poly(ADP-ribose). Nature, *296*: 271–272, 1982.
14. Müller, W.E.G. and Zahn, R. K. Poly(ADP-ribosylation) of DNA-dependent RNA polymerase I from quail oviduct. Dependence on progesterone stimulation. Mol. Cell. Biochem., *12*: 147–159, 1976.
15. Yukioka, M., Okai, Y., Hasuma, T., and Inoue, A. Non-preferential localization of poly(ADP-ribose) polymerase activity in transcriptionally active chromatin. FEBS Lett., *86*: 85–88, 1978.
16. Walker, J. and Pearson, C. K. NAD, ADP-ribosylation and transcription in permeabilized mammalian cells. Biochem. J., *199*: 813–817, 1981.
17. Mandel, P., Okazaki, H., and Niedergang, C. Purification and properties of calf thymus polyadenosine diphosphate ribose polymerase. FEBS Lett., *84*: 331–336, 1977.
18. Niedergang, C., Okazaki, H., and Mandel, P. Properties of purified calf thymus poly(adenosine diphosphate ribose) polymerase. Eur. J. Biochem., *102*: 43–57, 1979,
19. De Murcia, G., Jongstra-Bilen, J., Ittel, M. E., Mandel, P., and Delain, E. Poly (ADP-ribose) polymerase auto-modification and interaction with DNA: electron microscopic visualization. EMBO J., *2* (No. 5): 543–548, 1983.
20. Poirier, G. G., De Murcia, G., Jongstra-Bilen, J., Niedergang, C., and Mandel, P. Poly(ADP-ribosylation) of polynucleosomes causes relaxation of chromatin structure. Proc. Natl. Acad. Sci. U.S., *79*: 3423–3427, 1982.
21. Dubochet, J., Ducommun, M., Zollinger, M., and Kellenberger, E. A new preparation method for dark field electron microscopy of biomacromolecules. J. Ulstrastruct. Res., *35*: 147–167, 1971.
22. De Murcia, G. and Koller, Th. The electron microsocpic appearance of soluble rat liver chromatin mounted on different supports. Biol. Cell., *40*: 165–174, 1981.
23. Saragosti, S., Moyne, G., and Yaniv, M. Absence of nucleosomes in a fraction of SV 40 chromatin between the origin of replication and the region coding for the late leader RNA. Cell, *20*: 65–73, 1980.
24. Moyne, G., Freeman, R., Saragosti, S., and Yaniv, M. A high-resolution electron microscopy study of nucleosomes from simian virus 40 chromatin. J. Mol. Biol., *149*: 735–744, 1981.
25. Davis, R. W., Simon, M., and Davidson, N. D. Electron microscope heteroduplex methods for mapping regions of base sequence homology in nucleic acids. Methods Enzymol., *21B*: 413–428, 1971.
26. Hashida, T., Ohgushi, H., and Yoshihara, K. Highly effective DNA in stimulating

poly(ADP-ribose) polymerase reaction. Biochem. Biophys. Res. Commun., *88*: 305–311, 1979.

27. Kawaichi, M., Ueda, K., and Hayaishi, O. Multiple autopoly(ADP-ribosyl)ation of rat liver poly(ADP-ribose) Synthetase. J. Biol. Chem., *256*: 9483–9489, 1981.
28. Thoma, F., Koller, Th., and Klug, A. Involvement of histone H1 in the organization of the nucleosome and of the salt-dependent superstructures of chromatin. J. Cell Biol., *83*: 403–427, 1979.
29. Laemmli, U. K. Cleavage of structural proteins during the assembly of the head of bacteriophage $T_4$. Nature, *227*: 680–685, 1970.
30. Prell, B. and Vosberg, H-P. Analysis of covalent complexes formed between calf thymus DNA topoisomerase and single-stranded DNA. Eur. J. Biochem., *108*: 389–398, 1980.
31. Miwa, M., Ishihara, M., Takishima, S., Saikawa, N., Tanaka, M., Takasuka, N., Maeda, M., Yamaizumi, Z., Nishimura, S., Saito, H., Sugimura, T., Yokoyama, S., and Miyazawa, T. Novel structure of poly(ADP-ribose) with a branch linkage. *In;* M. E. Smulson and T. Sugimura (eds.), Novel ADP-ribosylations of Regulatory Enzymes and Proteins. (Developments in Cell Biology, Vol. 6), pp. 1–10, Elsevier/North-Holland Publisher Co., Amsterdam, 1980.
32. Gellert, M. DNA Topoisomerases. Annu. Rev. Biochem., *50*: 879–910, 1981.
33. Jongstra-Bilen, J., Ittel, M. E., Niedergang, C., Vosberg, H-P., and Mandel, P. DNA topoisomerase I from calf thymus is inhibited *in vitro* by poly ADP-ribosylation. Eur. J. Biochem., in press.

ADP-RIBOSYLATION, DNA REPAIR AND CANCER, M. MIWA ET AL. (EDS.), JAPAN SCI. SOC. PRESS, TOKYO/VNU SCIENCE PRESS, UTRECHT, PP. 83-91. 1983

# Nucleosomal Poly(ADP-ribose) Polymerase: Properties and Relaxation of the Chromatin Structure

R. AUBIN,*[1] A. FRÉCHETTE,*[1] G. de MURCIA,*[2] F. MALOUIN,*[1] A. LORD,*[1] P. MANDEL,*[3] and G. G. POIRIER*[1]

*Chromatin Research Unit, Centre de Recherche sur les Mécanismes de Sécrétion, Faculté des Sciences, Université de Sherbrooke, Sherbrooke, Québec, Canada J1K 2R1,*[1] Institut de Biologie Moléculaire et Cellulaire du CNRS, Laboratoire de Biophysique, 15 rue René Descartes,*[2] and Centre de Neurochimie du CNRS, 5 rue Blaise Pascal,*[3] 67084 Strasbourg Cédex, France*

*Abstract:* The properties of poly(ADP-ribose) polymerase associated to nucleosomes and nuclei were analyzed by measuring enzymatic activity, isolating poly(ADP-ribosylated) histones and by electron microscopy. The enzyme was found to be most active in tri- and tetranucleosomes. The presence of NaCl (0 to 200 mM) in the incubation medium showed a very small inhibitory effect on the enzymatic activity of isolated nucleosomes; however, the enzyme activity of incubated nuclei showed a gradual reduction with increasing salt concentrations leading to an almost complete inhibition at 200 mM NaCl. In parallel, the histones H1 extracted from isolated nucleosomes incubated at 200 mM NaCl were hyper(ADP-ribosylated), whereas the histones H1 extracted from nuclei chromatin showed no highly modified forms at 200 mM NaCl. The effect of ethidium bromide (EB) was also analyzed and was found to stimulate enzyme activity in tetranucleosomes and octanucleosomes but not in dinucleosomes and nucleosomes of 20 repeat units. Electron microscopic analysis of poly(ADP-ribosylated) nucleosomes does not show the presence of aggregates at low ionic strength. However at higher ionic strengths the presence of localized relaxed domains of chromatin is evidenced in condensed nucleosomes.

Poly(ADP-ribose) polymerase is a chromatin bound enzyme which, from the coenzyme NAD, will catalyze the covalent attachment of ADP-ribose units to various nuclear proteins (*1–3*). This enzyme, located in the nucleosomal architecture (*4*, *5*), has been shown to modulate architectural changes in chromatin (*6*, *7*).

Our studies have shown that, at concentrations above 10 $\mu$M NAD, histone H1 is the major histone acceptor of ADP-ribose subunits in pancreatic nucleosomes (*8*). Also, H1 ADP-ribosylation causes a change in the affinity of this histone for DNA as observed by Burzio *et al.* (*9*). These results prompted us to analyze the changes in chromatin structure caused by a selective histone H1 poly(ADP-ribosyl)ation. Thus, in order to have maximal effect of histone H1 ADP-ribosylation, we incubated nucleosomes in the presence of exogenous purified poly(ADP-ribose) polymerase and observed that under extensive hyper(ADP-ribosylation) of histone H1 there was a complete relaxation of nucleosomal architecture (*6*).

In this paper, we describe some properties of poly(ADP-ribose) polymerase associated to polynucleosomes and nuclei. We have analyzed the effect of NaCl and ethidium bromide (EB) on the enzymatic activity and the poly(ADP-ribosyl)ation of histone H1. We have also studied the effect of poly(ADP-ribosyl)ation on the morphology of polynucleosomes.

### *Materials*

Amylamine and micrococcal nuclease (*Staphylococcus aureus*) was purchased from Sigma Co., St. Louis, Missouri, U.S.A. $^{32}$P-labeled NAD (53 Ci/mmol (1 Ci=37 GBq)) was obtained from New England Nuclear.

### *Isolation and Purification of Pancreatic Nuclei and Nucleosomes*

Pancreatic nuclei were purified as elaborated elsewhere (*4*). The nuclei were digested with micrococcal nuclease to an acid solubility of 4–6% as described previously (*4, 8*). Enzymatic activity in nuclei and nucleosomes was assayed as described before and histones were isolated as described by Aubin *et al.* (*4*).

### *Properties of Nucleosomal Poly(ADP-ribose) Polymerase*

In order to localize poly(ADP-ribose) polymerase in pancreatic nucleosomes we have analyzed the distribution of the enzyme after an extensive micrococcal nuclease digestion. We have found that enzyme activity was present in the heavy monomer fraction and was most active in the tri-tetra nucleosomal region of the gradient (Fig. 1). This feature of the enzyme distribution is explained by the fact that mononucleo-

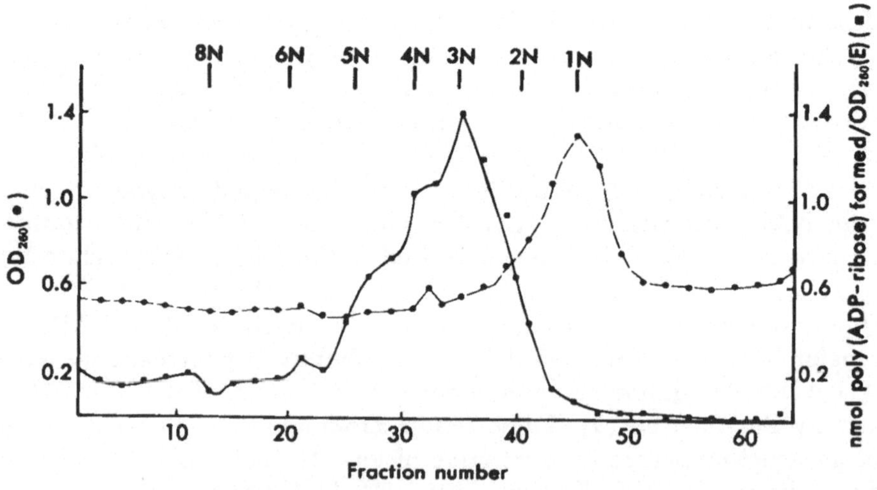


FIG. 1. Distribution of nucleosome-associated poly(ADP-ribose) polymerase within short chromatin subunits. Pancreatic nucleosomes were purified on 5–20% (w/v) linear sucrose gradient. Endogenous poly(ADP-ribose) polymerase activity was assessed in each fraction. (Taken from Aubin *et al.* (*4*)).

somes bearing poly(ADP-ribose) polymerase will sediment faster towards the dinucleosome peak. This suggests that poly(ADP-ribose) polymerase could bind mononucleosomes either in the core and/or linker region; thus, the selective binding of both core and linker DNA by poly(ADP-ribose) polymerase would most likely depend on the presence of single stranded breaks in core and linker DNA than on the conformation of the core or linker DNA. However, some shielding of the DNA strand breaks by nuclear proteins such as histones are more likely to happen in the core than in the linker DNA and might affect the binding of the enzyme and its activation.

One noteworthy feature of poly(ADP-ribose) polymerase is that it displays a maximal level of activity in tri- and tetranucleosomes (Fig. 1) generated by extensive or mild micrococcal nuclease digestion even under different conditions of digestion: *e.g.*, with and without NaCl (*4*, *8*). This enrichment of poly(ADP-ribose) polymerase activity in trimers and tetramers could be caused by a selective cleavage of localized relaxed regions (*6*) of chromatin or by a higher concentration of single strand breaks in these nucleosomes. Indeed Smulson *et al.* (*10*) have shown that poly (ADP-ribosylated) nucleosomes purified by affinity chromatography harboured numerous single strand breaks as compared to their control counterparts. Studies are in progress in our laboratory to verify these possibilities.

### *Effect of NaCl on Poly(ADP-ribose) Polymerase Activity in Nuclei and Nucleosomes*

Poly(ADP-ribose) polymerase has been shown to extensively poly(ADP-ribosylate) histone H1 at high NAD concentrations in nuclei and nucleosomes (*8*, *11*). The enzyme has also the ability to poly(ADP-ribosylate) core histones at lower NAD concentrations (*12*).

The activity of poly(ADP-ribose) polymerase will be modulated by the ADP-ribosylation levels of the enzyme and by the structure of the chromatin on which it is located. In order to evaluate some of the changes brought about by condensation of

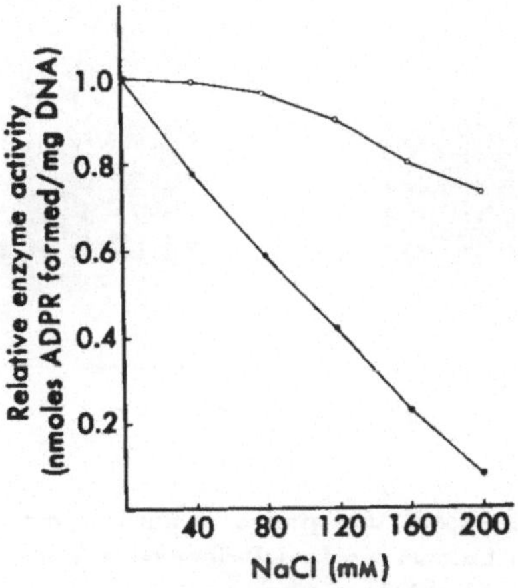


FIG. 2. Effect of increasing concentrations of NaCl ranging from 0 to 200 mM on the activity of poly(ADP-ribose) polymerase associated with nuclei (●) or nucleosomes (○).

the chromatin structure, we have analyzed the effect of NaCl on poly(ADP-ribose) polymerase activity and on histone modification in nuclei and nucleosomes. This study was performed to enable us to have some insight as to the responsiveness of the enzyme to various states of nuclear and nucleosomal structure condensation. Indeed the folding ratio of the DNA is much higher in nuclei than in nucleosomes (*13*, *14*).

In nuclei the activity of poly(ADP-ribose) polymerase was significantly reduced by NaCl (Fig. 2). Furthermore 200 mM NaCl caused a 95% reduction of the enzyme

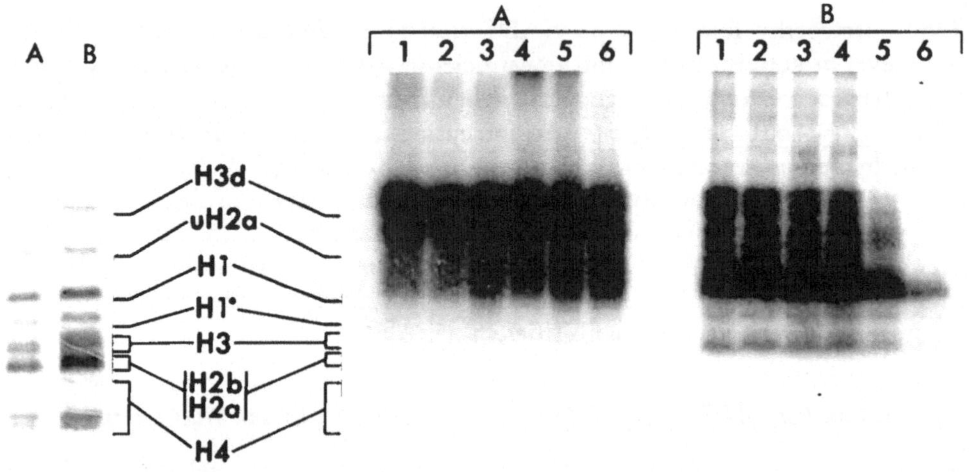


FIG. 3. Effect of NaCl on the poly(ADP-ribosyl)ation profile of nuclear and nucleosomal histones. Nucleosomal (A) and nuclear (B) histones were ADP-ribosylated in the presence of 100 μM [$^{32}$P] NAD for 30 min at 30°C, extracted and electrophoresed in 6 M urea–1 M acetic acid slab gels. Numbers 1 to 6 representing respectively 0, 40, 80, 120, 160, and 200 mM NaCl.

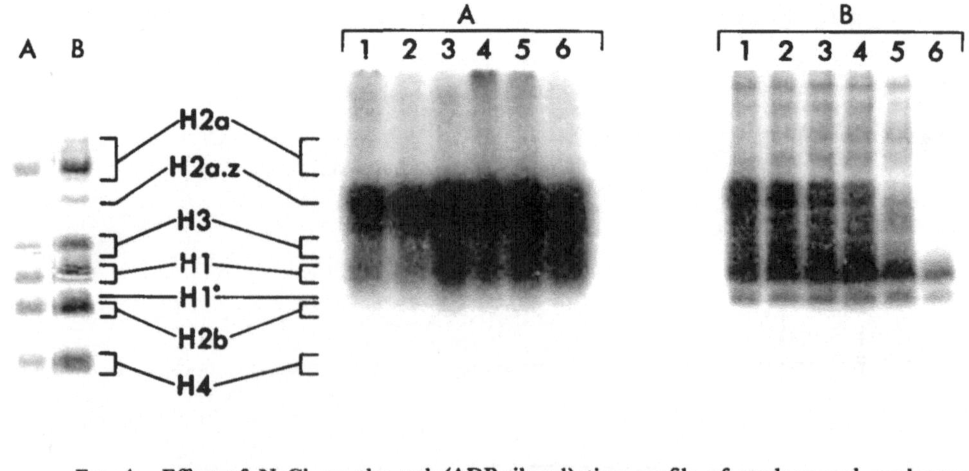


FIG. 4. Effect of NaCl on the poly(ADP-ribosyl)ation profile of nuclear and nucleosomal histones. Nucleosomal (A) and nuclear (B) histones were ADP-ribosylated in the presence of 100 μM [$^{32}$P] NAD for 30 min at 30°C, extracted and electrophoresed in 6 M urea, 1 M acetic acid, 6 mM Triton X-100 slab gels. Numbers 1 to 6 representing respectively 0, 40, 80, 120, 160, and 200 mM NaCl.

activity. However in nucleosomes, where the degree of folding is much smaller, the inhibition by NaCl was only observed at concentration higher than 80 mM and an inhibition of 25% was produced by 200 mM NaCl.

Thus these differential effects of NaCl on the nucleosomal and nuclear preparations suggest that the enzyme will respond differently to various degrees of DNA supercoiling. Sodium chloride has a much stronger inhibitory effect on the enzyme in nuclei in which the supercoiling or packing ratio is higher.

To further substantiate this differential effect of NaCl we have performed labeling experiments of nuclear and nucleosomal histones under the same experimental conditions (Figs. 3, 4). In nuclei we have found that the generation of hyper(ADP-ribosylated) histone H1 (*8*) was reduced at all concentrations of NaCl and its formation was abolished completely at 160 and 200 mM NaCl. The same modification profile was observed in the presence or absence of Triton X-100. The intensity of the intermediates between the stained band and the hyper(ADP-ribosylated) form of histone H1 was not decreased markedly until a concentration of 160 mM NaCl was reached.

In nucleosomes however, it was found that NaCl has no inhibitory effect but instead showed an enhancing effect on ADP-ribosylated intermediates of histone H1 and also on the formation of hyper(ADP-ribosylated) histone H1. This stimulation by NaCl of the ADP-ribosylation of histone H1 could be caused either by a direct effect of NaCl on the interaction between the enzyme and histone H1 or by an enhancement of an exchange (*15*) between ADP-ribosylated histone H1 molecules and unmodified histone H1 thereby favoring the ADP-ribosylation of histone H1. NaCl did not alter ADP-ribosylation of histone H2B in nuclei or nucleosomes.

The conclusion to be drawn from these results is that, even though NaCl causes changes in poly(ADP-ribose) polymerase activity in nuclei and nucleosomes, the level of poly(ADP-ribosyl)ation of histone H1 is unaltered in nucleosomes. Thus, in

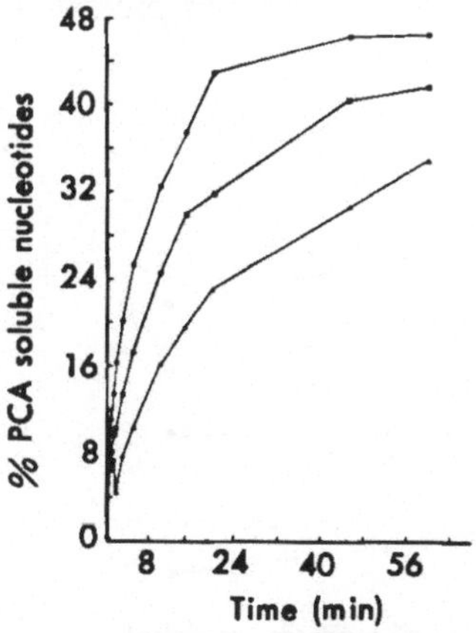


FIG. 5. Micrococcal nuclease digestion of pancreatic nuclei at different concentrations of NaCl. Pancreatic nuclei were digested in the presence of NaCl at concentrations of 0 mM (●), 80 mM NaCl (■), 150 mM NaCl (▲) in a buffer containing 0.25 M sucrose, 10 mM Tris-HCl pH 7.5, 5 mM $MgCl_2$, 1 mM $CaCl_2$, and micrococcal nuclease at a concentration of 1 unit/mg DNA at 30°C. At different times, samples were taken and acid solubility was measured in 0.7 N PCA.

order to evaluate the possible changes caused by poly(ADP-ribose) polymerase, a study of the poly (ADP-ribosylated) acceptors should be performed.

To characterize further the fact that high concentrations of NaCl induce condensation of nuclear DNA, we have carried out micrococcal nuclease digestion of nuclei utilizing 1 unit of nuclease/mg DNA. It was observed that nuclei incubated at lower salt concentrations digested much faster than the nuclei which had been incubated at higher salt concentrations (Fig. 5).

### *Effect of EB on Nucleosomal Poly(ADP-ribose) Polymerase Activity*

Since the enzyme seems to respond differently to NaCl in nuclei and nucleosomes we have used EB to further characterize this property of the enzyme. EB is an intercalating agent which will unwind the double helix and will permit to study various degrees of unwinding of the nucleosomal architecture and can unfold the nucleosomal architecture at a D/P ratio above 0.3 (*16*). We have treated nucleosomes of different repeat numbers with EB and we have found that this intercalating agent displayed a stimulatory effect of 40–50% on enzyme activity on tetranucleosomes and octanucleosomes at concentrations of 1–15 μg/ml (Fig. 6). On shorter nucleosomes (1–3N) no significant effect was observed while a slight stimulatory effect was noticed in 15–30N nucleosomes. The activation of the enzyme was detected at D/P ratio above 0.3. Thus, it is most likely that this stimulation is caused by a greater accessibility to poly(ADP-ribose) polymerase of core histone or other proteins since these ratio of dye unfold partially or completely the nucleosome structure.

Figure 7 shows control and poly(ADP-ribosylated) nucleosomes which were fixed at low ionic strength according to the method of de Murcia and Koller (*17*). The poly(ADP-ribosylated) nucleosomes were incubated with 1 mM NAD for 30 min at 30°C. Under these incubation conditions hyper(ADP-ribosylation) of histone H1 takes place (*8*). The poly(ADP-ribosylated) nucleosomes do not show any region of aggregation thus suggesting that the formation of histone H1 complex or hyper poly

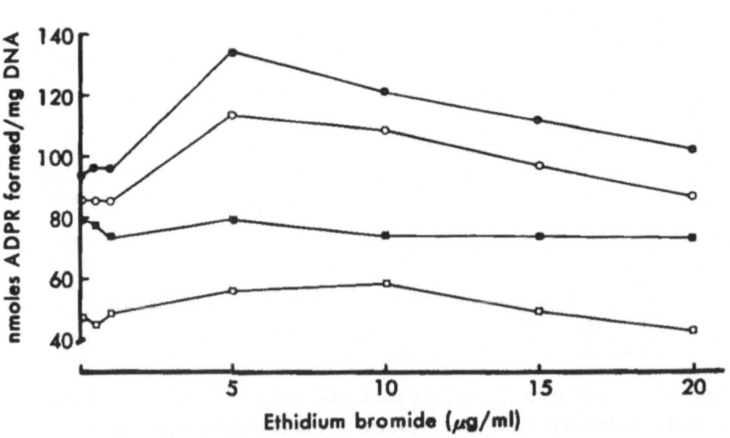


FIG. 6. Effect of various concentrations of EB on poly(ADP-ribose) polymerase activity in pancreatic nucleosomes. Nucleosomes were assayed at 100 μM NAD as described previously (*4*) at a concentration of DNA of $3OD_{260}$/ml. ● 4-6 N; ○ 8-10 N; ■ 1-3 N; □ 15-30 N.

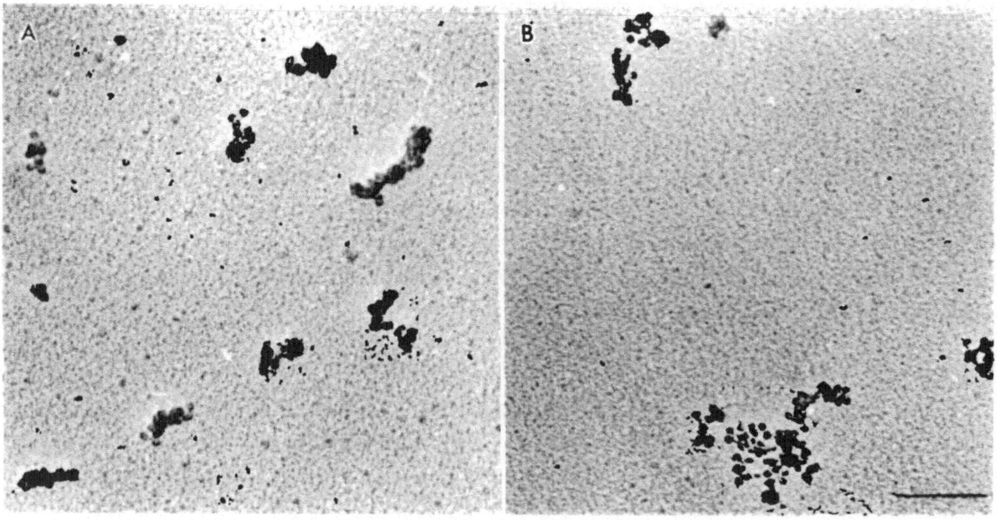


FIG. 7. Effect of poly(ADP-ribosyl)ation on the structure of chromatin at 30 mM NaCl. (A) native chromatin, (B) poly(ADP-ribosylated) chromatin. Polynucleosomes were poly(ADP-ribosylated) in the presence of 1.0 mM NAD for 30 min at 30°C and they were treated as previously described (*6*). The bar indicates 2,000 Å.

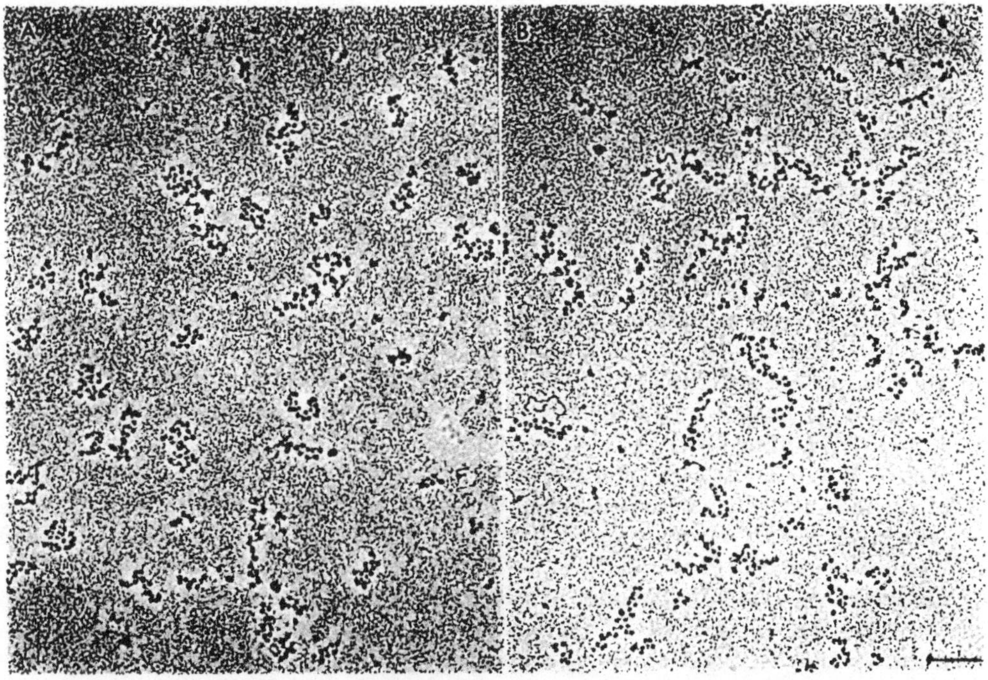


FIG. 8. Effect of poly(ADP-ribosyl)ation on the structure of chromatin at low ionic strength. (A) native chromatin, (B) poly(ADP-ribosylated) chromatin. The nucleosomes were incubated and treated as described in Fig. 7. The bar indicates 2,000 Å.

(ADP-ribosylated) histone H1 does not induce chromatin condensation or aggregation in pancreatic nucleosomes.

When very long chains of poly(ADP-ribosylated) nucleosomes were examined under the electron microscope at 30 mM NaCl we found that these nucleosomes stayed in a quite relaxed conformation as compared to control nucleosomes (Fig. 8). Thus poly(ADP-ribosyl)ation of polynucleosomes could cause a localized relaxation of the chromatin structure. This relaxation is most likely related to the hyper (ADP-ribosylation) of histone H1 as suggested by Poirier *et al.* (*6*). This type of relaxation is similar to what has been observed by Thomas and Koller for chromatin which was partially depleted of histone H1 (*18*).

ACKNOWLEDGMENT

The authors would like to thank A. Gaudreau, G. Grondin, and M. Martin for excellent technical assistance and C. Rancourt for typing the manuscript.

REFERENCES

1. Hayaishi, O. and Ueda, K. Poly(ADP-ribose) and ADP-ribosylation of proteins. Annu. Rev. Biochem., *46*: 95–116, 1977.
2. Mandel, P., Okazaki, H., and Niedergang, C. Poly(adenosine diphosphate ribose). Prog. Nucleic Acid Res. Mol. Biol., *27*: 1–51, 1982.
3. Wong, N.C.W., Poirier, G. G., and Dixon, G. H. Adenosine diphosphoribosylation of certain basic chromosomal proteins in isolated trout testis nuclei. Eur. J. Biochem., *77*: 11–12, 1977.
4. Aubin, R. J., Dam, V. T., Miclette, J., Brousseau, Y., and Poirier, G. G. Chromosomal protein poly(ADP-ribosyl)ation in pancreatic nucleosomes. Can. J. Biochem., *60*: 295–305, 1982.
5. Butt, T. R., Brothers, J. F., Giri, C. P., and Smulson, M. A nuclear protein-modifying enzyme is responsive to ordered chromatin structure. Nucleic Acids Res., *5*: 2775–2788, 1978.
6. Poirier, G. G., de Murcia, G., Jongstra-Bilen, J., Niedergang, C., and Mandel, P. Poly(ADP-ribosyl)ation of polynucleosomes causes relaxation of chromatin structure. Proc. Natl. Acad. Sci. U.S., *79*: 3423–3427, 1982.
7. Butt, T. R. and Smulson, M. Relationship between nicotinamide adenine dinucleotide concentration and *in vitro* synthesis of poly(adenosine diphosphate ribose) on purified nucleosomes. Biochemistry, *19*: 5235–5242, 1980.
8. Aubin, R. J., Dam, V. T., Huletsky, A., Brousseau, Y., Miclette, J., and Poirier, G. G. Hyper (ADP-ribosylation) of histone H1. Can. J. Biochem., *60*: 1085–1094, 1982.
9. Burzio, L. O., Koide, S. S., Puigdomenech, P., and Ruiz-Carillo, A. ADP-ribosylation of rat liver H1 and nucleosomal core histones. *In;* M. E. Smulson and T. Sugimura, (eds.), Novel ADP-ribosylation of Regulatory Enzymes and Proteins, Vol. 6, pp. 344–353, Elsevier/North-Holland, New York, 1981.
10. Smulson, M., Malik, N., Wong, M., Pomato, N., and Thraves, P. Fine resolution of the poly ADP-ribosylated domains of polynucleosomal chromatin: DNA gene and integrity analysis; mechanism of histone H1 modification. This volume, pp. 49–70.
11. Poirier, G. G., Savard, P., Rajotte, D., Morisset, J., and Lord, A. Properties of poly (ADP-ribose) synthestase from rat pancereas and poly(ADP-ribosylation) of basic nuclear proteins. Can. J. Biochem., *56*: 784–790, 1978.

12. Giri, C. P., West, M.H.P., and Smulson, M. Nuclear protein modification and chromatin substructure. I. Differential poly(adenosine diphosphate) ribosylation of chromosomal proteins in nuclei *versus* isolated nucleosomes. Biochemistry, *17*: 3495–3500, 1978.
13. Laemmli, U. K. Levels of organization of the DNA in eucaryotic chromosomes. Pharmacol. Rev. *30*: 469–476, 1979.
14. Lebkowski, J. S. and Lemmli, U. K. Evidence for two levels of DNA folding in histone-depleted HeLa interphase nuclei. J. Mol. Biol. *156*: 309–324, 1982.
15. Caron, F. and Thomas, J. O. Exchange of histone H1 between segments of chromatin. J. Mol. Biol., *146*: 513–537, 1981.
16. Erard, M., Das, G. C., de Murcia, G., Mazen, A., Pouyet, J., Champagne, M., and Daune, M. Ethidium bromide binding to core particle: comparison with native chromatin. Nucleic Acids Res., *6*: 3231–3253, 1979.
17. de Murcia, G. and Koller, T. The electron microscopic appearance of soluble rat liver chromatin mounted on different supports. Biol. Cell., *40*: 165–174, 1981.
18. Thoma, F. and Koller, T. Unravelled nucleosomes, nucleosome beads and higher order structures of chromatin: influence of non-histone components and histone H1. J. Mol. Biol., *149*: 709–735, 1981.

ADP-RIBOSYLATION, DNA REPAIR AND CANCER, M. MIWA ET AL. (EDS.),
JAPAN SCI. SOC. PRESS, TOKYO/VNU SCIENCE PRESS, UTRECHT, PP. 93-100, 1983

# Profiles of Antibodies to Poly(ADP-ribose), Left-handed Z-DNA, and Other Nuclear Constituents in Systemic Lupus Erythematosus and Progressive Systemic Sclerosis

Yoshiyuki KANAI,*[1] Yasuoki MOROI,*[2] Ryuichi YOKOHARI,*[3] and Sachio GOTO*[4]

*Department of Molecular Oncology, Institute of Medical Science, University of Tokyo, Tokyo 108,*[1] Department of Internal Medicine and Physical Therapy, Faculty of Medicine, University of Tokyo, Tokyo 113,*[2] Clinical Research Institute, National Medical Center, Tokyo 162,*[3] and Department of Biochemistry Faculty of Medicine, Kinki University, Osaka 589,*[4] Japan*

*Abstract:* A simple and sensitive method was developed for detection of antibodies to histones using a modification of the enzyme-linked immunosorbent assay (ELISA) for antibodies to nucleic acids. The antigens used for antibody assay were poly(ADP-ribose), left-handed Z-DNA, double-stranded DNA (dsDNA), single-stranded DNA (ssDNA), and calf thymus histones. These five antigen-antibody systems were used to examine the antibodies in patients with systemic lupus erythematosus (SLE) and progressive systemic sclerosis (PSS). A positive correlation was found between the antibody titers in patients with PSS. However, weak correlation was found between the titers of antibodies to these nuclear antigens in the SLE patients, except for correlations between the titers of antibodies to left-handed Z-DNA and histones, and those to ssDNA and dsDNA. From these data, it is suggested that presentation of modified chromatin to immune systems is involved in antinuclear antibody formation in PSS patients, whereas spontaneous expansion of clones for antinuclear antibodies may occur in SLE patients under certain conditions.

In 1977 we first reported the presence of antibodies to poly(ADP-ribose) in patients with systemic lupus erythematosus (SLE) (*1*). Later we were able to demonstrate the presence of poly(ADP-ribose) in M phase chromosomes of HeLa cells by the fluorescent antibody technique (*2*). Therefore, it is now probable that poly(ADP-ribose) can serve as an antigen in humans. The characteristics of naturally occurring antibodies to poly(ADP-ribose) in SLE patients have been described by us (*3, 4*), and Okolie and Shall (*5*). The data presented so far were obtained by radioimmunoassay with [$^{14}$C]poly(ADP-ribose) as tracer. Recently we developed a simple and sensitive enzyme-linked immunosorbent assay (ELISA) (*6*). This method has made it possible to measure various kinds of antinuclear antibodies without using radiolabeled antigens. It is unknown whether antigens are essential for triggering antinuclear antibody formation *in vivo* or whether such formation is a result of spontaneous expression of B cell clones under certain conditions *in vivo* (*7*). Recent *in vitro* studies indicated that ADP-ribosylation of chromatin occurs mainly on histone subgroups H1 and H2b, and also suggested that non-histone proteins undergo ADP-ribosylation (*8–10*). On the other hand, it is known that even self-components, once modified by drugs or virus infections, act as immunogens (*11*). The ADP-ribosylation reaction of histones should be involved in one of the modifications of self-components in a broad sense. Thus, we examined the effect of oligo(ADP-ribosyl)ation of histones on antibody production to histones by immunizing rabbits with oligo(ADP-ribosyl)ated histones. As expected, we found that oligo(ADP-ribosyl)ation of histones induces antibodies to poly(ADP-ribose) (*3*) and renders histones antigenic (*4*). This result suggests that modification of chromatin under certain *in vivo* conditions renders the chromatin more antigenic. If modified nuclear antigens are required for triggering the relevant antibody expansion *in vivo*, there should be some relation between the patterns of appearance of antinuclear antibodies. To investigate this possibility, we examined five nuclear antigen-antibody systems by an ELISA in patients with SLE and progressive systemic sclerosis (PSS), diseases which are considered to be a prototype of autoimmune disease characterized by the presence of antinuclear antibodies.

## *ADP-ribosylation of Histones and Histone Antigenicity*

Anti-histone antibodies are known to be present in patients with the idiopathic form of SLE (*12*) and with drug-induced lupus, and drug-induced lupus antibodies to histone were found in almost all the patients examined (*13*). Experimental induction of antibodies to histones in rabbits was achieved first by Stollar and Ward (*14*)

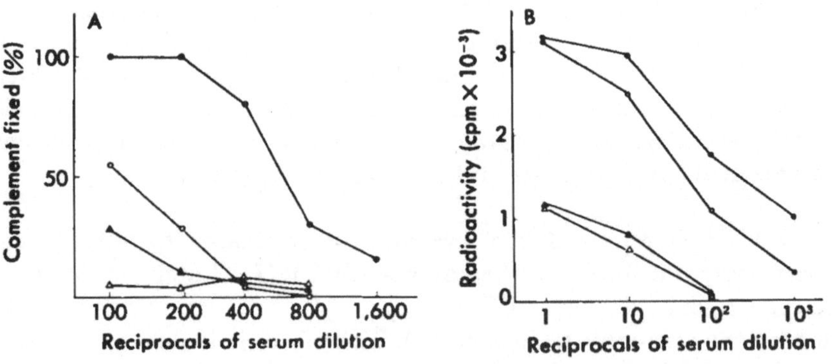


FIG. 1. A. Complement fixing antibody to rat histones in sera of different rabbits immunized with oligo(ADP-ribosyl)ated rat histones or unmodified rat histones. Oligo(ADP-ribosyl)ated rat histones and unmodified rat histones were prepared as described (*3*). Complement fixing antibody to rat histone was determined by the method of Mayer (*15*) with a slight modification (*4*). B. Poly(ADP-ribose) binding activity in sera of different rabbits immunized with oligo (ADP-ribosyl)ated rat histones or unmodified rat histones as prepared above. Poly(ADP-ribose) binding activity was measured by radioimmunoassay using [$^{14}$C]poly(ADP-ribose) as described (*1*). ○, ● rabbits immunized with oligo(ADP-ribosyl)ated rat histones; △, ▲ rabbits immunized with unmodified rat histones.

using complexes of whole histone with DNA or of histone subfraction with RNA: free calf thymus histones were not immunogenic.

The recent finding that ADP-ribosylation of chromatin occurs mainly on histone subgroups H1 and H2b (*8–10*) prompted us to study whether ADP-ribosylated histones alone are able to elicit antibodies to histones in rabbits when injected as an emulsion in Freund's complete adjuvant. Actually, examination of the antisera by a microcomplement fixation assay (*4*) originally developed by Mayer (*15*) showed that oligo(ADP-ribosyl)ated histones with two ADP-ribosyl units were immunogenic (Fig. 1A). Oligo(ADP-ribosyl)ated histones also induced antibodies to poly(ADP-ribose) (Fig. 1B). Thus oligo(ADP-ribosyl)ation of histones rendered the histones antigenic and induced antibodies reactive with poly(ADP-ribose). This procedure avoided the use of methylated bovine serum albumin (MBSA), which is an essential carrier for the experimental production of antibodies to DNA (*16*) and for the enhancement of antibodies to poly(ADP-ribose) (*17*). Chromatin is mainly, if not completely, composed of nucleosomes, repeating histone-DNA subunits (*18*), and poly-(ADP-ribosyl)ation of nucleosomes may induce alteration of the chromatin structure (*19*). Furthermore, it has been reported that conversion of poly(dG-dC)·poly(dG-dC), B-form DNA, to left-handed Z-DNA occurs in the presence of high salt (*20*) and that the presence of Z-DNA prevents nucleosome formation (*21*). The latter finding suggests that left-handed Z-DNA prevents formation of the normal chromatin structure. Since ADP-ribosylation of histones renders them antigenic, as mentioned above (Fig. 1A), we thought that these conformational changes or modification of chromatin would render chromatin antigenic. If such chromatin could be served as an antigen *in vivo*, results might demonstrate some relation between the patterns of appearance of antinuclear antibodies in autoimmune disease such as SLE and PSS. To investigate this possibility, we developed an ELISA for the detection of antibodies to nuclear materials.

### *ELISA for Antibodies to Nucleic Acids*

The procedures used for assay of antibodies to nucleic acids were essentially the same as those for assay of antibodies to poly(ADP-ribose), as described elsewhere (*6*), being as follows.

#### *1. Preparation of antigen-coated microplates*

Microplates were treated with 50 μl of poly L-lysine (PLL) solution (50 μg/ml). Then aliquots of a solution containing 100 ng of nucleic acid (50 μl) were introduced onto the microplates and evaporated at 37°C overnight to allow adherence of nucleic acids to the microplates. Residual antigen binding sites were blocked by treatment with Tris-buffered saline (TBS) (25 mM Tris, 140 mM NaCl, pH 7.4) containing 0.01% bovine serum albumin (BSA) (TBS-BSA) for 1 hr.

#### *2. Preparation of conjugate of alkaline phosphatase and protein A*

Equimolar solutions of alkaline phosphatase and protein A were combined and dialyzed against phosphate-buffered saline (PBS) (10 mM phosphate, 140 mM NaCl, pH 7.4) for 5 hr. Conjugation was achieved in the presence of 0.2% glutaraldehyde by gentle stirring for 2 hr at room temperature. The specific enzyme activity of the conjugate was 200–300 U/mg protein, as determined by measuring decomposition of substrate *p*-nitrophenyl-phosphate.

*3. Preparation of poly(ADP-ribose) and nucleic acid antigens*

a) Poly(ADP-ribose)—poly(ADP-ribose) was synthesized and purified as described previously (*22*). Briefly, after incubation of [$^{14}$C]NAD (2 mM) with calf thymus nuclei (equivalent to 5 mg DNA per ml reaction mixture) for 30 min, the ethanol-acetate precipitable fraction was extensively digested with pronase. The DNA fraction, which became flocculant on addition of 1 M NaCl and 2 volumes of ethanol, was rolled onto a glass rod and dissolved in 0.1 M Tris-HCl buffer, pH 8.0. The DNA solution was mixed with DNase I, RNase, and nuclease P1 and incubated overnight at 4°C. Oligo- and poly(ADP-ribose) were extracted from the mixture with phenol. Finally, the poly(ADP-ribose) fraction was eluted from a column of hydroxylapatite (HA) with 0.25 to 0.35 M Na-phosphate buffer (pH 6.8).

b) Double-stranded(ds)- and single stranded(ss) DNA—dsDNA was purified from a commercial preparation of calf thymus DNA by the method of Marmur (*23*), treated with nuclease S1 by the method of Stollar and Papalian (*24*), and finally purified on a column of HA. This purified DNA showed no more increase of $A_{260}$ even in the presence of nuclease S1. ssDNA was obtained by heating purified dsDNA at 100°C for 8 min, and then rapidly cooling the mixture to 0°C.

c) Left-handed Z-DNA—Left-handed Z-DNA was formed from poly(dG-dC)·poly(dG-dC) at high salt concentration and stabilized by bromination by the method of Lafer *et al.* (*25*).

*4. Determination of serum antibodies to nucleic acids*

Samples of 50 μl of test serum diluted 1:20 with TBS-BSA were introduced onto microplates coated with nucleic acid. After incubation at room temperature for 1 hr, they were washed with TBS and incubated with 50 μl of conjugate (100 ng) for 1 hr at room temperature. Then the microplates were washed and incubated with 100 μl of substrate, *p*-nitrophenyl-phosphate at 37°C for 30 min. The reaction was stopped with 100 μl of 0.1N NaOH, and then $A_{410}$ or $A_{405}$ was measured with a conventional spectrophotometer or an autoreader, which measures the absorbances in a 96-well microplate in 1 min. The $A_{410}$ or $A_{405}$ values obtained by subtracting the absorbance of control microplates from that of antigen-coated microplates were used as measures of antibody activity. All the test sera were heat-inactivated at 37°C for 30 min before assay.

## *ELISA for Antibodies to Histones*

Antibodies to histones have usually been measured by the complement fixation or immunofluorescent antibody technique (*12, 26*). Very recently, an ELISA with antigen-coated beads instead of antigen-coated microplates was reported (*27*). We developed an ELISA for anti-histone antibodies by slightly modifying the anti-nucleic acid antibody assay as described above. The assay procedure was essentially the same as for antinucleic acid antibody assay except that the microplates were treated with poly L-glutamic acid (PLG) to facilitate their binding of histone. Histones were added at 500 ng per well and after incubation for 2 hr at room temperature, the microplates were washed with TBS. Then residual antigen binding sites were blocked by overnight treatment with 0.1% BSA in TBS at 4°C and the microplates were washed with TBS. The subsequent procedure for assay of anti-histone antibodies was the same as that for antinucleic acid antibodies.

*Profiles of Antibodies to Nuclear Constituents in SLE and PSS*

Since the discovery of "LE" cells in SLE patients by Hargraves *et al.* (*28*), several antibodies to nuclear constituents have been reported (*7*, *30*). We measured five antinuclear antibodies—anti-poly(ADP-ribose), anti-dsDNA, anti-ssDNA, anti-Z-DNA, and anti-histone antibodies—in 37 patients with SLE, 21 patients with PSS, and 17 normal individuals (Fig. 2). Statistical evaluation of the titers of these antibodies in patients with SLE and PSS showed a positive correlation between the titers in patients with PSS ($r$=0.634–0.840), but weak correlation in those with SLE ($r$≦0.559), except for the titers of antibodies to Z-DNA and histones and those to ssDNA and dsDNA (Table 1). The positive correlation between the titers of

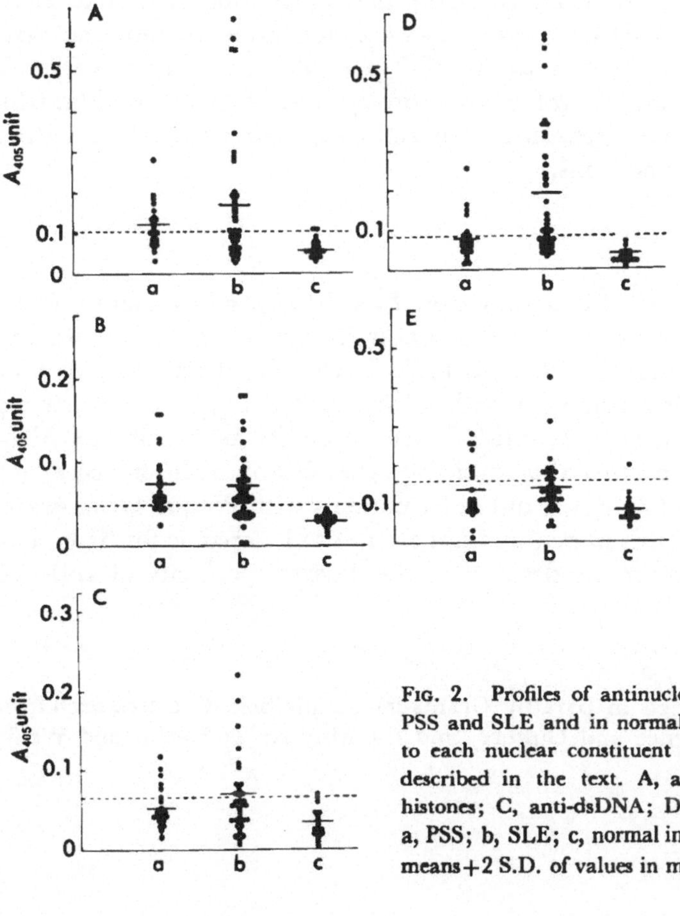


FIG. 2. Profiles of antinuclear antibodies in patients with PSS and SLE and in normal individuals. Antibody activity to each nuclear constituent was determined by ELISA as described in the text. A, anti-poly(ADP-ribose); B, anti-histones; C, anti-dsDNA; D, anti-ssDNA; E, anti-Z-DNA. a, PSS; b, SLE; c, normal individuals. Dotted lines indicate means+2 S.D. of values in mormal individuals.

TABLE 1. Correlation Coefficient ($r$) between Antibodies to A and B

| A | B | PSS $r$ | SLE $r$ |
|---|---|---|---|
| Poly(ADPR[a]) | Histone | 0.701 | 0.399 |
| Poly(ADPR) | Z-DNA | 0.731 | 0.359 |
| Poly(ADPR) | dsDNA | 0.717 | 0.294 |
| Poly(ADPR) | ssDNA | 0.634 | 0.492 |
| Histone | Z-DNA | 0.796 | 0.731 |
| Histone | dsDNA | 0.840 | 0.474 |
| Histone | ssDNA | 0.776 | 0.478 |
| dsDNA | ssDNA | 0.773 | 0.737 |
| dsDNA | Z-DNA | 0.724 | 0.559 |
| ssDNA | Z-DNA | 0.833 | 0.520 |

[a] ADPR, ADP-ribose.

antibodies to ssDNA and dsDNA in both PSS and SLE seems reasonable because of cross-reaction of anti-dsDNA antibodies with ssDNA (*29*). However, the positive correlation between the antibody titers to histones and Z-DNA in both PSS and SLE is of interest, because it has been reported that the presence of Z-DNA prevents nucleosome formation (*21*). It seems probable that prevention of formation of the nucleosomal structure by Z-DNA makes histones accessible to immune systems. These statistical findings suggest that in PSS patients presentation of modified chromatin to immune systems is involved in antinuclear antibody formation, whereas in SLE patients spontaneous expansion of B cell clones for antinuclear antibodies may occur under certain conditions.

### *Added in Proof*

After we have submitted this manuscript, Lafer *et al.*, independently from our studies, have first reported the presence of naturally occurring antibodies to left-handed Z-DNA in the patients with SLE (*31*). They have described two types of these antibodies. One type reacts with both ssDNA and Z-DNA. The other type is specific for Z-DNA. They stress that anti-Z-DNA antibodies are specific for SLE and do not appear frequently in other rheumatic deseases. However, in our case, anti-Z-DNA antibodies were also found in 8 out of 21 patients with PSS, so far as examined by an ELISA described in this paper, and appear not to be specific for SLE. Further studies should be required for the decision of the desease specificity of anti-Z-DNA antibodies.

ACKNOWLEDGMENTS

This work was supported in part by Grants-in-Aid for Scientific Research from the Ministry of Education, Science and Culture, and the Ministry of Health and Welfare of Japan.

## REFERENCES


1. Kanai, Y., Kawaminami, Y., Miwa, M., Matsushima, T., Sugimura, T., Moroi, Y., and Yokohari, R. Naturally-occurring antibodies to poly(ADP-ribose) in patients with systemic lupus erythematosus. Nature, *265*: 175–177, 1977.
2. Kanai, Y., Tanuma, S., and Sugimura, T. Immunofluorescent staining of poly(ADP-ribose) *in situ* in HeLa cell chromosomes in the M phase. Proc. Natl. Acad. Sci. U.S., *78*: 2801–2804, 1981.
3. Kanai, Y. and Sugimura, T. Comparative studies on antibodies to poly(ADP-ribose) in rabbits and patients with systemic lupus erythematosus. Immunology, *43*: 101–110, 1981.
4. Kanai, Y. and Sugimura, T. Systemic lupus erythematosus. *In;* O. Hayaishi and K. Ueda (eds.), ADP-ribosylation Reactions, pp. 533–546, Academic Press, New York, 1982.
5. Okolie, E. E. and Shall, S. The significance of antibodies to poly(adenosine diphosphate-ribose) in systemic lupus erythematosus. Clin. Exp. Immunol., *36*: 151–164, 1979.
6. Kanai, Y., Tauchi, M., Aotsuka, S., and Yokohari, R. A simple and rapid micro-enzyme-linked immunosorbent assay for antibodies to poly(ADP-ribose) in systemic lupus erythematosus. J. Immunol. Methods, *53*: 355–365, 1982.
7. Stollar, B. D. Anti-DNA antibodies. Clin. Immunol. Allergy, *1*: 243–260, 1981.
8. Okayama, H., Ueda, K., and Hayaishi, O. Purification of ADP-ribosylated nuclear proteins by covalent chromatography on dihydroxyboryl polyacrylamide beads and their characterization. Proc. Natl. Acad. Sci. U.S., *75*: 1111–1115, 1978.
9. Riquelme, P. T., Bruzio, L. O., and Koide, S. S. ADP-ribosylation of rat liver lysine-rich histone *in vitro*. J. Biol. Chem., *254*: 3018–3028, 1979.
10. Burzio, L. O., Riquelme, P. T., and Koide, S. S. ADP-ribosylation of rat liver nucleosomal core histones. J. Biol. Chem. *254*: 3029–3037, 1979.
11. Allison, A. C. Autoimmune diseases: Concepts of pathogenesis and control. *In;* N. Talal (ed.), Autoimmunity, Genetic, Immunologic, Virologic, and Clinical Aspects, pp. 91–139, Academic Press, New York, 1977.
12. Stollar, B. D. Serological analyses of histones. Methods Cell Biol., *18*: 105–122, 1978.
13. Tan, E. M. and Portanova, J. P. The role of histones as nuclear autoantigens in drug-related lupus erythematosus. Arthritis Rheum., *24*: 1064–1069, 1981.
14. Stollar, B. D. and Ward, M. Rabbit antibodies to histone fractions as specific reagents for preparative and comparative studies. J. Biol. Chem., *245*: 1261–1266, 1970.
15. Mayer, M. M. Complement and complement fixation. *In;* E. A. Kabat and M. M. Mayer (eds.), Kabat and Mayer's Experimental Immuno-Chemistry, pp. 133–240, Thomas, Springfield, 1961.
16. Plescia, O., Braun, W., and Palczuk, N. C. Production of antibodies to denatured deoxyribonucleic acid (DNA). Proc. Natl. Acad. Sci. U.S., *52*: 279–285, 1964.
17. Kanai, Y., Miwa, M., Matsushima, T., and Sugimura, T. Comparative studies on antibody and antibody production to poly(ADP-ribose) in mice. Immunology, *34*: 501–508, 1978.
18. McGhee, J. D. and Felsenfeld, G. Nucleosome structure. Annu. Rev. Biochem., *49*: 1115–1155, 1980.
19. Poirier, G. G., de Murcia, G., Jongstra-Bilen, J., Niedergang, C., and Mandel, P. Poly(ADP-ribosyl)ation of polynucleosomes causes relaxation of chromatin structure. Proc. Natl. Acad. Sci. U.S., *79*: 3423–3427, 1982.

20. Thamann, T. J., Lord, R. C., Wang, AH-J., and Rich, A. The high salt form of poly (dG-dC)·poly(dG-dC) is left-handed Z-DNA: Raman spectra of crystals and solutions. Nucleic Acids Res., *9*: 5443–5457, 1981.
21. Nickol, J., Behe, M., and Felsenfeld, G. Effect of the B-Z transition in poly(dG-m$^5$dc)· poly(dG-m$^5$dC) on nucleosome formation. Proc. Natl. Acad. Sci. U.S., *79*: 1771–1775, 1982.
22. Kanai, Y., Kawamitsu, H., Tanaka, M., Matsushima, T., and Miwa, M. A novel method for purification of poly(ADP-ribose). J. Biochem., *88*: 917–920, 1980.
23. Marmur, J. A procedure for the isolation of deoxyribonucleic acid from microorganisms. J. Mol. Biol., *3*: 208–218, 1961.
24. Stollar, B. D. and Papalian, M. Secondary structure in denatured DNA is responsible for its reaction with antinative DNA antibodies of systemic lupus erythematosus sera. J. Clin. Invest., *66*: 210–219, 1980.
25. Lafer, E. M., Möller, A., Nordheim, A., Stollar, B. D., and Rich, A. Antibodies specific for left-handed Z-DNA. Proc. Natl. Acad. Sci. U.S., *78*: 3546–3550, 1981.
26. Tan, E. M., Robinson. J., and Robitaille, P. Studies on antibodies to histones by immunofluorescence. Scand. J. Immunol., *5*: 811–818, 1976.
27. Gioud, M., Kaci, M. A., and Monier, J. C. Histone antibodies in systemic lupus erythematosus. A possible diagnostic tool. Arthritis Rheum., *25*: 407–413, 1982.
28. Hargraves, M. M., Richmond, H., and Morton, R. Presentation of two bone marrow elements: the "tart" cell and the "L.E" cell. Proc. Mayo Clin., *23*: 25–28, 1948.
29. Koffler, D., Carr, R. I., Agnello, V., Fiezi, T., and Kunkel, H. G. Antibodies to polynucleotides: distribution in human serums. Science, *166*: 1648–1649, 1969.
30. Norman, D. D., Kurata, N., and Tan, E. M. Profiles of antinuclear antibodies in systemic rheumatic diseases. Ann. Intern. Med., *83*: 464–469, 1975.
31. Lafer, E. M., Valle, R.P.C., Möller, A., Nordheim, A., Schur, P. H., Rich, A., and Stollar, B. D. Z-DNA-specific antibodies in human systemic lupus erythematosus. J. Clin. Invest., *71*: 314–321, 1983.

# ENZYMATIC MONO(ADP-RIBOSYL)ATION

ADP-RIBOSYLATION, DNA REPAIR AND CANCER, M. MIWA ET AL. (EDS.), JAPAN SCI. SOC. PRESS, TOKYO/VNU SCIENCE PRESS, UTRECHT, PP. 103-109, 1983

# Characterization of NAD: Arginine ADP-ribosyltransferases in Animal Tissues

Joel Moss, Paul A. Watkins, and David A. Yost

*Laboratory of Cellular Metabolism, National Heart, Lung, and Blood Institute, National Institutes of Health, Bethesda, Maryland 20205, U.S.A.*

*Abstract:* Two NAD: arginine ADP-ribosyltransferases (transferase "A" and "B") were identified in turkey erythrocytes and purified to homogeneity. Both transferases in the presence of NAD catalyzed the ADP-ribosylation of arginine, other low molecular weight guanidino compounds and proteins. ADP-ribosyltransferase A was activated by chaotropic salt or histone. Activation was associated with the disaggregation of an inactive, rapidly sedimenting, high molecular weight species to a protomeric form of ~28,000 daltons; this protomer ↔ aggregate transition was rapidly reversible. In the presence of salt, the $K_m$'s for NAD and arginine methyl ester were 15 μM and 1.3 mM, respectively; the turnover number for the purified enzyme was ~9,900 mol·min$^{-1}$·mol enzyme$^{-1}$. ADP-ribosyltransferase B exhibited a substrate specificity clearly distinct from that of transferase A. Transferase B had a $M_r$ of 32,000, slightly larger than that of the transferase A protomer. The activity of transferase B was unaffected by histone and inhibited by chaotropic salts; its $K_m$'s for NAD and arginine methyl ester of 36 μM and 3 mM, respectively, were similar to those obtained with transferase A. These studies are consistent with the presence of two different NAD: arginine ADP-ribosyltransferases in turkey erythrocytes exhibiting distinct kinetic, regulatory, and physical properties.

ADP-ribosylation of proteins, resulting from the transfer of the ADP-ribose moiety of NAD to an amino acid in a protein acceptor, is critical to the regulation of several key metabolic pathways. Two types of ADP-ribosyltransferases have been described: a) poly(ADP-ribose) synthetases catalyze the formation of sometimes branched homopolymers of ADP-ribose moieties covalently linked to protein (*1, 2*); b) mono-ADP-ribosyltransferases attach a single ADP-ribose moiety to a specific amino acid in a protein (*3*). Animal cells contain both poly- and mono-ADP-ribosyltransferases (*1–3*). Poly(ADP-ribosyl)ation, primarily a nuclear process, appears to be involved in DNA excision/repair, cellular proliferation and differentiation (*1, 2*); in contrast, the function of endogenous mono(ADP-ribosyl)ation in animal cells is not known.

Certain bacterial toxins which act on animal cells do so by catalyzing the mono-(ADP-ribosyl)ation of regulatory enzymes or proteins in the host cell (*4–7*). One set of bacterial toxins acts on the hormone-sensitive adenylate cyclase system. Choleragen (cholera toxin) and *Escherichia coli* (*E. coli*) heat-labile enterotoxin, agents responsible in part for cholera and traveler's diarrhea, respectively, activate adenylate cyclase by ADP-ribosylating a regulatory protein of the cyclase system (*4*); a *Bordetella pertussis* toxin also modifies a component of the cyclase system, leading to the loss of responsiveness to inhibitory hormones and an enhanced response to stimulatory agents (*5*). Diphtheria toxin and Pseudomonas Exotoxin A belong to a second set of toxins which inhibit protein synthesis in animal cells by ADP-ribosylating elongation factor II (*6, 7*); this modification leads to the death of susceptible cells. In addition to the bacterial toxins, mono-ADP-ribosyltransferases have been identified in turkey erythrocytes (*8*); these enzymes have been purified to homogeneity (*9*), and their kinetic, physical and regulatory properties examined.

### *Identification and Purification of NAD: Arginine ADP-ribosyltransferase A from Erythrocytes*

An NAD: arginine ADP-ribosyltransferase from the soluble fraction of turkey erythrocytes was purified >500,000-fold by successive chromatography on phenyl-Sepharose, carboxymethyl cellulose, NAD-agarose, and concanavalin A agarose (*9*). The specific activity of the purified transferase, approximately 350 $\mu$mol$\cdot$min$^{-1}\cdot$mg$^{-1}$ (*9*), was comparable to or higher than that of other enzymes, such as NAD glycohydrolase (*10*) or poly (ADP-ribose) synthetase (*11, 12*), that utilize NAD as an ADP-ribose donor. The purified transferase appeared as a single protein band of $M_r$=28,300 on sodium dodecyl sulfate-polyacrylamide gel electrophoresis (*9*). Based on a monomeric molecular weight of 28,300, the specific activity corresponds to a turnover number of ~9,900 mol of ADP-ribose transferred to arginine methyl ester per min per mol of enzyme. On gel permeation chromatography (Ultrogel AcA 54), only one protein peak with a $K_{av}$ slightly less than that of chymotrypsinogen (25,300 daltons) was identified (*9*). This protein cochromatographed with ADP-ribosyltransferase activity; a constant specific activity was observed across the protein peak (*9*). The purified transferase used both proteins and low molecular weight guanidino compounds such as arginine and agmatine as ADP-ribose acceptors (*13*). The preference for guanidino compounds was similar to that observed previously with choleragen and *E. coli* heat-labile enterotoxin, with agmatine$\geqslant$arginine methyl ester> D- and L-arginine>guanidinopropionate (*14*). Compounds with a positive charge in the vicinity of the guanidino group (agmatine, arginine methyl ester) were clearly more effective than those with a negative charge (arginine, guanidinopropionate). The reaction catalyzed by the transferase was stereospecific; the enzyme used $\beta$-NAD as substrate and in the presence of L-arginine formed $\alpha$-ADP-ribose-L-arginine, which rapidly anomerized to give $\alpha$- and $\beta$-ADP-ribose-L-arginine (*13*). The enzyme used both NAD and NADP as donors of ADP-ribose and phosphoribosyl-ADP-ribose, respectively; NAD was clearly more effective than NADP (*13*). The ability of the avian enzyme to use NADP contrasts markedly with the inability of the bacterial toxin choleragen to use this nucleotide (*15, 16*).

In addition to catalyzing the ADP-ribosylation of low molecular weight guanidino compounds, the transferase modified various purified proteins, such as lysozyme, histones, and polyarginine (*8*). Several particulate and soluble erythrocyte proteins were also ADP-ribosylated by the transferase (Moss and Stanley, unpublished observations). The substrate specificity for both purified proteins and crude mixtures differed from that of choleragen.

The product of the transferase reaction in all cases appeared to be a mono-ADP-ribosylated protein. Snake venom phosphodiesterase digestion of proteins, previously [$^{14}$C] ADP-ribosylated with transferase and [adenine-U-$^{14}$C] NAD, released [$^{14}$C]-5′-AMP (*8*); [$^{14}$C] phosphoribosyl-5′-AMP, a major product of phosphodiesterase digestion of poly-ADP-ribosylated proteins, was not found. ADP-ribosylation of these proteins was inhibited by guanidino compounds. This finding is consistent with the proposal that both substrates compete for the same catalytic site (*8*).

### *Activation of NAD: Arginine ADP-ribosyltransferase A by Histones and Chaotropic Salts*

The activity of the NAD: arginine ADP-ribosyltransferase was enhanced 10- to 20-fold in the presence of salts (*17*, *18*); the potency of salts with $SCN^- > Cl^- > PO_4^{3-}$, corresponded to the Hofmeister series for the salting out of englobulins (*18*). Maximal activity in NaCl was observed at concentrations between 200–300 mM; an equivalent concentration for NaSCN was ~70 mM. Activation of the transferase by salt was rapid and reversible and appeared to be independent of effects on enzyme stability (*18*). Gel permeation chromatography and sedimentation velocity analysis revealed that, in the absence of salt, the transferase existed as a high molecular weight species (Fig. 1) (*18*). Salt dissociated the transferase to a protomeric form whose molecular weight was similar to the monomeric species observed on sodium dodecyl sulfate-polyacrylamide gels (Fig. 1) (*8*).

The transferase was also activated by histones at micromolar concentrations

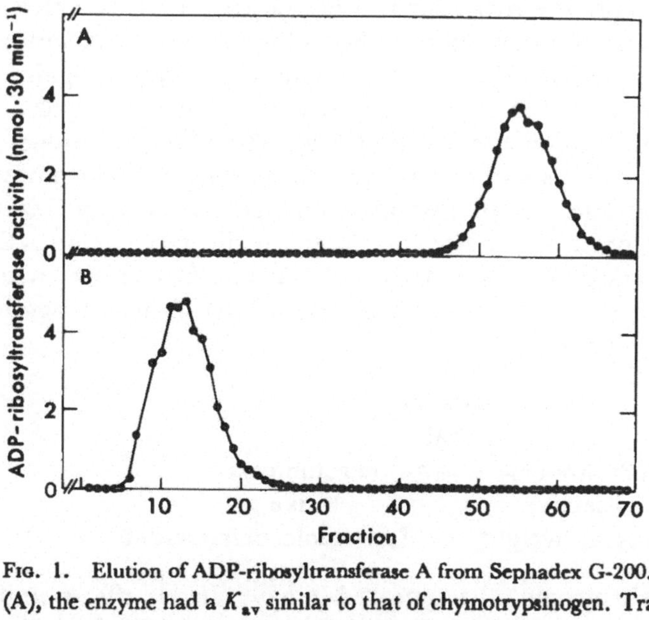


FIG. 1. Elution of ADP-ribosyltransferase A from Sephadex G-200. In the presence of 1 M NaCl (A), the enzyme had a $K_{av}$ similar to that of chymotrypsinogen. Transferase A eluted in the void volume when NaCl was omitted (B). (Data from ref. *18*).

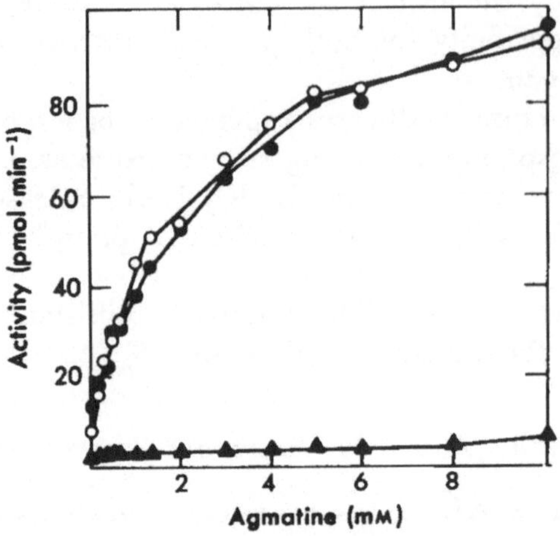


FIG. 2. Effect of NaCl and histone on the activity of ADP-ribosyltransferase A, assayed with the indicated concentration of agmatine and no addition (▲), 200 mM NaCl (○), or histone (●), 20 $\mu$g/ml. Maximal activation was observed with these concentrations of additions. (Data from ref. *19*).

(*17*, *19*). Histone accelerated the ADP-ribosylation of agmatine and other low molecular weight guanidino compounds; the extent of activation observed with histone was identical to that obtained with NaCl (Fig. 2). Histones were effective at concentrations where they did not serve as substrates for the transferase (*17*); maximal stimulation of the ADP-ribosyltransferase under standard assay conditions was observed with 30 $\mu$g/ml histone, whereas histones served as ADP-ribose acceptor at ~3 mg/ml. Both lysine- and arginine-rich histones were effective in stimulating transferase activity, but arginine-rich histone were clearly more effective as ADP-ribose acceptors (Moss and Stanley, unpublished observations). Activation by histone was rapid and reversible (*19*). Histone also stabilized the enzyme against thermal denaturation at concentrations that were clearly related to those required to activate rather than to serve as an ADP-ribose acceptor (*19*). The effect of histone on transferase activity was specific and was not observed with other purified proteins (*17*); in particular, polycations such as ployarginine and polylysine were inactive. Although polyarginine did not activate the transferase, it did serve as an ADP-ribose acceptor in a reaction catalyzed by the activated enzyme. Concurrent with activation of transferase A by histone, the enzyme was dissociated from the aggregated to the protomeric form (*19*). Thus,

$$\underset{\text{High molecular weight}}{\underset{\text{Inactive}}{(\text{Protomer})_n}} \;\underset{}{\overset{\text{Histone}}{\underset{\text{Salt}}{\rightleftharpoons}}}\; \underset{\text{Low molecular weight}}{\underset{\text{Active}}{(\text{Protomer})_n}} \qquad (1)$$

activation by histone or chaotropic salts appeared to have similar effects on the physical and kinetic properties of the enzyme (Reaction 1).

### *Isolation of a Histone- and Salt-independent NAD: Arginine ADP-ribosyltransferase*

A second NAD: arginine ADP-ribosyltransferase (ADP-ribosyltransferase B) was identified and purified from turkey erythrocytes. This enzyme differed from transferase A in that the activity was independent of salt and histone (*20*). Both transferase A and transferase B were present in the soluble fraction from turkey erythrocytes. Unlike transferase A, transferase B did not specifically bind to phenyl-Sepharose. Transferase B was further purified from the breakthrough of a phenyl-Sepharose column by chromatography on concanavalin A-Sepharose, carboxy-methyl cellulose, Procion Red-agarose, and Ultrogel AcA 54. The purified transferase exhibited one band on sodium dodecyl sulfate-polyacrylamide gel electrophoresis;

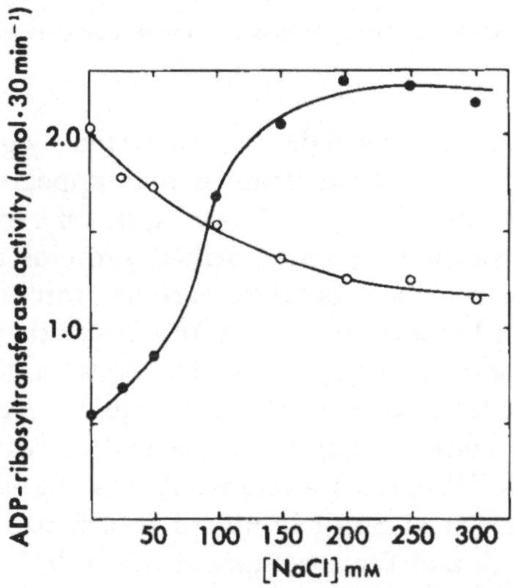


FIG. 3. Effect of NaCl on transferases A (●) and B (○) activity using agmatine as substrate. Whereas NaCl activated transferase A, transferase B was inhibited. (Data from ref. *20*).

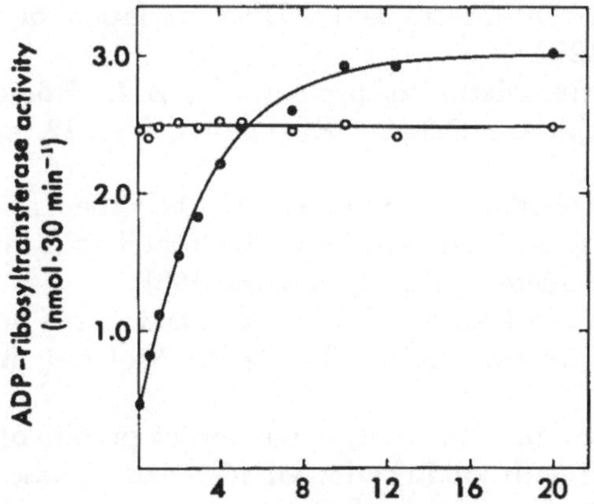


FIG. 4. Effect of histone on activity of transferases A (●) and B (○) using agmatine as substrate. Histone activated transferase A but had no effect on transferase B. (Data from ref. *20*).

TABLE 1. Comparison of the Kinetic and Physical Properties of NAD: Arginine ADP-ribosyltransferases A and B

| Property | Transferase A | Transferase B |
|---|---|---|
| $K_m$ NAD | ~15 $\mu$M | ~36 $\mu$M |
| $K_m$ arginine methyl ester | 1.3 mM[b] | 3.0 mM |
| $V_{max}$ ($\mu$mol·min$^{-1}$·mg$^{-1}$) | 350 | 60 |
| Effect of histone | | |
| activity | ↑↓ | None |
| size | ↓↓ | None |
| Effect of chotropic salts | | |
| activity | ↑↑ | ↓↓ |
| size | ↓↓ | None |
| Protomeric molecular weight[a] | ~28,000 | ~32,000 |

[a] Estimated by sodium dodecyl sulfate-polyacrylamide electrophoresis and gel permeation chromatography.
[b] Determined in the presence of NaCl.

the apparent $M_r$ was 32,000. A similar molecular weight was estimated by gel filtration on Ultrogel AcA 54 and TSK 3,000 SW. Transferase B thus appeared to be slightly larger than the transferase A monomer. Like transferase A, transferase B used both low molecular weight guanidino compounds and purified proteins as ADP-ribose acceptors. The substrate specificity of transferase B for various proteins clearly differed from that of transferase A. As with transferase A, ADP-ribosylation of proteins by transferase B was inhibited by arginine methyl ester. The $K_m$'s for NAD and arginine methyl ester were 36 $\mu$M and 3 mM, respectively, with the $V_{max}$ being 60 $\mu$mol·min$^{-1}$·mg$^{-1}$. The major difference between transferases A and B was in their responses to histones and chaotropic salts. Whereas transferase A was stimulated by both histone and salts, histone had no effect and salt inhibited transferase B (Figs. 3, 4). Some characteristics of transferases A and B are compared in Table 1.

## REFERENCES


1. Hayaishi, O. and Ueda, K. Poly(ADP-ribose) and ADP-ribosylation of proteins. Annu. Rev. Biochem., *46*: 95–116, 1977.
2. Pekala, P. and Moss, J. Poly ADP-ribosylation of proteins. *In;* B. L. Horecker and E. R. Stadtman (eds.), Current Topics in Cellular Regulation, Vol. 22, pp. 1–240, Academic Press, New York, 1983.
3. Vaughan, M. and Moss, J. Mono(ADP-ribosyl)transferases and their effects on cellular metabolism. *In;* B. L. Horecker and E. R. Stadtman (eds.), Current Topics in Cellular Regulation, Vol. 20, pp. 205–246, Academic Press, New York, 1981.
4. Moss, J. and Vaughan, M. Mechanism of action of choleragen and *E. coli* heat-labile enterotoxin: Activation of adenylate cyclase by ADP-ribosylation. Mol. Cell. Biochem., *37*: 75–90, 1981.
5. Katada, T. and Ui, M. ADP ribosylation of the specific membrane protein of C6 cells by islet-activating protein associated with modification of adenylate cyclase activity. J. Biol. Chem., *257*: 7210–7216, 1982.
6. Pappenheimer, A. M., Jr. Diphtheria toxin. Annu. Rev. Biochem., *46*: 69–94, 1977.
7. Iglewski, B. H. and Kabat, D. NAD-dependent inhibition of protein synthesis by *Pseudomonas aeruginosa* toxin. Proc. Natl. Acad. Sci. U.S., *72*: 2284–2288, 1975.

8. Moss, J. and Vaughan, M. Isolation of an avian erythrocyte protein possessing ADP-ribosyltransferase activity and capable of activating adenylate cyclase. Proc. Natl. Acad. Sci. U.S., *75*: 3621–3624, 1978.
9. Moss, J., Stanley, S. J., and Watkins, P. A. Isolation and properties of an NAD- and guanidine-dependent ADP-ribosyltransferase from turkey erythrocytes. J. Biol. Chem., *255*: 5838–5840, 1980.
10. Swislocki, N. I. and Kaplan, N. O. Purification and characterization of diphosphopyridine nucleosidase from pig brain. J. Biol. Chem., *242*: 1083–1088, 1967.
11. Mandel, P., Okazaki, H., and Niedergang, C. Purification and properties of calf thymus polyadenosine diphosphate ribose polymerase. FEBS Lett., *84*: 331–336, 1977.
12. Ito, S., Shizuta, Y., and Hayaishi, O. Purification and characterization of poly(ADP-ribose) synthetase from calf thymus. J. Biol. Chem., *254*: 3647–3651, 1979.
13. Moss, J., Stanley, S. J., and Oppenheimer, N. J. Substrate specificity and partial purification of a stereospecific NAD- and guanidine-dependent ADP-ribosyltransferase from avian erythrocytes. J. Biol. Chem., *250*: 8891–8894, 1979.
14. Moss, J. and Vaughan, M. Specificity of ADP-ribosylation by choleragen (cholera toxin), *E. coli* heat-labile enterotoxin and an arginine- and NAD-dependent avian erythrocyte transferase. *In;* M. E. Smulson and T. Sugimura (eds.), Developments in Cell Biology, Vol. 6, Novel ADP-Ribosylations of Regulatory Enzymes, and Proteins, pp. 391–399, Elsevier/North-Holland, New York, 1980.
15. Gill, D. M. Involvement of nicotinamide adenine dinucleotide in the action of cholera toxin *in vitro*. Proc. Natl. Acad. Sci. U.S. *72*: 2064–2068, 1975.
16. Moss, J., Osborne, J. C., Jr., Fishman, P. H., Brewer, H. B., Jr., Vaughan, M., and Brady, R. O. Effect of gangliosides and substrate analogues on the hydrolysis of nicotinamide adenine dinucleotide by choleragen. Proc. Natl. Acad. Sci. U.S., *74*: 74–78, 1977.
17. Moss, J. and Stanley, S. J. Histone-dependent and histone-independent forms of an ADP-ribosyltransferase from human and turkey erythrocytes. Proc. Natl. Acad. Sci. U.S., *78*: 4809–4812, 1981.
18. Moss, J., Stanley, S. J., and Osborne, J. C., Jr. Effect of self-association on activity of an ADP-ribosyltransferase from turkey erythrocytes: Conversion of inactive oligomers to active protomers by chaotropic salts. J. Biol. Chem., *256*: 11452–11456, 1981.
19. Moss, J., Stanley, S. J., and Osborne, J. C., Jr. Activation of an NAD: arginine ADP-ribosyltransferase by histone. J. Biol. Chem., *257*: 1660–1663, 1982.
20. Yost, D. A. and Moss, J. Amino acid-specific ADP-ribosylation: Evidence for two distinct NAD: arginine ADP-ribosyltransferases in turkey erythrocytes. J. Biol. Chem., *258*: 4926–4929, 1983.

ADP-RIBOSYLATION, DNA REPAIR AND CANCER, M. MIWA ET AL. (EDS.),
JAPAN SCI. SOC. PRESS, TOKYO /VNU SCIENCE PRESS, UTRECHT, PP. 111-118, 1983

# *Xenopus* Mono(Adenosine Diphosphate Ribosyl) Transferase: Purification, Assay, and Properties

Francois GODEAU and S. S. KOIDE
*The Population Council, The Rockefeller University, New York, New York 10021, U.S.A.*

*Abstract:* Various tissues of *Xenopus laevis* possess mono(ADP-ribosyl) transferase activity. The *Xenopus* liver enzyme was purified approximately 40,000-fold by sequential chromatography on phenyl Sepharose, DEAE-Sephadex, $NAD^+$-Agarose, and Con A-Sepharose. The enzyme is relatively heat-stable, has no cation requirements, transfers monomers of ADP-ribose from $NAD^+$ to arginine-rich histones H3 and H4, and is not dependent upon DNA for activity. Its $M_r$ was estimated to be 30 kilodaltons (KD). The amino acid acceptor of the ADP-ribose moiety is probably arginine, and the bond formed is resistant to the hydrolytic action of neutral hydroxylamine. The transferase activity was neither inhibited nor precipitated by antisera directed against the A fragment of cholera or diphtheria toxins. Theophylline, thymidine, nicotinamide, and some of its analogues blocked the enzymatic activity, while nicotinic acid was ineffective. When a DNA-free extract of HeLa cells was incubated with the transferase system, several proteins were found to be ADP-ribosylated. A band at position 90 KD showed high specific radioactivity.

A zymographic method for assaying transferase activity in crude tissue extracts was developed. In the initial step, the extracts were analyzed by sodium dodecyl sulfate (SDS)-polyacrylamide gel electrophoresis. Using this method, a histone-dependent mono(ADP-ribosyl) transferase corresponding to an $M_r$ of about 30 KD was demonstrated in every tissue examined. Treatment with reducing agents practically abolished transferase activity in contrast to the enhancement of choleragen A fragment activity. A variant of the transferase at about 80 KD was detected. It differed from the 30 KD enzyme in that histones inhibited its activity while agmatine had no influence.

The post-translational modification of proteins by ADP-ribosylation is mediated by two distinct enzymatic systems, *i.e.*, the poly(ADP-ribosyl) synthetase and mono (ADP-ribosyl) transferase. The former enzyme is located in the cell nucleus, is dependent upon DNA for activity and links covalently the ADP-ribose moiety of $NAD^+$ to proteins by an ester bond to form chains of varying lengths. The latter enzyme is located in the cytoplasm and affects mono(ADP-ribosyl)ation of acceptor proteins.

Poly(ADP-ribose) formation was reported to be involved in DNA synthesis and/or repair (*1–4*), in chromatin structure (*5, 6*), and cellular differentiation (*7, 8*).

Mono(ADP-ribosyl) transferase activity has been demonstrated in the following extranuclear sites: mitochondria (*9–11*); ribosomes (*12*); *Xenopus* oocytes (*13*); and mouse testicular and Sertoli cells (*14*). The enzyme from turkey erythrocytes has been purified and the monomeric form has an estimated $M_r$ of 28 KD (*15–18*). Both $NAD^+$ and $NADP^+$ are substrates. The $K_m$ of $NAD^+$ is about 30 $\mu$M. The purified enzyme stimulates adenylate cyclase obtained from bovine brain, while that from turkey erythrocytes and other tissues does not (*18*).

Thus, a function of the transferase is to stimulate cAMP formation. The site of modification on the cyclase by ADP-ribosylation may be the regulatory subunit as demonstrated for choleragen (*18, 19*). The following observations support this contention. Rat hepatocytes and thyroid cells possess a membrane-bound ADP-ribosyl transferase activity (*20*) which is stimulated by isoproterenol (*21*) and thyrotropin (*22, 23*), respectively, suggesting that the increase in cAMP formation induced by the hormone is dependent upon ADP-ribosylation of the cyclase probably by modification of the guanine nucleotide regulatory subunit.

### *Mono ADP-ribosylated Proteins*

The presence of a distinct mono ADP-ribosylating system is supported by the demonstration of the existence of mono ADP-ribosylated proteins in cells and subcellular fractions. These modified proteins were found in mitochondria, plasma membranes, microsomes, and cytosol fractions of rat liver (*24, 25*). Various proteins act as acceptors and vary widely in their sizes. The covalent bond between the ADP-ribose and proteins isolated from mitochondria and plasma membrane is re-

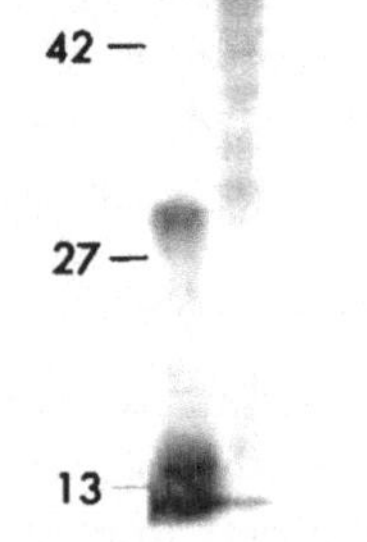


FIG. 1. Radioautogram of mono ADP-ribosylated proteins in extract from HeLa cells. Extract was prepared according to Manley *et al.* (*26*) and was incubated with purified *Xenopus* liver ADP-ribosyl transferase. The reaction mixture was analyzed by SDS-polyacrylamide gel electrophoresis; a, calf thymus histones used as control; b, HeLa cell extract. Arrow denotes 90 KD position.

sistant to hydroxylamine treatment (*9, 24, 25*), suggesting a non-ester bond possibly to the guanidine group of arginine.

A cell-free system prepared from HeLa cells was used to assay for transcriptional capacity or RNA synthesis as described by Manley (*26*). This extract does not contain DNA and has no endogenous mono(ADP-ribosyl) transferase activity. Poly-(ADP-ribosyl) synthesis activity can be demonstrated in the extract by the addition of DNA and radiolabeled $NAD^+$. When purified *Xenopus* liver transferase was added to the HeLa cell extract and incubated with radiolabeled $NAD^+$ and the reaction mixture analyzed by SDS-polyacrylamide gel electrophoresis, a major band with high incorporated radioactivity was detected at an estimated position of 90 KD (Fig. 1). Properties of this acceptor substance are being determined.

### *Assay for ADP-ribosyl Transferase Activity*

The previously reported method for the determination of transferase activity is to measure the amount of nicotinamide liberated from radiolabeled $NAD^+$ (*15, 16*). It is not feasible to measure transferase activity in crude tissue extract by this method because of the presence of competing enzymes such as $NAD^+$ glycohydrolase, ADP-ribose protein hydrolase, and phosphodiesterase. To overcome these drawbacks, a new zymographic method was developed. Interference by competing enzymes was

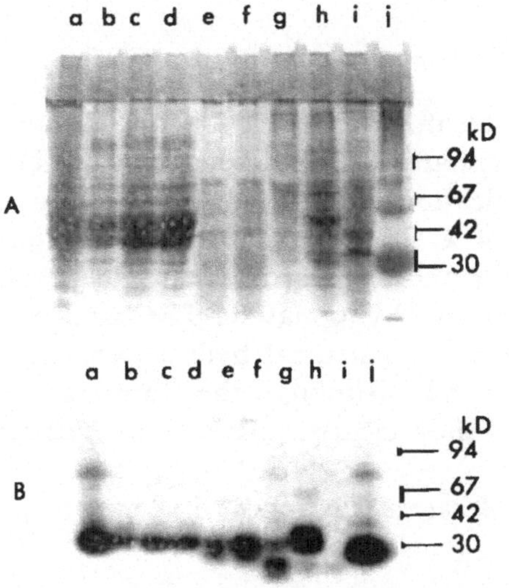


FIG. 2. Zymographic assay for ADP-ribosyl transferase activity in crude extracts of *Xenopus laevis* tissues. Tissues were homogenized in medium containing 50 mM potassium phosphate buffer, pH 7.0, 1 mM 2-mercaptoethanol, 2 mM EDTA, 2 mM phenylmethylsulfonylfluoride, and 0.2% Nonidet-P40 centrifuged at 2,000 × *g* for 10 min. The supernatant sample was suspended in a medium (1 : 4 v/v) containing 40% glycerol, 0.01% bromophenol blue, 2 mM Tris-HCl, pH 6.8, 4% SDS, and no reducing agent for 5 min at 30°C. Electrophoresis was performed as described by Laemmli (*30*). Gel was immersed in 25% Triton X-100 solution, washed and incubated with [$^{32}$P]NAD for 4 hr at 30°C. Gel was fixed in a solution containing 5% trichloroacetic acid (TCA), 1 mM NAD, and 0.25% Coomassie Blue, washed and processed for radioautography. A. Coomassie Blue staining. B. Radioautogram.

minimized, if not eliminated, by an initial step of separation by electrophoresis and subsequent detection and measurement by radioautography.

In brief, crude tissue extracts were subjected to SDS-polyacrylamide gel electrophoresis under non-reducing conditions in the presence of Triton X-100. The enzymatic activity was detected by incubating the gel in a medium containing [$^{32}$P]-NAD$^+$ and histones. The gel was immersed in 50% trichloroacetic acid, thereby fixing the radiolabeled ADP-ribosylated proteins. The gel was then washed and developed by an radioautographic procedure (Fig. 2). The density of the spots on the film was measured. Transferase activity in tissue extracts containing 10 to 80 $\mu$g of proteins was measured.

*Transferase Activity in Various Tissues*

ADP-ribosylating activity was found in all *Xenopus* tissues examined. The activity was highest in skeletal muscle, heart, liver, and erythrocytes. The radioactivity was incorporated into a major protein band corresponding to 30 KD. When histones were omitted from the reaction mixture, there was a slight residual activity, suggesting auto-ADP-ribosylation of the enzymes. Agmatine, at a concentration of 100 mM, blocked the transferase activity by about 90%. The enzymatic activity was extremely sensitive to treatment with a reducing agent. Less than 1% of the activity remained after treatment with 2-mercaptoethanol.

*Comparison of the Properties of Xenopus Hepatic Transferase and Cholera Toxin*

Mono(ADP-ribosyl) transferase was purified from the soluble fraction of *Xenopus* liver by chromatography on phenyl-Sepharose, DEAE-Sephadex, NAD$^+$-Agarose and Con A-Sepharose (Table 1). A purification of about 40,000-fold was achieved. The transferase was stable to heating at 56°C for 30 min and did not require DNA for activity. It had no cation requirements and utilized histones and protamines as acceptors. The enzymatic activity was inhibited by agmatine. The bond between the ADP-ribose moiety and histone was resistant to neutral hydroxylamine. The estimated $M_r$ of the transferase was about 30 KD. Although the transferase and choleragen have certain common properties, they clearly possess other discriminating features.

The ADP-ribosylating activity of choleragen was weak compared to that of the

TABLE 1. Purification of Mono(ADP-ribosyl) Transferase from *Xenopus* Liver

| Purification steps | Total units (pmol·min$^{-1}$) | Specific activity (pmol·min$^{-1}$·mg$^{-1}$) | Protein yield purification (mg) | (%) | (folds) |
|---|---|---|---|---|---|
| Cytosol (100,000×*g* superatant) | 390 | 0.4 | 900 | 100 | 1 |
| Phenyl-Sepharose | 2,440 | 14 | 166 | 620 | 34 |
| DEAE-Sephadex | 3,930 | 187 | 21 | 1,000 | 435 |
| NAD$^+$-Agarose | 1,685 | 887 | 1.9 | 432 | 2,062 |
| Con A-Sepharose | 400 | 17,400 | 0.02 | 102 | 40,000 |

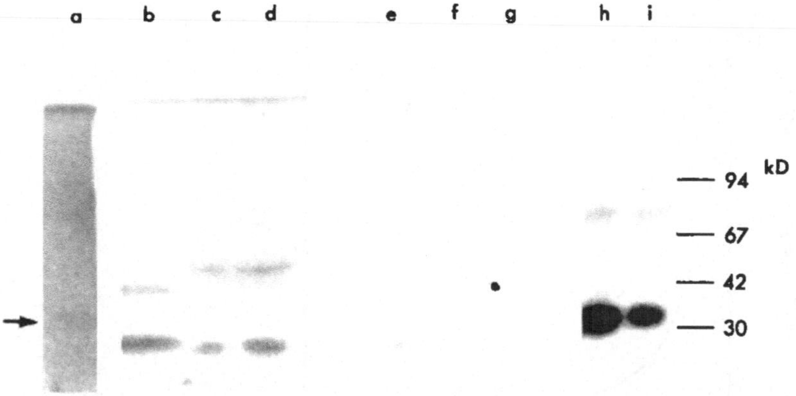


FIG. 3. Effects of reducing agents on choleragen and *Xenopus* muscle mono(ADP-ribosyl) transferase activities determined by zymographic analysis. Replicate samples of cholera toxin and crude extract of *Xenopus laevis* skeletal muscle were treated with and without 5% 2-mercaptoethanol and subjected to zymographic analysis (see Fig. 2). The reaction mixture for ADP-ribosylation contained [$^{32}$P]NAD and calf thymus histones (2 mg/ml). a, unreduced cholera toxin, 10 $\mu$g (arrow); b, e, reduced cholera toxin, 10 $\mu$g; c, f, reduced *Xenopus* muscle extract, 20 $\mu$g; d, g, reduced *Xenopus* muscle extract, 40 $\mu$g; h, unreduced *Xenopus* muscle extract, 40 $\mu$g; i, unreduced *Xenopus* muscle extract, 20 $\mu$g. a–d, 12 days exposure; e–i, 1 day exposure.

transferase when determined by the described zymographic method (Fig. 3). Treatment with 2-mercaptoethanol markedly stimulated the ADP-ribosylating activity of choleragen by forming the active Al fragment while that of the transferase was drastically reduced. The reduced transferase migrated with an apparent mobility of 20 KD while that of the active cholera toxin Al fragment migrated at about 22 KD. These findings suggest that choleragen and transferase are composed of distinctly different subunits linked by disulfide bonds. Antiserum to choleragen A fragment did not block transferase activity nor immunoprecipitated the enzyme.

### *Variants of ADP-ribosyl Transferase*

An active band corresponding to 80 KD was found on analysis of *Xenopus* tissue extracts by the above described zymographic method (Fig. 3). This enzymatic activity was inhibited by histones but not by other guanine compounds, suggesting a distinctly different ADP-ribosylating system from the 30 KD enzyme. Moreover, treatment with 2-mercaptoethanol yielded reduced fragments of about 45 KD. Although the identity of this band is not known, it might be NAD+ glycohydrolase, poly(ADP-ribosyl) synthetase or a variant of ADP-ribosyl transferase.

### *Functional Role of Transferase*

Potent inhibitors of the transferase were tested for their capacity to block maturation in *Xenopus* oocytes induced with progesterone. It has been demonstrated that all the agents increasing cAMP levels inhibit germinal vesicle breakdown (GVBD) (*27*). Inhibitors of transferase activity also blocked progesterone-induced matura-

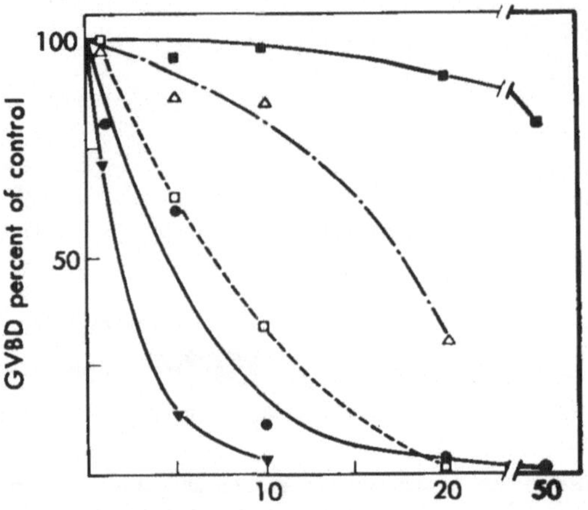


FIG. 4. Effect of varying concentrations of inhibitors of ADP-ribosyl transferase activities on progesterone-induced maturation of *Xenopus* oocyte. ■ nicotinic acid; △ 7-methyl nicotinamide; □ 3-aminobenzamide; ● nicotinamide; ▼ methyl nicotinate.

tion of *Xenopus* oocytes (Fig. 4). Nicotinamide did not influence protein synthesis in *Xenopus* oocytes not alter intracellular NAD+ levels; however, cAMP content increased significantly. The burst of protein phosphorylation and GVBD induced by microinjection with maturation promoting factor (*28*) were blocked by nicotinamide treatment (*29*). These present results suggest that mono(ADP-ribosyl) transferase may be involved in the modulation of adenylate cyclase activity in intact cells and that inhibition of ADP-ribosylation may not be sufficient to inactivate the cyclase.

## CONCLUDING REMARKS

Choleragen A fragment and ADP-ribosyl transferase share common features, suggesting that the animal enzyme may simulate the toxin by activating adenylate cyclase (*18–23*). The $M_r$ of the unreduced (30 KD) and the reduced (22 KD) forms of these two factors are comparable. They also mediate the same biochemical reaction; namely, transfer of monomers of ADP-ribose from NAD+ to the guanidine group of arginine or arginine-containing proteins (*13–15, 17–23*). In spite of their difference in turnover number and substrate specificity (*15–17*), it is reasonable to propose that they promote, influence or serve similar cellular functions. We have explored this thesis and have uncovered additional discriminating characteristics as follows: no immunologic homology between choleragen A fragment and *Xenopus* transferase was found; effects of reducing agents on these two factors were markedly different, indicating basic differences in their structure; inhibition of oocyte maturation by inhibitors of ADP-ribosyl transferase suggests that in addition to adenylate cyclase other key enzymes may be modulated by ADP-ribosylation in intact cells.

## ACKNOWLEDGMENTS

This study was supported by PHS NICHD Grant No. RO1-HD-13184 and Rockefeller Foundation Grant No. GA PS 8115 and 8208. The authors gratefully thank Dr. M. Gill for the gift of antiserum against cholera toxin A fragment.

## REFERENCES


1. Burzio, L. and Koide, S. S. A functional role of poly ADPR in DNA synthesis. Biochem. Biophys. Res. Commun., *40*: 1013–1019, 1970.
2. Burzio, L. and Koide, S. S. Mode of inhibition of DNA synthesis induced by adenosine diphosphoribosylation of nuclear protein. Biochem. Biophys. Res. Commun., *42*: 1185–1190, 1971.
3. Miller, E. G. Stimulation of nuclear poly(adenosine diphosphate-ribose) polymerase activity from HeLa cells by endonucleases. Biochim. Biophys. Acta, *395*: 191–200, 1975.
4. Berger, N. A., Sikorski, G. W., Petzold, S. J., and Kurohara, K. K. Association of poly(adenosine diphosphoribose) synthesis with DNA damage and repair in normal human lymphocytes. J. Clin. Invest., *63*: 1164–1171, 1979.
5. Burzio, L. O., Puigdomench, P., Ruiz-Carrillo, A., and Koide, S. S. Adenosine diphosphate ribosylation and regulation of nucleic acid synthesis. *In;* M. C. Niu and H. H. Chuang (eds.), The Role of RNA in Development and Reproduction, pp. 114–131, Science Press, Beijing, 1981.
6. Poirier, G. G., de Murcia, G., Jongstra-Bilen, J., Niedergang, C., and Mandel, P. Poly(ADP-ribosyl)ation of polynucleosomes causes relaxation of chromatin structure. Proc. Natl. Acad. Sci. U.S., *79*: 3423–3427, 1982.
7. Caplan, A. I. and Rosenberg, M. J. Interrelationship between poly(ADP-Rib) synthesis, intracellular NAD levels and muscles or cartilage differentiation from mesodermal cells of embryonic chick limb. Proc. Natl. Acad. Sci. U.S., *72*: 1852–1857, 1975.
8. Pekala, P. H., Lane, M. D., Walkins, P. A., and Moss, J. On the mechanism of preadipocyte differentiation. Masking of poly(ADP-ribose) synthesis activity during differentiation of 3T3-L1 preadipocytes. J. Biol. Chem., *256*: 4871–4876, 1981.
9. Kun, E., Zimber, P. H., Chang, A. C., Puschendorf, B., and Grunicke, H. Macromolecular enzymatic product of $NAD^+$ in liver mitochondria. Proc. Natl. Acad. Sci. U.S., *72*:1436–1440, 1975.
10. Burzio, L. O., Luke, M., and Koide, S. S. Poly(adenosine diphosphate ribose) synthase activity in mitochondria of *Xenopus* oocytes. Fed. Proc., *38*: 618, 1979.
11. Burzio, L., Concha, I., Saez, L., and Koide, S. Adenosine diphosphate ribosylation of rat testis mitochondrial proteins. Fed. Proc., *41*: 678, 1982.
12. Roberts, J. H., Stark, P., Giri, C. P., and Smulson, M. Cytoplasmic poly(ADP-ribose) polymers during the HeLa cell cycle. Arch. Biochem. Biophys., *171*: 305–315, 1975.
13. Boquet, P., Godeau, F., and Duflot, E. The existence of an NAD dependent ADP-ribosyl transferase in *Xenopus laevis* oocytes. Eur. J. Cell. Biol., *22*: 112, 1980.
14. Momii, A. and Koide, S. S. Adenosine diphosphate-ribosyl transferase activity in permeabilized mouse testicular cells. Arch. Biochem. Biophys., *214*: 628–633, 1982.
15. Moss, J., Stanley, S. J., and Oppenheimer, N. J. Substrate specificity and partial purification of a stereospecific NAD and guanidine dependent ADP ribosyl transferase from avian erythrocytes. J. Biol. Chem., *254*: 8891–8894, 1979.
16. Moss, J., Stanley, S. J., and Watkins, T. Isolation and properties of an NAD and guanidine dependent ADP ribosyl transferase from turkey erythrocytes. J. Biol. Chem., *255*: 5838–5840, 1980.
17. Moss, J. and Stanley, S. J. Histone dependent and histone independent forms of an ADP ribosyl transferase from human and turkey erythrocytes. Proc. Natl. Acad. Sci. U.S., *78*: 4809–4812, 1981.
18. Moss, J. and Vaughan, M. Isolation of an avian erythrocyte protein possessing ADP-

ribosyl transferase activity and capable of activating adenylate cyclase. Proc. Natl. Acad. Sci. U.S., *75*: 3621–3624, 1978.

19. Moss, J. and Vaughan, M. Mechanism of action of choleragen: Evidence for ADP-ribosyl transferase activity with arginine as an acceptor. J. Biol. Chem., *252*: 2455–2457, 1977.
20. Beckner, S. K. and Blecher, M. Regulatory states of adenylate cyclase in RL-PR-C cloned rat hepatocytes. Biochim. Biophys. Acta, *673*: 467–476, 1980.
21. Reilly, T. M., Beckner, S., McHugh, E. M., and Blecher, M. Isoproterenol-induced ADP-ribosylation of a single plasma membrane protein of cultured differentiated RL-PR-C hepatocytes. Biochem. Biophys. Res. Commun., *98*: 1115–1120, 1981.
22. DeWolf, M.J.S., Vitti, P., Ambesi-Impiombato, F. S., and Kohn, L. D. Thyroid membrane ADP ribosyl transferase activity. Stimulation by thyrotropin and activity in functioning and nonfunctioning rat thyroid cells in culture. J. Biol. Chem., *256*: 12287–12296, 1981.
23. Vitti, P., DeWolf, M.J.S., Acquaviva, A. M., Epstein, M., and Kohn, L. D. Thyrotropin stimulation of the ADP-ribosyl transferase activity of bovine thyroid membrane. Proc. Natl. Acad. Sci. U.S., *79*: 1525–1529, 1982.
24. Wielckens, K., Bredehorst, R., Adamietz, P., and Hilz, H. Protein-bound polymerase and monomeric ADP-ribose residues in hepatic tissues. Comparative analyses using a new procedure for the purification of poly(ADP-ribose). Eur. J. Biochem., *117*: 69–74, 1981.
25. Adamietz, P. Acceptor proteins of poly(ADP-ribose). *In;* O. Hayaishi and K. Ueda (eds.), ADP-Ribosylation Reactions, pp. 77–101, Academic Press, New York, 1982.
26. Manley, J. L., Fire, A., Cano, A., Sharp, P. A., and Gefter, M. L. DNA-dependent transcription of adenovirus genes in a soluble whole cell extract. Proc. Natl. Acad. Sci. U.S., *77*: 3855–3859, 1980.
27. Godeau, F., Schorderet-Slatkine, S., Boquet, P., Schoderet, M., Younglai, E. V. and Baulieu, E.-E. Involvement of cAMP in the cytoplasmic control of meiotic cell division in *Xenopus laevis* oocytes. Adv. Cyclic. Nucl. Res., *14*: 421–428, 1981.
28. Maller, J. L. and Krebs, E. G. Regulation of oocyte maturation. Curr. Top. Cell. Regul., *16*: 271–311, 1980.
29. Godeau, F., Boquet, P., and Koide, S. S. Nicotinamide inhibition of *Xenopus laevis* oocyte maturation, in preparation.
30. Laemmli, U. K. Cleavage of structural proteins during the assembly of the head of bacteriophage T4. Nature, *227*: 680–685, 1970.

# ENZYMATIC POLY(ADP-RIBOSYL)ATION

ADP-RIBOSYLATION, DNA REPAIR AND CANCER, M. MIWA ET AL. (EDS.),
JAPAN SCI. SOC. PRESS, TOKYO/VNU SCIENCE PRESS, UTRECHT, PP. 121-128, 1983

# Properties of Poly(ADP-ribose) Synthetase and ADP-ribosyl Histone Splitting Enzyme

Masashi KAWAICHI, Jun OKA, Jing-yuan ZHANG, Kunihiro UEDA, and Osamu HAYAISHI

*Department of Medical Chemistry, Kyoto University, Faculty of Medicine, Kyoto 606, Japan*

*Abstract:* Poly(ADP-ribose) synthetase catalyzes initiation, elongation, and branching of poly(ADP-ribose) on various acceptor proteins including itself. The automodification took place at about 15 sites on the synthetase molecule of rat liver; at each site was bound poly(ADP-ribose) as large as >80 ADP-ribose units long with branched structures. Such extensive automodification accompanied a marked increase in the apparent $M_r$ (to >500,000) of the synthetase. Based on this unique property, the synthetase was identified as the main acceptor of poly(ADP-ribose) in isolated nuclei and permeabilized cells. The enzyme-bound poly(ADP-ribose) was not transferred to histone under the various conditions tested. These results suggested that the automodified enzyme does not serve as an intermediate in the modification of other proteins, but plays some biological role(s) as a structural element.

ADP-ribosyl histone splitting enzyme, an enzyme responsible for the cleavage of ADP-ribose-protein bonds, was purified 5,000-fold to apparent homogeneity from rat liver. The purified enzyme was composed of a single polypeptide of $M_r \approx 80,000$. The substrate specificity was broad for the protein portion, but strict for the ADP-ribose portion; only mono(ADP-ribosyl) proteins served as the substrates. The splitting reaction appeared to proceed by a non-hydrolytic mechanism producing a derivative of ADP-ribose with an altered terminal ribose.

Synthesis of poly(ADP-ribosyl) proteins involves three different reactions, *i.e.*, initial attachment of ADP-ribose to an acceptor protein (*1*), subsequent elongation of the ADP-ribose chain to a polymer (*2*), and branching of the polymer (*3*). The former reaction is the formation of carboxylesters of ribose (*4–7*), and the latter two are that of ribosyl-ribosides (*3, 8*). A single enzyme, poly(ADP-ribose) synthetase, catalyzes these three reactions (*1, 2, 9*). In addition to these functions as a catalyst, poly(ADP-ribose) synthetase serves as an acceptor of the polymer (*9, 10*).

Degradation of poly(ADP-ribosyl) proteins is carried out by two enzymes, poly(ADP-ribose) glycohydrolase (*11, 12*) and ADP-ribosyl histone splitting enzyme (*13*). The former cleaves the ribosyl-riboside bond, and the latter splits the carboxylester bond between ADP-ribose and protein. Another enzyme, phosphodiesterase, has been shown to play a minor role in the degradation of poly(ADP-ribose) (*14*).

In spite of extensive studies on biosynthesis and degradation of poly(ADP-ribosyl) proteins by many researchers, the mechanism and regulation of these enzyme reactions are not yet well understood. In this articles, we review our recent studies on automodification of poly(ADP-ribose) synthetase and novel properties of the ADP-ribosyl histone splitting enzyme.

*Multiple Autopoly(ADP-ribosyl)ation of Poly(ADP-ribose) Synthetase*

Yoshihara *et al.* (*10*) observed that automodification of calf thymus poly(ADP-ribose) synthetase accompanied an increase in the $M_r$ of the enzyme. A similar result was obtained with rat liver poly(ADP-ribose) synthetase (Fig. 1). When the purified synthetase was incubated for a short time (1 sec) with a low concentration (1 $\mu$M) of [$^{32}$P]NAD in the presence of DNA alone, the automodified enzyme comigrated with the unmodified enzyme ($M_r$=110,000) in sodium dodecyl sulfate(SDS)-gel electrophoresis at pH 8.5. When the incubation time was prolonged keeping the NAD concentration at 100 $\mu$M, the $M_r$ of the automodified enzyme increased markedly with time to >500,000. Such a remarkable increase in the $M_r$ was suggestive of automodification with very large polymers at, probably, multiple sites on the enzyme molecule.

When similar experiments were carried out using SDS-gels at neutral pH or lithium dodecyl sulfate-gels at acidic pH, no significant radioactivity migrated faster than the enzyme protein band (*9*), indicating that the synthetase produced little amount of free poly(ADP-ribose).

Changes in size, number, and branching of enzyme-bound poly(ADP-ribose)

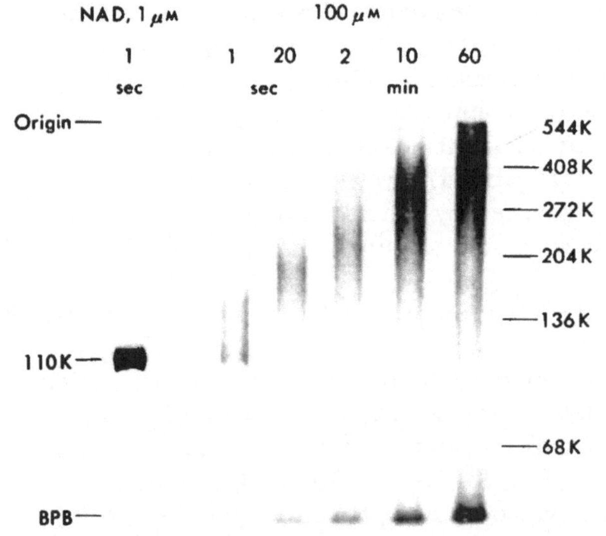


FIG. 1. SDS-gel electrophoresis of variously automodified poly(ADP-ribose) synthetase. The enzyme was incubated with [$^{32}$P]NAD under indicated conditions and electrophoresed in an SDS-(3%)polyacrylamide gel. The gel was subjected to autoradiography. BPB, bromphenol blue.

were then analyzed along the time course of automodification using the phosphodiesterase digestion method (Fig. 2). When the synthetase was incubated with 100 μM NAD at 37°C, the number of polymers on the enzyme reached a maximum value of about 15 polymers/enzyme molecule in 10 min, and remained constant thereafter. On the other hand, the chain length continued to increase to at least 80 ADP-ribose units in 60 min. The frequency of branching was constant at a level of once every 50 ADP-ribose residues. These results suggested that the enzyme had approximately 15 acceptor sites, and at each site was bound very long poly(ADP-ribose) having branched structures.

Based on these data and the result obtained from incubations at 0°C (data not shown), the progression of automodification was depicted schematically as shown in Fig. 3. When the synthetase was incubated at 0°C, a single enzyme molecule produced about 3 polymers with the average chain length of 2.5 ADP-ribose units in

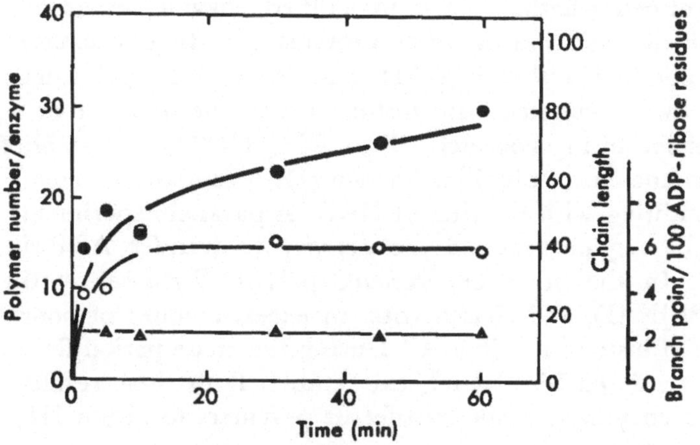


FIG. 2. Changes in chain number, chain length, and branching frequency during automodification of poly(ADP-ribose) synthetase. Incubation was carried out at 37°C with 100 μM NAD. ● chain length; ○ polymer number; ▲ branching. (From Kawaichi *et al.* (*9*)).

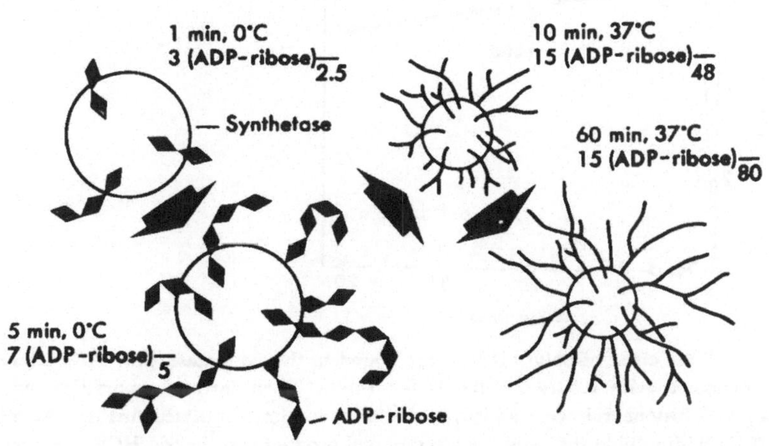


FIG. 3. Progression of multiple autopoly(ADP-ribosyl)ation of poly(ADP-ribose) synthetase.

1 min. After a 5-min incubation, both the number and the average chain lengtł doubled. In this manner, the number and the size of polymers increased almost ir parallel until all 15 sites were occupied by polymers. The automodification, there· fore, seems to proceed in a quasidistributive manner in the early stage of the reac· tion. Then, only the size continued to increase up to >80 ADP-ribose units. Branche: were produced at a constant frequency of once every 50 ADP-ribose residues ir· respective of chain length. After extensive modification, for example, a 60-min incu· bation with 100 $\mu$M NAD, the $M_r$ of the automodified enzyme was calculated tc exceed 600,000. This figure agreed well with the value estimated from SDS-gel elec· trophoresis (Fig. 1).

*No Transfer of Enzyme-bound Poly(ADP-ribose) to Histones*

Poly(ADP-ribose) synthetase utilizes various nuclear proteins as acceptors o ADP-ribose. We examined whether the automodified enzyme served as an inter· mediate in poly(ADP-ribosyl)ation of other proteins (*9*). In one experiment, thı synthetase was automodified with [$^{32}$P]NAD, and isolated by gel filtration. Thı isolated enzyme was then incubated with histone H1 in the presence or absence o NAD, and the conversion of incorporated $^{32}$P to a 5% $HClO_4$-extractable form wa: measured; this acid is known to solubilize histone H1. Essentially no conversion wa: observed in either condition with or without NAD. A parallel experiment indicatec that the automodified enzyme retained the activity to transfer ADP-ribose fron NAD to the histone. In the other experiment, poly(ADP-ribose) synthetase wa: pulse-labeled with [$^{32}$P]NAD, and chased with an excess amount of nonradioactivı NAD in the presence of histone H1 (Fig. 4). During the chase period for 20 min, es· sentially no $^{32}$P was rendered 5% $HClO_4$-extractable. Both these results indicatec that the automodified enzyme did not transfer its polymers to histon H1, and thu:

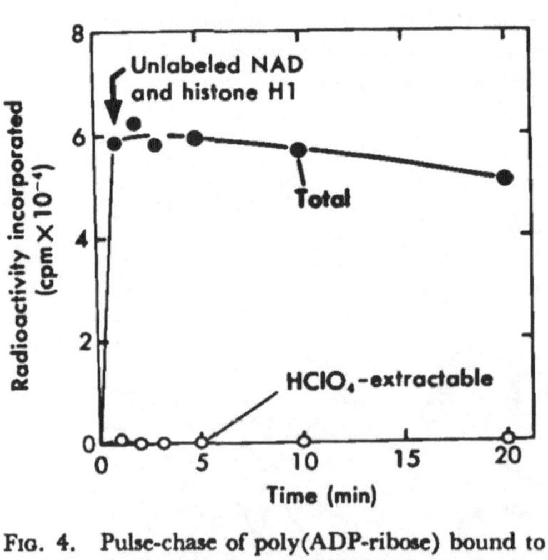


FIG. 4. Pulse-chase of poly(ADP-ribose) bound to the synthetase. Poly(ADP-ribose) synthetası was incubated with 0.5 $\mu$M [$^{32}$P]NAD for 1 min. Then (arrow), nonradioactive NAD (fina 7 mM) and histone H1 were added, and the incubation continued further. At intervals, 20% $Cl_3CCOOH$-insoluble radioactivities in the total product and the 5% $HClO_4$-extractable produc were determined.

excluded the possibility of the automodified enzyme serving as an intermediate in poly(ADP-ribosyl)ation of other proteins.

*Poly(ADP-ribose) Synthetase as the Main Acceptor of Poly(ADP-ribose)*

When human peripheral lymphocytes were hypotonically permeabilized and incubated for 20 sec with 1 $\mu$M [$^{32}$P]NAD, the major product migrated in a broad band having the smallest $M_r$=110,000 in SDS-gel electrophoresis (Fig. 5) (*15*). When lymphocytes were incubated with increasing NAD concentrations or for longer times, a tremendous increase in the $M_r$ of the main product was observed. Based on the smallest $M_r$ and the behavior under these various conditions, the product was identified as the synthetase-bound poly(ADP-ribose). The synthetase also proved to be the main acceptor in nuclei isolated from rat liver (*16*). These results indicated that poly(ADP-ribose) synthetase plays biological roles not only as a catalyst modifying other proteins but also as a structural element in a poly(ADP-ribosyl)ated form.

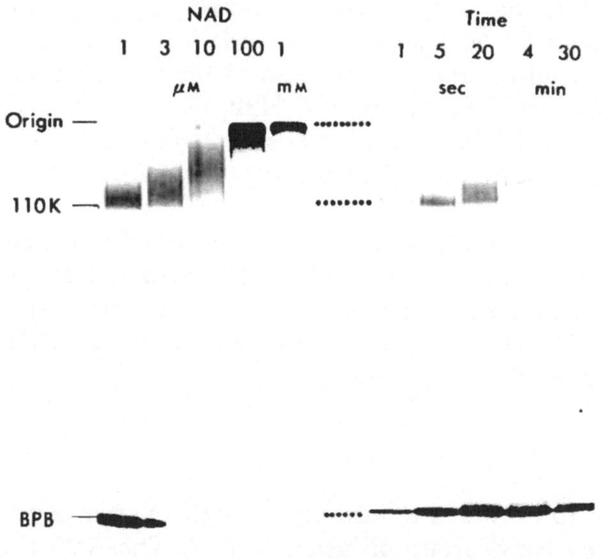


FIG. 5. SDS-gel electrophoresis of poly(ADP-ribosyl)ated lymphocyte proteins. Human lymphocytes hypotonically permeabilized were incubated for 20 sec with various concentrations of [$^{32}$P]NAD (left) or for various times with 3 $\mu$M [$^{32}$P]NAD (right). The materials solubilized by SDS from the lymphocytes were electrophoresed in SDS-(7.5%)polyacrylamide gels, and the gels were subjected to autoradiography. BPB, bromphenol blue. (From Ogata *et al.* (*15*)).

*Purification and Properties of ADP-ribosyl Histone Splitting Enzyme*

The enzyme which splits ADP-ribosyl protein bonds was recently purified 5,000-fold to apparent homogeneity from rat liver (*17*). Upon conventional fractionation using aqueous media, about 80% of the enzyme activity was recovered in the post-nuclear supernatant. Because of this abundance, the supernatant was used as the enzyme source. The purified enzyme exhibited a single protein band at the position

TABLE 1. Chain Length Distribution of ADP-ribose Chains in ADP-ribosyl Histone HI and Split Products

| ADP-ribose chain | $^{14}C$ in substrate (a) (cpm) | $^{14}C$ in split products | |
|---|---|---|---|
| | | + Enzyme (b) (cpm) | – Enzyme (c) (cpm) |
| Monomer | 5,850 | 4,540 | 252 |
| Dimer | 2,760 | 155 | 119 |
| Trimer | 2,260 | 125 | 99 |
| Tetramer | 1,240 | 70 | 49 |
| Pentamer | 614 | 25 | 26 |

A mixture of mono- and oligo([$^{14}C$]ADP-ribosyl)ated histone HI was incubated for 40 min in 1 M $NH_4OH$ (a) or for 10 min with (b) or without (c) enzyme at pH 6.0. The released materials were separated according to chain length by DEAE-cellulose column chromatography (*18*), and each ADP-ribose chain class was examined for radioactivity.

of $M_r$=83,000 upon SDS-polyacrylamide gel electrophoresis. The Stokes radius was 5.2 nm and the sedimentation coefficient was 3.4 S. From these values, $M_r$=75,000 was calculated. Taken together, this enzyme was composed of a single and asymmetrical polypeptide of approximate $M_r$=80,000 with a frictional ratio of 1.8.

This enzyme had a broad substrate specificity toward the protein portion. Various ADP-ribosylated proteins including histone H1, H2B, nonhistone proteins, and the N-terminal pentapeptide of H2B (*6*) all served as substrates of this enzyme. The $K_m$ value for ADP-ribosyl H2B was estimated as 8.5 $\mu$M, and $V_{max}$ as 970 nmol/min/mg protein.

In contrast, this enzyme was highly, if not absolutely, specific to mono(ADP-ribose) residues. When a mixture of mono- and oligo(ADP-ribosyl) histone H1 was treated with this enzyme, only mono(ADP-ribose) was released from histone H1 in a significant amount (Table 1). This enzyme did not act on deadenylylated substrates, for example, phosphoribosyl pentapeptide of H2B.

### *Analysis of Split Product*

ADP-ribose was reported to be linked to histones through an ester bond between the terminal ribose and the carboxyl group of histones (*4–7*). The ADP-ribosyl histone splitting enzyme, therefore, had been tacitly assumed to catalyze hydrolytic cleavage of the ester bond. Our recent analyses, however, threw doubt on this assumption.

Our previous studies revealed that the split product was not identical with ADP-ribose, although it contained all moieties of ADP-ribose molecule and no peptide fragment (*13*). The treatment of the product with snake venom phosphodiesterase yielded AMP from the adenine half, but not ribose 5-phosphate from the other half. The "ribose" portion obtained by further treatment with alkaline phosphatase migrated apart from authentic ribose or any other pentoses upon paper chromatography (Fig. 6). This was in contrast to the fact that the nonenzymatic product obtained from the same substrate by alkaline treatment coincided exactly with ADP-ribose in chromatographies and was convertible to ribose by similar enzyme treatments.

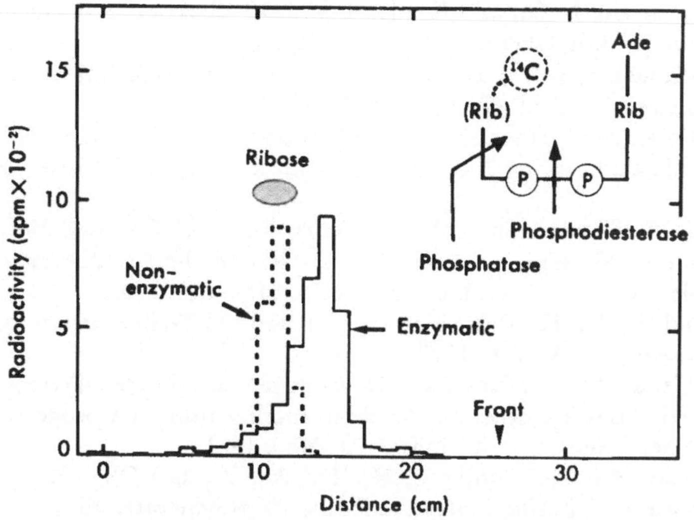


FIG. 6. Paper chromatography of split products treated with phosphodiesterase and phosphatase. ADP-[$^{14}$C]ribosyl histone H2B was treated for 10 min with ADP-ribosyl histone splitting enzyme (enzymatic) or 0.1 M NaOH (nonenzymatic). The split products were digested with snake venom phosphodiesterase and bacterial alkaline phosphatase (inset). The digests were analyzed by paper chromatography.

These results raised a number of possibilities. (i) The enzymatic reaction may be hydrolysis, but the product quickly undergoes rearrangement(s) within the molecule or reacts with other molecule(s) in the reaction mixture. (ii) The reaction may not be hydrolysis but, for example, elimination producing an unsaturated derivative of the terminal ribose. Several enzymes (eliminases or lyases) that degrade polysaccharides are known to catalyze this type of reaction (*19*). Further characterization of the split product as well as the substrate is necessary before obtaining the final conclusion on the mechanism of action of the ADP-ribosyl histone splitting enzyme.

ACKNOWLEDGMENT

This work was supported in part by grants-in-aid for Scientific Research and Cancer Research from the Ministry of Education, Science and Culture, Japan, and by research grants from the Princess Takamatsu Cancer Research Fund and the Vitamin B Research Committee.

REFERENCES

1. Kawaichi, M., Ueda, K., and Hayaishi, O. Initiation of poly(ADP-ribosyl) histone synthesis by poly(ADP-ribose) synthetase. J. Biol. Chem., *255*: 816–819, 1980.
2. Ueda, K., Kawaichi, M., Okayama, H., and Hayaishi, O. Poly(ADP-ribosyl)ation of nuclear proteins: Enzymatic elongation of chemically synthesized ADP-ribose· histone adducts. J. Biol. Chem., *254*: 679–687, 1979.
3. Miwa, M., Saikawa, N., Yamaizumi, Z., Nishimura, S., and Sugimura, T. Structure of poly(adenosine diphosphate ribose): Identification of 2′-[1″-ribosyl-2″-(or 3″-) (1‴-ribosyl)]adenosine-5′,5″,5‴-tris(phosphate) as a branch linkage. Proc. Natl. Acad. Sci. U.S., *76*: 595–599, 1979.

4. Riquelme, P. T., Burzio, L. O., and Koide, S. S. ADP-ribosylation of rat liver lysine-rich histone *in vitro*. J. Biol. Chem., *254*: 3018–3028, 1979.
5. Burzio, L. O., Riquelme, P. T., and Koide, S. S. ADP-ribosylation of rat liver nucleosomal core histones. J. Biol. Chem., *254*: 3029–3037, 1979.
6. Ogata, N., Ueda, K., and Hayaishi, O. ADP-ribosylation of histone H2B: Identification of glutamic acid residue 2 as the modification site. J. Biol. Chem., *255*: 7610–7615, 1980.
7. Ogata, N., Ueda, K., Kagamiyama, H., and Hayaishi, O. ADP-ribosylation of histone H1: Identification of glutamic acid residues 2, 14, and the COOH-terminal lysine residue as modification sites. J. Biol. Chem., *255*: 7616–7620, 1980.
8. Hayaishi, O. and Ueda, K. Poly(ADP-ribose) and ADP-ribosylation of proteins. Annu. Rev. Biochem., *46*: 95–116, 1977.
9. Kawaichi, M., Ueda, K., and Hayaishi, O. Multiple autopoly(ADP-ribosyl)ation of rat liver poly(ADP-ribose) synthetase: Mode of modification and properties of automodified synthetase. J. Biol. Chem., *256*: 9483–9489, 1981.
10. Yoshihara, K., Hashida, T., Yoshihara, H., Tanaka, Y., and Ohgushi, H. Enzyme-bound early product of purified poly(ADP-ribose) polymerase. Biochem. Biophys. Res. Commun., *78*: 1281–1288, 1977.
11. Miwa, M. and Sugimura, T. Splitting of the ribose-ribose linkage of poly(adenosine diphosphate-ribose) by a calf thymus extract. J. Biol. Chem., *246*: 6362–6364, 1971.
12. Ueda, K., Oka, J., Narumiya, S., Miyakawa, N., and Hayaishi, O. Poly ADP-ribose glycohydrolase from rat liver nuclei, a novel enzyme degrading the polymer. Biochem. Biophys. Res. Commun., *46*: 516–523, 1972.
13. Okayama, H., Honda, M., and Hayaishi, O. Novel enzyme from rat liver that cleaves an ADP-ribosyl histone linkage. Proc. Natl. Acad. Sci. U.S., *75*: 2254–2257, 1978.
14. Miwa, M., Nakatsugawa, K., Hara, K., Matsushima, T., and Sugimura, T. Degradation of poly(adenosine diphosphate ribose) by homogenates of various normal tissues and tumors of rats. Arch. Biochem. Biophys., *167*: 54–60, 1975.
15. Ogata, N., Kawaichi, M., Ueda, K., and Hayaishi, O. Poly(ADP-ribosyl)ation of 110,000 dalton protein in human lymphocytes treated with *N*-methyl-*N'*-nitro-*N*-nitrosoguanidine. Biochem. Int., *1*: 229–236, 1980.
16. Ogata, N., Ueda, K., Kawaichi, M., and Hayaishi, O. Poly(ADP-ribose) synthetase, a main acceptor of poly(ADP-ribose) in isolated nuclei. J. Biol. Chem., *256*: 4135–4137, 1981.
17. Oka, J., Ueda, K., and Hayaishi, O. Purification and properties of ADP-ribosyl histone splitting enzyme. Seikagaku, *52*: 755, 1980 (in Japanese).
18. Kawaichi, M., Oka, J., Ueda, K., and Hayaishi, O. A new method for oligo(ADP-ribose) fractionation according to chain length. Biochem. Biophys. Res. Commun., *101*: 672–679, 1981.
19 Meyer, K. Hyaluronidases. *In;* P. D. Boyer (ed.), The Enzymes, Vol. 5, pp. 307–320, Academic Press, New York, 1971.

ADP-RIBOSYLATION, DNA REPAIR AND CANCER, M. MIWA ET AL. (EDS.),
JAPAN SCI. SOC. PRESS, TOKYO/VNU SCIENCE PRESS, UTRECHT, PP. 129-140, 1983

# The Life History of Poly(ADP-ribose)

Miyoko IKEJIMA,*[1] Yosuke SAWADA,*[2] An Quoc DANG,*[2] Robert J. SUHADOLNIK,*[2] and D. Michael GILL*[1]

*Department of Molecular Biology and Microbiology, Tufts University Schools of Medicine, Veterinary Medicine and Dental Medicine, Boston, Massachusetts 02111,*[1] and Department of Biochemistry, Temple University School of Medicine, Philadelphia, Pennsylvania 19122, U.S.A.*[2]*

*Abstract:* We present arguments and data in support of the following sequence of events in ghost cells. 1) Poly(ADP-ribose, ADPR) synthetase, activated by binding to DNA at a break or other anomaly, synthesizes chains of polymer upon itself, adding new residues at the proximal 1″ terminus of the growing chain. 2) Free chains of poly(ADPR) are produced by release from the active site or by internal glycohydrolysis of the growing chains without hindering the continued elongation. Subsequent glycohydrolysis cuts the free chains down to monomers. 3) ADP-ribosylation of core histones may involve soluble intermediates of some form. Glycohydrolysis generates a limit digest with only short histone-bound chains. 4) Most ADP-ribosylation of histone H1 occurs by the addition of single free ADPR residues independently of poly(ADPR) synthetase.

*Rationale*

It is clear that poly(ADP-ribose, ADPR) has something to do with the repair of DNA breaks caused by mutagens or radiation and may have something similar to do with events in differentiation but we have remarkably little idea how the formation of poly(ADPR) fits into these processes. We are far being able to state in molecular terms what spatial arrangement exists between the broken ends of DNA strands, poly(ADPR) synthetase, the polymer itself, and neighboring proteins. It is our aim to try and design such a model.

A good starting point is the fact that poly(ADPR) is made efficiently only in the presence of broken DNA (*1–5*). The bulk of the DNA remains unmodified during the construction of poly(ADPR) itself: other systems that are permitted by the polymer synthesis apparently repair the damage. It is fair to assume that the poly(ADPR) synthetase recognizes a DNA break by binding to it and it is the binding that causes the increase in the enzyme's activity. Different types of DNA damage activate the enzyme to different extents correlating well with the difficulty of repairing the damage. Flush-ended double-stranded breaks without terminal 5′ phosphate groups are the most effective activators yet identified: for these a few hundred NAD residues per minute are assembled into poly(ADPR) in the immediate vicinity of the break. How long does it survive? How is it arranged in space? The sort of role we imagine for it in DNA repair depends rather critically on the answers.

In our present state of ignorance we can construct three kinds of plausible model.

a) There might be no particular structure to the polymer. The appearance of a lot of an acidic macromolecule (considerably more acidic than DNA) in the immediate vicinity of the DNA break would certainly provide a new binding substrate for local histones and might promote the disassembly of neighboring nucleosomes. Perhaps this undoes the chromatin structure sufficiently to allow the repair system access to the damaged DNA. The repair enzymes would have to dislodge poly(ADPR) synthetase molecules from the DNA (how otherwise could they modify the DNA?) thereby preventing any further polymer synthesis. While the repair was being effected, the mass of pre-formed polymer would be digested (by poly(ADPR) glycohydrolase) and the histones could once more assemble into nucleosomes.

b) We construct a rather different picture if we focus on the covalently-modified histones that acquire short chains of poly(ADPR). Many experts have believed that such modification is the primary event, *i.e.*, that poly(ADPR) synthetase transfers ADPR directly from NAD to certain side chains on the histones and then to histone-bound growing chains. If this were true, we would have to consider how the synthetase, while bound to the broken DNA, is able to react in a precise orientation with specific groups on specific nucleosomal histones which in the first instance must be sited at unpredictable and random distances from the DNA damage sites. We would also have to explain how a synthetase molecule was able to move away from the histone as synthesis of the chain proceeded. There is no easy solution to this puzzle unless one of the proteins dissociates. We would need to assume either that the target histone comes away from its nucleosome and assumes an appropriate orientation in free solution, or that the synthetase comes loose from the damaged DNA and remains active long enough to interact with local nucleosomes. Neither supposition would be necessary, however, if the modification of histones were not a primary event; if instead there were a soluble and chemically reactive derivative of ADPR or of poly(ADPR) that was made by the enzyme and that could react with any protein within a short range. These would include, of course, histones, non histone proteins, any other abundant nuclear protein such as T antigen after SV40 infection (*6*) and perhaps the synthetase itself. In fact all of these proteins are ADP-ribosylated by one means or another but it is an entirely different matter to decide which of the modifications are functionally relevant.

There is another reason to suspect that histone-bound chains are not formed primarily: they are not long enough. Every report finds that the chains on histones are very short; rarely over 4 residues long and frequently averaging barely more than one. Yet the system initially makes much longer chains. *In vitro* one can easily detect free chains with 60 residues. It hardly seems likely that long free chains all come from very short bound chains, but short chains could arise from the transfer or attachment of long chains followed by glycohydrolysis. The end result may alter the histone's function but it does not seem wise to attribute the whole raison d'être of poly(ADPR) to the modification of histones by short chains or single residues; nature could surely achieve this result more economically.

c) A third possible model is that poly(ADPR) plays a structural role, forming some kind of scaffold around the DNA break. Could it indeed interact with itself or with other molecules in such a way that it prevented the two ends of a double-strand break from drifting apart? In this situation there would be one enzyme molecule on each side of the break synthesizing poly(ADPR) in opposite directions. Given their high negative charge, the two strands could easily interact with nucleosomes and make some kind of local mesh. Could the two strands also interact with each other? Model building suggests so (*7*) but the absence of physical evidence for "double-stranded" polymer now makes this proposal less likely (*8*).

To distinguish between these possibilities we decided to focus on two related questions. In our view the crucial discriminator is the order of formation of the various products. In particular is the enzyme-bound polymer all a product or is some of it, at least, an intermediate of the synthesis?

One always sees poly(ADPR) synthetase gathering labeled ADPR upon itself (*2*, *9–15*). Is this bound polymer the consequence of secondary addition away from the active site (ADPR derivatives react slowly with any protein) or is it polymer in the process of elongation? The scale of the enzyme-bound material suggests the second interpretation, for *all* of the poly(ADPR) produced by purified synthetase seems to be attached to the enzyme (*10*, *13*, *15*)—hardly the result expected from a chance interaction of chains with some secondary sites on the enzyme.

The second question follows directly from the first: in which direction are poly (ADPR) chains elongated? Is it by addition of new ADPR residues to the 2′OH terminus (mechanism 1), as most have assumed, or is it by addition at the 1″ terminus (mechanism 2) which can happen only if the growing chain remains attached to the enzyme and retains some of the energy of the nicotinamide-ribose bond? Thus if we could prove that the ADPR residues are inserted at the 1″ end, it would automatically follow that elongation takes place on the enzyme. Correspondingly, extension of chains on the active site of the synthetase would almost certainly have to occur at the 1″ ends.

Mechanism 1

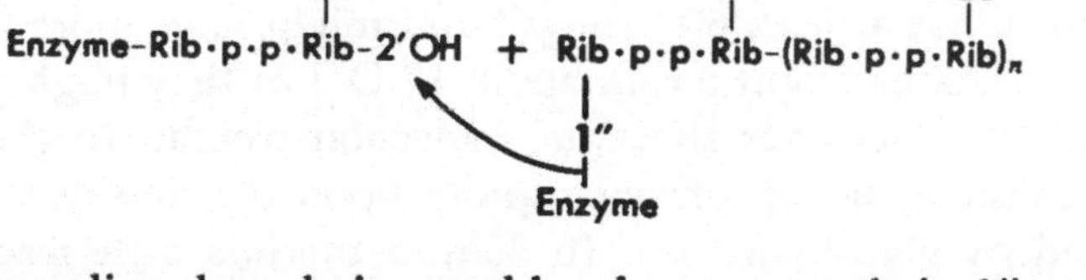


Mechanism 2

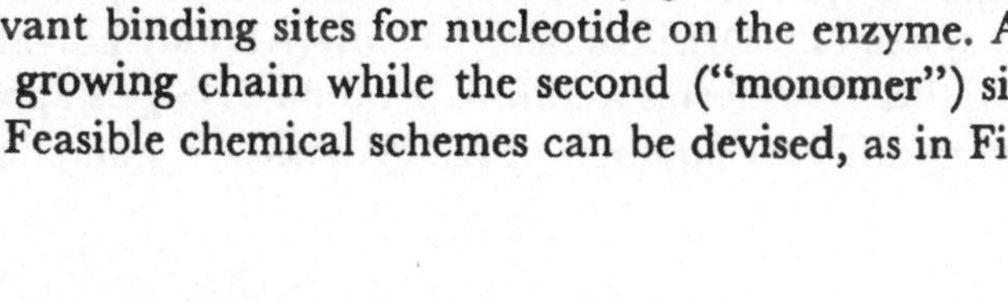


It is important to realize that chains could only grow at their 1″ ends if there were at least two relevant binding sites for nucleotide on the enzyme. A "polymer" site must contain the growing chain while the second ("monomer") site binds the next NAD or ADPR. Feasible chemical schemes can be devised, as in Fig. 1.

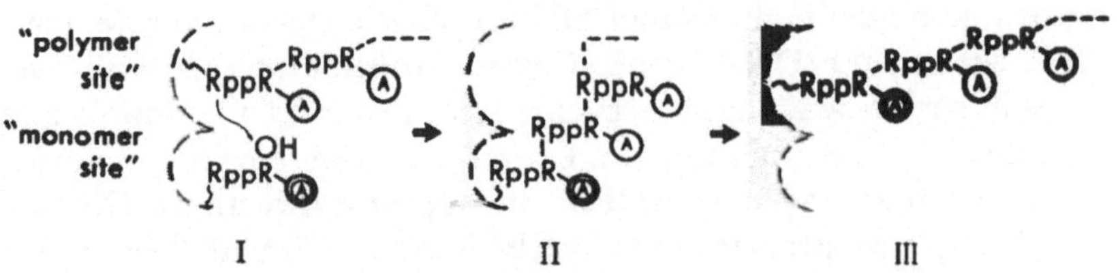


FIG. 1. One possible two-site model for poly(ADPR) synthesis. This shows how chains might be constructed by adding new residues at the 1″ end, proximal to the synthetase. Here we represent that the incoming ADPR residue is attached to the enzyme covalently (I) but it is also conceivable that the proximal residue in stages I and II could be NAD bound at the "monomer" site non-covalently.

## *Incorporation of ADPR by DNase I—Treated HeLa Cell Ghosts*

We collect actively-growing HeLa cells and place them at $1\text{–}2\times10^7$ cells/ml (5–10% v/v) in ice-cold "lysis buffer" (0.05% Trition X-100, 5% dextran T-110, 10 mM $MgCl_2$, 0.1 mM $CaCl_2$, and 40 mM tris-HCl, pH 8.0). To the permeabilized cells ("ghosts") we add, together, 10 $\mu$g/ml DNase I and 10 $\mu$M [$^{32}$P]NAD and incubate at 37°C for 3 to 30 min. Although soluble components equilibrate with the medium, this system has the advantage of resembling the situation *in vivo* to a reasonable extent. With it we can investigate certain properties of the entire integrated poly(ADPR) system, and not just those properties that survive nuclear isolation, for example.

The DNase is sufficient to suddenly create a large number of breaks in the nuclear DNA that activate, we calculate, every molecule of the synthetase. Continued digestion reduces the DNA to products that no longer activate poly(ADPR) synthetase, and degradation takes over. In effect we generate a pulse of poly(ADPR) that peaks at about 3 min and is degraded during the following 30 min or so. Such events act as a model of events *in vivo* when a single synthetase molecule is activated for a short while by a DNA break and creates a transient pulse of polymer.

We have deliberately not fractionated the products but have subjected the entire reaction mixture to sodium dodecyl sulfate (SDS) gel analysis and autoradiography in order to study all the large NAD-derived products at once. Over 75% of the trichloroacetic acid (TCA)-insoluble products are accounted for on the gels. We may lose a small amount of material if some products are unstable in SDS. Upon fixation we lose small products, including short free chains. The principle products seen are a) free chains of poly(ADPR) of all sorts of lengths. These form a "ladder" up the gel which has at least 60 "rungs" and might have more if the resolution were better. The free chains might terminate in 1″ OH or they might be attached to some small molecule that does not affect the molecular weight. In ghosts these free chains are largely transient, being formed rapidly upon the addition of DNase but being soon degraded by glycohydrolase. In 30 min there is little free polymer left but it represents the greater part of the early synthesis and may be most relevant to the function of the system in DNA repair; b) poly(ADPR) polymerase MW 115,000, with attached ADPR. This forms a diffuse band on SDS gels apparently representing molecules carrying different numbers of ADPR residues. Sometimes this diffuse band is replaced by an agglomeration at the top of the main gel; c) diffuse protein bands of smaller sizes, some of which may represent proteolytic fragments of the enzyme; d) core histone with single residues and short chains attached. Histone H1 is labeled by a different process in this system, as described later.

### *Is 2′Deoxy ADPR Incorporated into Free and Bound Chains?*

We tried to determine the direction of chain growth using [$^{32}$P] 2′deoxy NAD as a substrate for the polymerase. This was synthesized from [$^{32}$P] 2′deoxy ATP as described (*16*). It had already been shown to be incorporated into some products, albeit inefficiently (*17*).

The reason was that 2′d ADPR can only be present at the 2′OH terminus of chains. If chains grew by addition of residues at the 2′OH terminus (mechanism 1) 2′d ADPR could be incorporated efficiently into the terminal positions. All chains, of

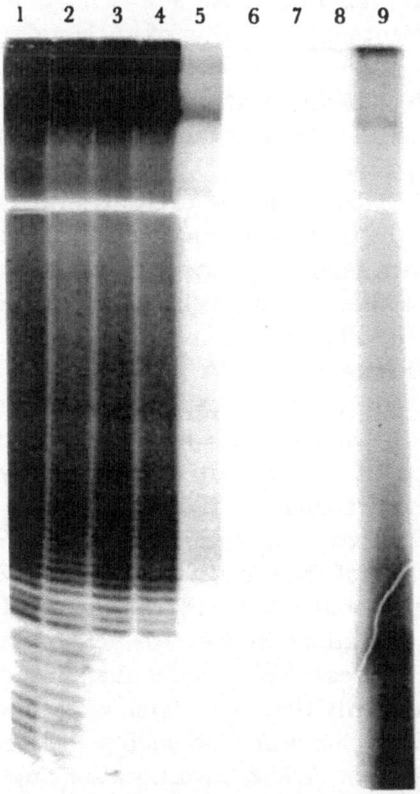


FIG. 2. Incorporation from NAD and 2′deoxy NAD compared. HeLa cells at $2\times10^7$/ml were incubated 3 min 37 C in lysis buffer with DNase I and different combinations of NAD and 2′deoxy NAD. Tracks 1–5: [$^{32}$P]NAD 22 μM, 3.9 Ci/mmole (this includes 12 μM unlabeled NAD supplied by the cells) and various concentrations of unlabeled 2′dNAD. 1) none; 2) 10 μM; 3) 100 μM; 4) 1 mM; 5) 5 mM. Tracks 6–8: [$^{32}$P] 2′d NAD, 10 μM, 8.7 Ci/mmol, and various concentrations of unlabeld NAD. 6) 12 μM; 7) 22 μM; 8) 112 μM. Track 9: [$^{32}$P] 2′dNAD 10 μM, 135 Ci/mmole, and 22 μM unlabeled NAD. In each case incorporation was terminated by adding SDS and boiling. The samples were electrophoresed for 9.5 hr at 450 volts on a long gradient gel.

all sizes, could be capped by one residue of 2′d ADPR. If, however, chains grew at the 1″ terminus (mechanism 2) it would be difficult or impossible (depending on the precise chemistry) for them to be labeled with 2′dADPR.

We wanted to look first at free polymer, for at the strat we could expect the synthetase to be labeled under either mechanism, and had therefore to use systems that produced free polymer: this excluded pure synthetase (*10–15*). We used HeLa ghosts that contained endogenous NAD, so we were sure that free chains were being produced, and supplied either [$^{32}$P]NAD or [$^{32}$P] 2′dNAD.

Some results are shown in Fig. 2. The incorporation of ADPR is totally inhibited at high levels of 2′dNAD. With less 2′dNAD chains were formed that contained ADPR (necessarily so!) but were not visibly labeled from 2′dNAD: compare tracks 2 and 7 each of which had 22 $\mu$M NAD and 10 $\mu$M 2′dNAD, with the concentration difference balanced by a difference in specific activity.

To search harder for labeling from 2′dNAD we increased its specific activity by a factor of 15.5 and reduced the NAD concentration further. The higher specific activity resulted in an increased background labeling of the autoradiogram which is seen also without tissue and seems to represent an interaction of nucleotides with the polyacrylamide itself (track 9: compare to track 2). On the film, but probably not in the reproduction, we can see some slight labeling of the core histones, but in the region below the core histones, where we expect the maximum sensitivity of the experiment, there are no labeled free chains visible.

By laying the autoradiogram of track 9 over various exposures of track 2 we determined that about 10% of the label in track 2 would have been detectable through the background, had it been present. Combining this with the ratio of specific activities (135 *vs.* 3.9 Ci/mmole) we find that the amount of 2′dADPR in free chains was less than 1/350th of the amount of ADPR. This difference is not as large as it may appear a) because the affinity of the synthetase for 2′dNAD seems to be only one fourth of that for NAD ($K_i$ and $K_m$ repsectively); and b) "capped" chains necessarily have less 2′dADPR than ADPR. Thus a ten-residue chain, which is among the smallest on a fixed gel (*cf. 18*), could have 9 or 10 ADPRs but only 1 2′deoxy ADPR. Despite these considerations, the labeling of decamers by 2′dADPR is still less than one-ninth ($350/(4\times10)$) of that expected if chains grew by addition at the 2′OH end.

We must now consider the slight labeling of the synthetase itself from 2′dNAD. In different experiments this amounted to about 0.1 to 1% of the labeling from NAD (0.14% in Fig. 2). However the modifications by the two precursors seem to be similar in nature. Equal proportions were released by post-incubation with venom phosphodiesterase and the proteolytic fragments described later were also similar (see Fig. 4). We must consider the possibility therefore that such a low level of incorporation, not distinguished in any way from ADPR incorporation, may in fact represent the product of a small amount of [$^{32}$P]NAD in the [$^{32}$P] 2′dNAD. Indeed we find that some of the synthetase-associated label derived from the 2′dNAD preparation can be displaced in a postincubation with an excess of unlabeled NAD but not by 2′dNAD.

We can certainly conclude from these experiments that the inhibition of chain growth by 2′dNAD is not solely the result of capping growing chains with 2′dADPR. If this was the only effect we would have detected labeled free chains easily.

We are left with two possibilities.

One is that 2′dNAD is a very poor substrate which reacts with the enzyme and substitutes directly for NAD to a finite but limited extent. It would be incorporated into the termini of chains to the same limited extent which would be below the limit of detection of our experiment. This would be exceptional behavior for 2′dNAD, for we find that it is a reasonable substrate for diphtheria toxin and cholera toxin. It also substitutes reasonably well for NAD in dehydrogenations (*16*).

The second possibility is that the apparent incorporation into the synthetase and core histones is due to a very slight contamination with NAD and that 2′dNAD is not a complete substrate at all. We know that 2′dNAD must bind reversibly to the synthetase, for it inhibits ADPR incorporation with competitive kinetics, but if a 2′OH group were required for the next step, it would be unable to progress from the reversible site to any acid-insoluble product.

### *The Binding of ADPR and Poly(ADPR) to the Synthetase*

There appear to be at least two ways in which ADPR is attached covalently to the synthetase. On SDS gels of HeLa cell ghosts the early polymeric product runs more slowly than the synthetase and forms a smear extending upwards from 115,000 daltons, the molecular weight of the enzyme. Sometimes this material runs even more slowly forming a band on the top of the main gel or even remains in the stacking gel (it always enters 5% polyacrylamide).*

After synthesis of polymer stops, the association changes and we see more of a sharply-resolved radioactive band at the 115,000 dalton position. It becomes more prominent and sharper with time (compare tracks 1 and 3 in Fig. 3). Its sharpness suggests that it may consist only of monomers. Indeed by direct measurement of the gel band we find that it reaches a limit of about 0.5 ADPR/synthetase molecule.**

Both the monomeric and the polymeric material are heterogeneous with regard to alkaline stability (as reported in ref. *13*). However, the polymer tends to be attached by alkali-labile bonds whereas half or more of the monomer units tend to be attached by alkali-stable bonds. Some of the apparent "monomer" may in fact be "short polymer" attached in the alkali-labile manner. It is notable that early in an incubation when polymer synthesis is rapid, little alkali-stable monomer exists, but

* Such behavior suggests a size of 500,000 daltons or more, which is larger than the synthetic capacity of the synthetase can allow. Its $V_{max}$ has been variously estimated at 150–300 residues per min and similar values can be calculated, from data in ref. *2, per DNA end*, under conditions where broken DNA is limiting: it cannot therefore be thought that the $V_{max}$ was underestimated because few molecules were active. After purification all the stainable enzyme suffers from this aggregation (*10, 15*). Consonant with this argument, alkali releases chains that are much too short to account for the implied mass. We must, therefore, infer that some kind of aggregation occurs. We proved that this is possible by taking the result of a first SDS gel, rotating it 90°C and running it into a second SDS gel before fixation. Radioactive material that had run at or near the 115,000 dalton position in the first dimension formed a smear in the second dimension that extened upwards to the gel top.

** This value could easily be in error by a factor of two or more. We derived the enzyme content of HeLa cells from the data of Jump and Smulson (*9*) who reported a 2,200-fold increase in specific activity purifying synthetase from HeLa cells. This implies that 1 $\mu$l of packed cells contains $3.8 \times 10^{11}$ synthetase molecules, or $1.9 \times 10^{9}$ in each of $2 \times 10^{5}$ cells.

1 2 3 4 5 6 7 8 9 10

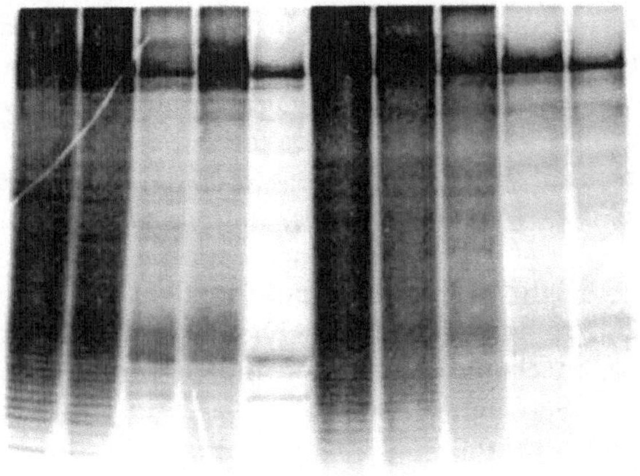

FIG. 3. Time course illustrating labeling of core histones. HeLa cell ghosts, as in Fig. 2, were incubated with [$^{32}$P]NAD for 5 min (1), 10 min (2), 30 min (3), 10 min (4) with poly(ADPR) glycohydrolase present for the last 5 min, 30 min with glycohydrolase present for the last 25 min (5). Tracks 6–10 duplicate 1–5 except that the ghosts were washed once after lysis and before the incorporation. The glycohydrolase was a gift from Dr. M. Miwa. Histones 2b/3 (top), 2a, 4 (bottom) are seen most clearly in track 5.

1 2 3 4 5 6 7 8 9 10

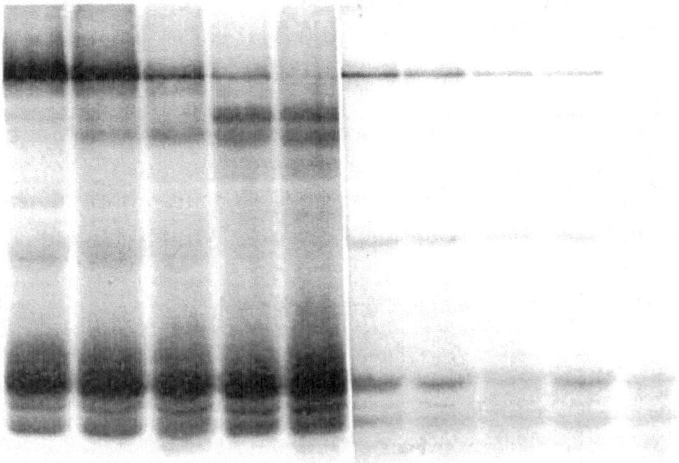

FIG. 4. Partial proteolysis of labeled synthetase. HeLa cell ghosts were incubated with [$^{32}$P]-NAD (tracks 1–5) or [$^{32}$P] 2′dNAD (tracks 6–10) for 20 min. Tracks 1 and 6 are controls. The other samples were postincubated with proteases for 15 sec before boiling in SDS. 2 and 7, chymotrypsin 50 $\mu$g/ml; 3 and 8, chymotrypsin 150 $\mu$g/ml; 4 and 9, elastase 50 $\mu$g/ml; 5 and 10, 150 $\mu$g/ml. Tracks 1-5 were exposed to X-ray film for 1 day but 6–10 were exposed for 13 days. The synthetase is the most prominent band near the gel top.

later, when polymer synthesis has stopped, the residue of monomer is alkali stable. An attractive interpretation is that there are two relevant sites on the enzyme that are both involved in polymer synthesis. Perhaps the alkali-labile site may be entered only in the presence of damaged DNA but then is the usual locus of the growing polymer chain. The alkali-stable site may be used during the polymer synthesis and also holds a single residue when polymer synthesis has stopped. It looks as if individual synthetase molecules might have one or the other site filled, but not both at once.

### *Proteolytic Fragmentation of Labeled Synthetase*

The concept that there might be separable sites one of which might bind polymers and the other monomers encouraged a search for ways of separating the two. By partial proteolysis we expected to generate fragments that held only polymer, which would be heterogenous in size, and, like the starting material, would give broad fuzzy bands on SDS gels. We hoped to separate these from other fragments which held single ADPR residues and which would give sharp bands on autoradiography. This method would work only if the two sites were sufficiently distant to have a protease site between them. In the event we could find no such separation. We tried four proteases (chymotrypsin, trypsin, elastase, and the V8 protease) each of which generated several large fragments of the labeled enzyme. All of these bands were however diffuse, even when the starting material gave a sharp band (Fig. 4). Some fragments were similar to those described by Nishikimi *et al.* (*14*).

### *Origin of ADPR Residues Attached to Core Histones*

ADP-ribosylation of the core histones in HeLa cell ghosts seem to be coupled to the formation of free polymer and enzyme-bound polymer in that all of these are dependent on DNA damage and are inhibited by thymidine and nicotinamide and their analogs. When the synthetase carries short chains we find short chains on histones but when the enzyme carries only monomers we see only monomers on the histones.

The extent of labeling of the core histones is much reduced when HeLa cell ghosts are washed free of soluble components (Fig. 3, compare tracks 6–10 [washed] with tracks 1 to 5). This suggests that a soluble component is required for the transfer of ADPR chains to these histones. The experiment in Fig. 3 suggests that glycohydrolase (*19*) is not this soluble factor.

This situation in HeLa cell ghosts differs from some previous reports but the differences are resolved by the realization that there are two quite distinct routes of labeling of core histones with ADPR. The first route described here is dominant in freshly-lysed cells (and therefore *in vivo* perhaps?) and leads to modifications that are *stable* to alkali and hydroxylamine. All four histones seem to be labeled, with some preference to H2b/H3 and H4 over H2a. The first route is lost upon washing and is therefore deficient in isolated nuclei. There the second route dominates which yields alkali-labile residues attached to carboxyl groups particularly on H2b but not on H4 (*20, 21*).

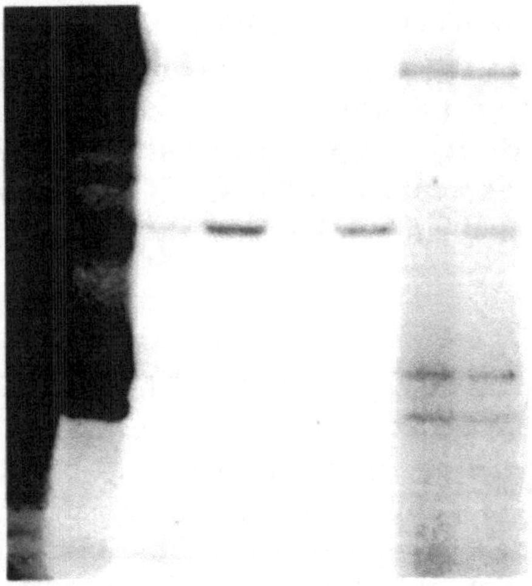


FIG. 5. Origin of the label on histone H1. Six pairs of samples are represented: the first of each pair represents a 5 min incubation of HeLa ghosts with DNase. The second a 30 min incubation. H1 labeling increases with incubation time while the other bands decrease in intensity. The precursor were: 1 and 2, [$^{32}$P] NAD, 22 $\mu$M; 3 and 4, [$^{32}$P]ADPR (NAD plus *Neurospora crassa* NADase); 5 and 6, [$^{32}$P]ADPR plus 10 mM AMP; 7 and 8, [$^{32}$P] 2′dNAD, 10 $\mu$M; 9 and 10, [$^{32}$P] 2′dADPR (2′dNAD plus *N. crassa* NADase); 11 and 12, [$^{32}$P] 2′dADPR plus 10 mM 2′dAMP.

Such heterogeneity of the histone product with respect alkaline stability has been reported for several years (*e.g.*, *22*).

*Labeling of Histone H1*

As reported (*23*, *24*) there is some thymidine-sensitive ADP-ribosylation of certain carboxyl residues on histone H1. This yields short chains that are released by alkali or hydroxylamine (*22*).

However, in cell ghosts the predominant modification of histone H1 seems to represent an independent reaction that has nothing to do with poly(ADPR) (Fig. 5). It has the following properties: 1) It is not inhibited by thymidine or similar inhibitors. In fact, with an appropriate inhibitor we can eliminate all other labeling and reveal histone H1 labeling. 2) Its labeling is not dependent upon DNA damage. Consequently it does not share the biphasic progress curve of the other products. Its labeling continues instead at a constant rate throughout an incubation of ghosts with NAD. At very long times it represents the dominant product. 3) It represents the addition of single residues: we cannot detect any chains. 4) 2′d NAD is as good a precursor as NAD. 5) Hydrolysis of NAD (2′dNAD) on storage or by added NAD glycohydrolase increases the H1 labeling in HeLa cell ghosts (whilst decreasing the labeling of all the other products). This suggests that ADPR (2′d ADPR) is the immediate precursor. The labeling is not affected by including an excess of non-radioactive AMP (2′d AMP) indicating that the ADPR is not first split to AMP or through AMP recycled to NAD (Fig. 5). 6) The adduct is alkali stable and is thereby quite distinct from the products characterized in refs. *23* and *24*.

7) The addition is not inhibited by ApppppA which had been reported to inhibit the modification of H1 by purified synthetase (*25*).

ACKNOWLEDGMENT

This work was supported by NIH grants GM28482 (to DMG) and GM26134 (to RJS).

REFERENCES

1. Benjamin, R. C. and Gill, D. M. ADP-ribosylation in mammalian cell ghosts. J. Biol. Chem., *255*: 10493–10501, 1980.
2. Benjamin, R. C. and Gill, D. M. Poly(ADP-ribose) synthesis *in vitro* programmed by damaged DNA. J. Biol. Chem., *255*: 10502–10508, 1980.
3. Miller, E. G. Stimulation of nuclear poly(adenosine diphosphate-ribose) polymerase activity from HeLa cells by endonucleases. Biochim. Biophys. Acta, *395*: 191–200, 1975.
4. Berger, N. A., Weber, G., Kaichi, A., and Petzold, S. J. Relation of poly(adenosine diphosphoribose) synthesis to DNA synthesis and cell growth. Biochim. Biophys. Acta, *509*: 105–117, 1978.
5. Ohgushi, H., Yoshihara, K., and Kamiya, T. Bovine thymus poly(adenosine diphosphate ribose) polymerase: Physical properties and binding to DNA. J. Biol. Chem., *255*: 6205–6211, 1980.
6. Goldman, N., Brown, M., and Khoury, G. Modification of SV40 T antigen by poly ADP-ribosylation. Cell, *24*: 567–572, 1981.
7. Gill, D. M., Caplan, R. A., Bredlau, C. E., and Meren, R. The structure and function of poly(ADP-ribose). *In:* M. Harris (ed.), Poly(ADP-ribose), pp. 85–99, DHEW Publication, Bethesda, 1973.
8. Miwa, M., Saitô, H., Sakura, H., Saikawa, N., Watanabe, F., Matsushima, T., and Sugimura, T. A $^{13}C$ NMR study of poly(adenosine diphosphate ribose) and its monomers: evidence of $\alpha$-(1″ → 2′) ribofuranosyl 1 ribofuranoside residue. Nucleic Acids Res., *4*: 3997–4005, 1977.
9. Jump, D. B. and Smulson, M. Purification and characterization of the major non-histone protein. Acceptor for poly(adenosine diphosphate ribose) in HeLa cell nuclei. Biochemistry, *19*: 1024–1030, 1980.
10. Ogata, N., Kawaichi, M., Ueda, K., and Hayaishi, O. Poly(ADP-ribosyl)ation of 110,000 dalton protein in human lymphocytes treated with N-methyl-N′-nitro-N-nitrosoguanidine. Biochem. Int., *1*: 229–236, 1980.
11. Kawaichi, M., Ueda, K., and Hayaishi, O. Multiple autopoly(ADP-ribosyl)ation of rat liver poly(ADP-ribose) synthetase. J. Biol. Chem., *256*: 9483–9489, 1982.
12. Yoshihara, K., Hashida, T., Yoshihara, H., Tanaka, H., and Ohgushi, H. Enzyme-bound early product of purified poly(ADP-ribose) polymerase. Biochem. Biophys. Res. Commun., *78*: 1281–1288, 1977.
13. Yoshihara, K., Hashida, T., Tanaka, H., Matsunami, N., Yamaguchi, A., and Kamiya, T. Mode of enzyme-bound poly(ADP-ribose) synthesis and histone modification by reconstituted poly(ADP-ribose) polymerase-DNA cellulose complex. J. Biol. Chem., *256*: 3471–3478, 1981.
14. Nishikimi, M., Ogasawara, K., Kameshita, I., Taniguchi, T., and Shizuta, Y. Poly-(ADP-ribose) synthetase, the DNA binding domain and automodification domain. J. Biol. Chem., *257*: 6102–6105, 1982.

15. Ferro, A. M. and Olivera, B. M. Poly(ADP-ribosylation) *in vitro*, reaction parameters and enzyme mechanism. J. Biol. Chem., *257*: 7808–7813, 1982.
16. Suhadolnik, R. J., Lennon, M. B., Uematsu, T., Monahan, J. E., and Baur, R. Role of adenine ring and adenine ribose of NAD in binding and catalysis with alcohol lactate and glyceraldehyde-3-phosphate dehydrogenases. J. Biol. Chem., *252*: 4125–4133, 1977.
17. Lichtenwalner, D. M. and Suhadolnik, R. J. ADP-ribosylation of histone and non-histone chromosomal proteins with NAD and 2′deoxyNAD using nuclei isolated from rat liver and HeLa cells. Biochemistry, *18*: 3749–3755, 1979.
18. Cleaver, J. E. and Boyer, H. W. Solubility and dialysis limit of DNA oligonucleotides. Biochim. Biophys. Acta, *262*: 116–124, 1972.
19. Miwa, M., Tanaka, M., Matsushima, T., and Sugimura, T. Purification and properties of a glycohydrolase from calf thymus splitting ribose-ribose linkages of poly-(adenosine diphosphate ribose). J. Biol. Chem., *249*: 3475–3482, 1974.
20. Burzio, L. O., Riquelme, P. T., and Koide, S. S. ADP-ribosylation of rat liver nucleosomal core histones. J. Biol. Chem., *254*: 3029–3037, 1979.
21. Ogata, N., Ueda, K., and Hayaishi, O. ADP-ribosylation of histone H2B. J. Biol. Chem., *255*: 7610–7615, 1980.
22. Bredehorst, R., Wielckens, K., Gartemann, A., Lengyel, H., Klapproth, K., and Hilz, H. Two different types of bonds linking single ADP-ribose residues covalently to proteins: Quantitation in eukaryotic cells. Eur. J. Biochem., *92*: 129–135, 1978.
23. Riquelme, P. T., Burzio, L. O., and Koide, S. S. ADP-ribosylation of rat liver lysine-rich histone *in vitro*. J. Biol. Chem., *254*: 3018–3028, 1979.
24. Ogata, N., Ueda, K., Kagamiyama, H., and Hayaishi, O. ADP-ribosylation of histone H1: Identification of glutamic acid residues 2, 14 and the COOH-terminal lysine residue as modification sites. J. Biol. Chem., *255*: 7616–7620, 1980.
25. Yoshihara, K. and Tanaka, Y. ADP-ribosylation of diadenosine 5′,5‴-$P^1$,$P^4$-tetraphosphate by poly(ADP-ribose) polymerase *in vitro*. J. Biol. Chem., *256*: 6756–6761, 1981.

ADP-RIBOSYLATION, DNA REPAIR AND CANCER, M. MIWA ET AL. (EDS.),
JAPAN SCI. SOC. PRESS, TOKYO/VNU SCIENCE PRESS, UTRECHT, PP. 141-152, 1983

# Poly(ADP-ribose) Synthetase Associated to Cytoplasmic Structures of Meiotic Cells

Luis O. BURZIO, Ilona I. CONCHA, Jaime FIGUEROA, and Margarita CONCHA

*Instituto de Bioquímica, Facultad de Ciencias, Universidad Austral de Chile, Casilla 567, Valdivia, Chile*

*Abstract:* An active poly(ADP-ribose) synthetase has been found to be associated to mitochondria of frog oocytes and of testis from various species. The enzyme was also found in association with the microsomal-ribosomes fraction of the testis. Both enzymes are strongly inhibited by nicotinamide, thymidine, and theophylline, and they synthesize oligomers of ADP-ribose of an average chain length of 2–6 residues. The apparent $K_m$ for $NAD^+$ of the mitochondrial enzyme is 22 $\mu$M which contrasts with the corresponding value of the chromatin-bound enzyme, 210 $\mu$M $NAD^+$. Treatment of the isolated mitochondria with Triton X-100 revealed that the enzyme is associated to a polydisperse structure. The higher specific activity of the enzyme associated to mitochondria was found in testis of 20 to 30 days old rats. This development study suggests that the maximal activity of the enzyme is associated to mitochondria belonging to primary spermatocytes.

Poly(ADP-ribose) synthetase was also found in association of the microsome-ribosome fraction of rat, mouse, carp, and bull testis. This activity is 20 to 30 times higher than in the same fraction of the liver. Contrary to the chromatin-bound enzyme, the microsome-ribosome enzyme activity is stimulated 3 to 4 times by DNA, but it is slightly inhibited by histone H1. The enzyme seems to be associated to ribonucleoproteins of a very polydisperse nature. The possible relationship of poly(ADP-ribose) synthetase with the intermitochondrial cement and with the chromatoid body will be discussed. These are very unique structures only present in germinal cells.

The biochemistry of poly(ADP-ribose) has been very well documented in previous work (*1–3*). A very fascinating finding has been the demonstration that poly(ADP-ribose) is also found as a ramified structure (*4, 5*). The synthesis of this polymer is catalyzed by the enzyme poly(ADP-ribose) synthetase, which has been purified to homogeneity from various sources (*6–8*). In simple terms, the reaction consists in the transfer of a first residue of ADP-ribose from $NAD^+$ to nuclear proteins, followed by succesive transfer of further residues. These oligomers and polymers of ADP-ribose have been found covalently attached to histones H1 and H2B (*9–12*) and to the enzyme itself (*13–15*).

Although the exact biological function of poly(ADP-ribose) synthetase is still unclear, experimental evidences published recently strongly suggest that the enzyme participates in modulation of chromatin structure (*16–18*) and in DNA repair (*19–21*).

The enzyme has been found ubiquitously in the nuclei of a great variety of eukaryotic cells, including plant (*22*) and human (*23–25*). The first report on the presence of a cytoplasmic poly(ADP-ribose) synthetase, was published by Roberts *et al.* (*26*) working with HeLa cells. The enzyme was associated with ribosomes and polysomes, and, to demonstrate the activity of the enzyme, the assay has to be performed in the presence of DNA and histones.

Contemporary, Kun *et al.* (*27*) reported a poly(ADP-ribose) synthetase activity associated to rat liver mitochondria. According to the authors, the enzyme was poorly inhibited by nicotinamide and insensitive to thymidine. Furthermore, they claim that the reaction was reversible (synthesis of $NAD^+$) in the presence of nicotinamide.

In the present work we would like to discuss our recent findings, which show that, in the germinal cell line, there is an active poly(ADP-ribose) synthetase associated to the mitochondrial fraction and to the microsomal-ribosomal fraction.

*Localization of Poly(ADP-ribose) Synthetase in Frog Oocytes*

We have developed a simple procedure to isolate large quantities of frog oocytes nuclei (*28*). These large nuclei, also known as germinal vesicles, contains a quite active poly(ADP-ribose) synthetase, together with DNA polymerases and RNA polymerases (*28*). Furthermore, it was shown that the enzyme was stimulated 3 to 4 times after induction of the first meiotic division with progesterone (*29*). So, in agreement with a large number of evidences, it was established that the frog oocyte nucleus also contains poly(ADP-ribose) synthetase, and an unknown functional correlation with the meiotic division was suggested (*29*).

Later studies however, showed that, after cell fractionation of the oocytes, an active poly(ADP-ribose) synthetase was also associated to the crude mitochondrial fraction. Centrifugation through a linear sucrose gradient (0.9 M to 2.1 M) yield a quite pure mitochondrial fraction (*30*). The distribution of the poly(ADP-ribose) synthetase through the gradient coincided with the distribution of the mitochondrial marker, succinate-cytochrome *c* reductase, strongly suggesting that the former enzyme was indeed associated with the mitochondria.*

However, a major critisism to this experimental approach, is the possible contamination of the mitochondrial fraction with the nuclear poly(ADP-ribose) synthetase. The frog oocyte nuclei is very large (about 300 μm) and therefore it is reasonable to think, that during homogenization of the ovary, the nuclei will be destroyed, and part of its content together with the enzyme will contaminate the mitochondrial fraction.

However, this problem can be avoided by a very simple procedure. Since the nuclei is so large, a single oocyte can be easily enucleated under low magnification

* Burzio *et al.* unpublished results.

TABLE 1. Poly(ADP-ribose) Synthetase in the Mitochondrial Fraction of Enucleated *C. caudiverbera* Oocytes

| Sample | Activity (pmol ADP-ribose/100 oocytes) |
|---|---|
| Experiment I (52) | 42.3 |
| Experiment II (74) | 63.0 |
| Experiment III (103) | 47.9 |
| Experiment IV (84) | 89.6 |

The isolated oocytes were enucleated as described in the text. The numbers in parenthesis indicated the number of enucleated oocytes employed to prepare the mitochondrial fraction. The assay was carried out as described (*34*).

(*31*). Furthermore, it is also possible to observe, if the large nucleus removed from the oocytes is intact or partially broken. Therefore, it is possible to enucleate a large number of oocytes (200 or more) discarding the ones in which the removed nuclei looks partially broken. These oocytes can be homogenized to prepare mitochondria and, if poly(ADP-ribose) synthetase is found in this fraction, it can not be explained as a contamination with the nuclear enzyme.

This experimental study was carried out with oocytes obtained from a Chilean frog, *Caudiverbera caudiverbera*. The ovary was removed and collected in Gurdon's medium (*32*), and a single oocytes were dissected out of pieces of ovary under low magnification. The isolated oocytes were transferred to a petri dish which contains Callant's saline medium (5 parts of 0.1 M KCl and 1 part of 0.1 M NaCl) (*33*). With a glass micro needle, the oocytes were punctured in the animal hemisphere, and the nucleus was squeezed out by applying gentle pressure to the cell with forceps. The oocytes in which their nucleus was broken or have leaked during the enucleation procedure, were discarded. The enucleated oocytes were homogenized in 0.25 M sucrose, 25 mM Tris-HCl pH 7.5, 50 mM KCl, and 5 mM $MgCl_2$, and the mitochondrial fraction was prepared as described (*34*). Poly(ADP-ribose) synthetase was assayed as described (*34*) and the results are shown in Table 1. The specific activity of the enzyme fluctuates between oocytes from different females. Furthermore, no major differences was found between the activity of enzyme associated to mitochondrial fraction of enucleated oocytes and intact oocytes.

TABLE 2. Poly(ADP-ribose) Synthetase Activity in Testis and Sperm Mitochondria

| Mitochondria | Activity (pmol ADP-ribose/mg protein) |
|---|---|
| Mouse testis | 217 |
| Rat testis | 443 |
| Hamster testis | 135 |
| Bull testis | 256 |
| Carp (*Cyprinus carpio*) testis | 411 |
| Bull sperm (treated with 1% Triton X-100) | 3.7 |
| Bull sperm (hypotonic treatment) | 2.0 |

Mitochondria were prepared and assayed as described (*34*).

*Poly(ADP-ribose) Synthetase Activity Associated to Mitochondria of Testicular Cells*

We have reported recently that, the mitochondrial fraction isolated from rat testis, contains an active poly(ADP-ribose) synthetase (*34*). As shown in Table 2, similar results were obtained with mitochondria isolated from mouse, carp (teleost fish, *Cyprinus carpio*), bull, and hamster testis. In contrast, when rat (*34*) or bull sperm (Table 2) were demembraned to expose the mitochondrial sheath (*34*), a negligible activity was found.

We also reported that mitochondria isolated from rat liver or brain shows a very low activity. These results are opposite to the paper of Kun *et al.* (*27*), who reported the presence of an active poly(ADP-ribose) synthetase in rat liver mitochondria. To detect this enzymatic activity they recommend sonication of the mitochondria. However, in our hand neither sonication nor treatment with 1% Triton X-100 revealed any significative activity. Moreover, we have prepared the protein soluble extract by sonication of the rat liver mitochondria according to their procedure (*27*), and we were unable to detect any significative activity, even in the presence of DNA. At the present time, it is not clear to us the reason for these different results.

Among the various criterions that we presented to discard the possibility of contamination of the mitochondrial fraction with the nuclear enzyme, was the affinity

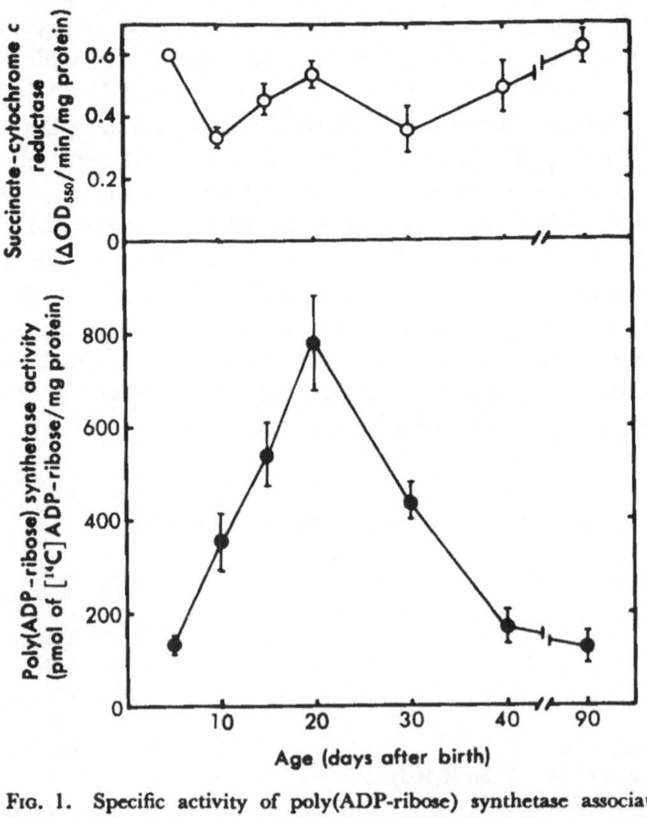


FIG. 1. Specific activity of poly(ADP-ribose) synthetase associated to mitochondria during development of the rat testis germinal epithelium. For comparison, the specific activity of succinate-cytochrome *c* reductase is also shown (top).

for $NAD^+$. While the apparent $K_m$ for $NAD^+$ of the nuclear enzyme was $2.1 \times 10^{-4}$ M, the $K_m$ for the mitochondrial-associated enzyme was $2.2 \times 10^{-5}$ M (*34*).

As discussed before, the enzyme was not detected in eyaculated sperm (bull) or in epididymal sperm (rat). Therefore, a reasonable question, is in which type of testicular cell, the mitochondria are associated to poly(ADP-ribose) synthetase. In the classical paper of Clermont and Perey (*35*), they show that, after birth, there is a chronological relationship between the age of the rat and the type of germinal cell present in the seminiferous tubules. This situation has been extensively used to correlate the appearance of a particular enzyme (*36*) or chromosomal protein (*37*) in the developing rat testis, with a particular type of spermatogenic cell. Hence, the mitochondrial fractions were prepared from testis of male rats at different ages (in days) after birth, and the activity of poly(ADP-ribose) synthetase was measured. As shown in Fig. 1, the activity of the enzyme is relatively low (160 pmoles/mg protein) at 5 days of age. Then, the specific activty of the enzyme rose rapidly up to reach a maximal value of about 700 pmoles/mg of protein at 25 days of age. Later on, the activity decreased and reached a constant level in the adult animals of approximately 150 pmoles/mg of protein. In contrast to this behaviour the activity of the mitochondrial succinate-cytochrome *c* reductase remains more or less constant during the development of the rat testis (Fig. 1).

At the age of 25 days, the germinal epithelium of the testis is rich in primary spermatocytes, specially pachytene spermatocytes (*35*). Since around this age, the isolated mitochondrial shows the highest specific activity of poly(ADP-ribose) synthetase, it is reasonable to conclude that mitochondria from primary spermatocytes (*e.g.*, pachytene) are associated with poly(ADP-ribose) synthetase. Of course, it is also possible that the activity of the enzyme is maximal in pachytene spermatocytes, and, that in other type of germinal cells the activity associated to mitochondria decrease progressively and disappears in sperm mitochondria. A definitive answer will came from mitochondria isolated from purified spermatogenic cells. Work is in progress to isolated highly pure pachytene spermatocytes and other germinal cells by sedimentation at unit gravity (*38*).

### *Poly(ADP-ribose) Synthetase Associated to the Microsome-ribosome Fraction of Testis*

If the rat testis post-mitochondrial supernatant is subjected to a series of centrifugation up to $27{,}000 \times g$ no activity of poly(ADP-ribose) synthetase was recovered in these pellets. However, if the same supernatant was centrifuged at $110{,}000 \times g$ for 2 hr a microsomal-ribosome (M-R) fraction was obtained containing a very active poly(ADP-ribose) synthetase. As shown in Table 3, similar levels of enzymatic activity were found in the same cellular fraction prepared from mouse and bull testis. On the other hand, the same fraction prepared from rat liver showed a quite low level of enzymatic activity (Table 3).

The apparent $K_m$ for $NAD^+$ of the poly(ADP-ribose) synthetase associated to the M-R, fraction was 105 $\mu$M, which is different from the $K_m$ of the mitochondrial-associated enzyme (22 $\mu$M) and the $K_m$ of the nuclear enzyme (210 $\mu$M). Furthermore, the enzyme was inhibited more than 90% by 1 mM nicotinamide, thymidine, and theophylline (data not shown).

TABLE 3. Poly(ADP-ribose) Synthetase Activity Associated to the Microsomal-ribosomal Fraction of Testis

| Microsomal-ribosomal fraction prepared from: | Activity (pmol ADP-ribose/mg protein) | |
|---|---|---|
| | Control | + DNA |
| Rat testis | 437 | 1,156 |
| Rat liver | 15 | — |
| Mouse testis | 240 | 794 |
| Bull testis | 175 | 253 |

The tissue was homogenized in S-TKM buffer (0.25 M sucrose, 25 mM Tris-HCl, pH 7.5, 50 mM KCl, and 5 mM $MgCl_2$). To obtain the M-R fraction, the postmitochondrial supernatant was centrifuged at 110,000 × *g* for 2 hr. The enzyme assay was carried out as described (*34*).

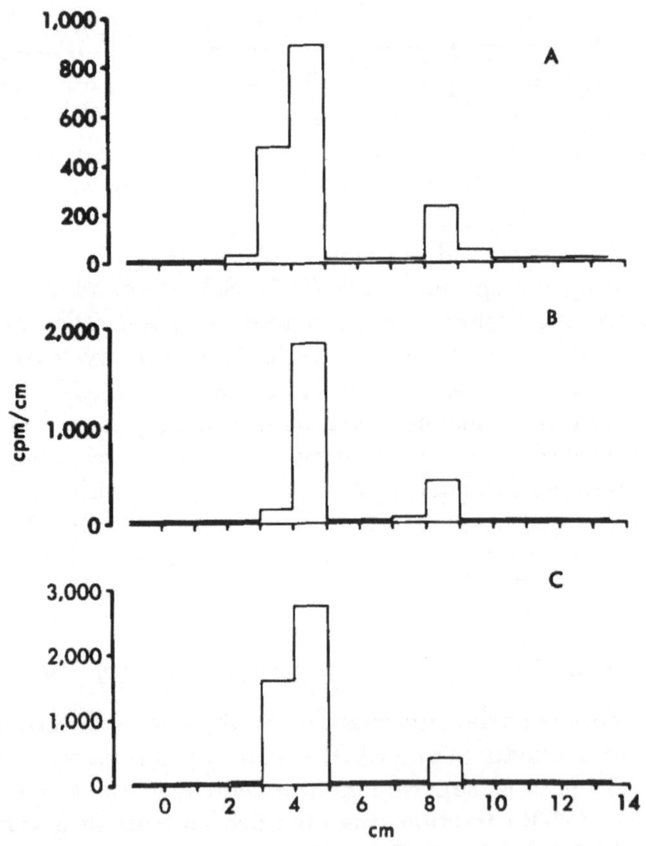


FIG. 2. Average chain length of oligomers of ADP-ribose synthesized by the rat testis M-R poly(ADP-ribose) synthetase. The hydrolysis products of venom phosphodiesterase were chromatographed on PEI cellulose. A and B show the results of the product synthesized by the M-R enzyme, without and with DNA respectively. C represents the chromatogram of the hydrolysis products of an oligomer of ADP-ribose with an average chain length of 11.

Roberts *et al.* (*26*) have already reported a poly(ADP-ribose) synthetase bound to free and polysomal ribosomes in HeLa cells. To demonstrate the enzymatic activity it was necessary to add DNA and histones to the assay. In contrast, the enzyme associated to the M-R fraction of the rat testis was active without DNA. However, if DNA was added to the assay, the incorporation of [$^{14}$C]NAD$^+$ was stimulated 2 to 4 times (Table 3). Histone H1 on the other hand, has a small inhibitory effect. It was also interesting to know whether the stimulatory effect of DNA induces an increase in the number of oligomeric chains of ADP-ribose or, was just an increase in the length of the synthesized polymers. As shown in Fig. 2, the average chain length measured as described (*9*), was 5.4 and 5.3 without and with DNA, respectively. So the effect of DNA is actually to stimulate an increase of the number of chains synthesized, rather than to increase the length of the polymers.

The oligomers of ADP-ribose are attached to proteins through a bond which is largely labile to NaOH, but partially stable to $NH_2OH$. If the M-R fraction previously incubated with [$^{14}$C]NAD$^+$, is treated with 1 M $NH_2OH$ for 60 min at 37°C, only about 20% of the incorporated radioactivity was released. On the other hand, incubation for 15 min with 0.1 N NaOH releases about 60% of the radioactivity and, after 1 hr, about the same amount of radioactivity (40%) still remains trichloroacetic acid (TCA)-insoluble. At the present time we do not know if, the 40% of radioactivity not released by the treatment with NaOH, is because some chemical bonds between the oligomers of ADP-ribose and proteins are alkali resistant, or, because all the bonds were actually cleaved by this treatment, but about 40% of the radioactivity are oligomers of ADP-ribose long enough to be TCA-precipitable.

When the rat testis post-mitochondrial supernatant was layered on top of a 2 M sucrose cushion (*39*) and centrifuged at 25,000 rpm for 2 hr, all the poly(ADP-

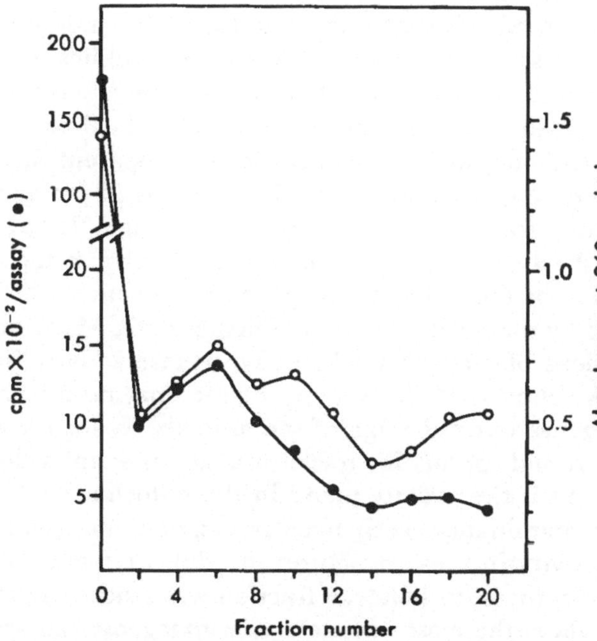


FIG. 3. Sedimentation of poly(ADP-ribose) synthetase in a sucrose gradient. The rat testis M-R fraction was treated with 1% Triton X-100 and sedimented through a linear sucrose gradient (15% to 40%) at 100,000×*g* for 2 hr. After fractionation of the gradient, the enzyme activity and the absorbance at 260 nm were determined. Sedimentation from right to lfet.

ribose) synthetase activity was recovered at the 0.25/2 M sucrose interphase. If previous to centrifugation, the supernatant was treated with 1% Triton X-100, once again most of the enzymatic activity was recovered in the 0.25/2 M sucrose interphase, but, about 20% was found in the pellet. This material that sedimented through the 2 M sucrose, has a typical spectra of ribonucleoproteins. Therefore, the pellet obtained after centrifugation of the rat testis post-mitochondrial supernatant at 110,000×*g* for 2 hr, was resuspended in homogenization buffer (0.25 M sucrose, 25 mM Tris-HCl, pH 7.5, 5 mM $MgCl_2$, and 50 mM KCl) and made 1% in Triton X-100. The clear homogenate was layered on top of a linear sucrose gradient (15% to 40%) and centrifuged at 25,000 rpm for 2 hr at 4°C. As shown in Fig. 3, the activity of poly(ADP-ribose) synthetase is distributed near the bottom of the tube, coinciding, more or less, with the absorbance at 260 nm. However, most of the enzymatic activity was recovered in the translucent pellet. Once again, the fractions with higher activity are associated to a material with the spectra of ribonucleoproteins.

## GENERAL DISCUSSION

The results presented here demonstrated that, in germinal cells there is a considerable amount of poly(ADP-ribose) synthetase associated to the mitochondrial fraction and to the microsomal-ribosomal fraction. The demonstration that the enzyme is really associated to these fractions, and it is not a contamination with the nuclear enzyme, is reminiscent of the demonstration of mitochondrial DNA. Since the amount of poly(ADP-ribose) synthetase in the nuclei is so large, it is reasonable to think on a contamination of the mitochondria and microsome fraction with the nuclear enzyme. However, in this study, as well as in a previous publication (*34*), we have shown that the enzyme associated to the cytoplasm structures have differences with the nuclear enzyme. First, in enucleated oocytes a poly(ADP-ribose) synthetase activity was found in the mitochondrial fraction. Second, the apparent $K_m$ for NAD of the mitochondrial-associated enzyme and the M-R enzyme is 22 μM and 105 μM, respectively, which are different from the $K_m$ of the nuclear enzyme (210 μM). Third, the chemical stability toward alkaline conditions and 1 M $NH_2OH$ of the bond between oligomers and proteins of the cytoplasmic structures, seems to be different from the product synthesized by the nuclear enzyme. Furthermore, Monii and Koide (*40*) reported on the variations of poly(ADP-ribose) synthetase activity in isolated testis nuclei of mouse. They found that the activity of the enzyme increased with the progression of spermatogenesis and the age of the animals, in such a way that, the total nuclear activity in round spermatids nuclei was about 4 times the activity in pachytene spermatocytes. So if the activity found in the mitochondrial and M-R fraction of the testis was contamination with nuclear enzyme, one should expect higher level of enzymatic activity in these structures in adult animals. In contrast, the maximal level of activity in the mitochondrial fraction was reached in testis from animals 20 to 30 days old, where the most advanced spermatogenic cell are the pachytene spermatocytes (*35*). Later on, the activity decreased and reached a constant lower level in the adult.

The enzyme associated to the rat testis M-R fraction was stimulated 2 to 4 times by DNA. However, the activity of the enzyme without DNA was important (Table

3). These assays were carried out in the presence of 50 $\mu$M $NAD^+$ which is below the apparent $K_m$ for $NAD^+$ of the enzyme (105 $\mu$M). More recent experiments done in the presence of 300 $\mu$M $NAD^+$ revealed a level of activity similar to the one found in the testis nuclei. Therefore, one pertinent question is, to which nucleic acid is the enzyme bound, in order to show considerable activity without added DNA. Although preliminar experiments have shown that the activity of the enzyme is partially affected by DNAase I but not by RNAase A, further studies are necessary to clear up this crucial point.

But, perhaps, the most important question is, to which cytoplasmic structure of the germinal cells is the poly(ADP-ribose) synthetase associated with. We would like to suggest here, that the enzyme is associated to the intermitochondrial cement (*41*) and to the chromatoid body and satellite (*42*). These structures are very enigmatic components of the germinal cell cytoplasm in all the species analyzed. During the meiotic prophase of spermatogenesis, and specially in pachytene spermatocytes, the mitochondria are found as cluster, or aggregated of three to a dozen or more of them (*41*, *42*). In the interstices between the mitochondria there is a dense and amorphous material, which has been referred to, as intermitochondrial cement (*41*). The chronological relationship between the appearance of these mitochondrial association in the germinal epithelium with the highest specific activity of poly(ADP-ribose) synthetase in the same cellular fraction (Fig. 1), strongly suggests that the enzyme is associated to the intermitochondrial cement.

Electron microscopy has shown that, the chromatoid body is a rather conspicuous structure of early spermatids (*42*). It is a pleomorphic structure composed of fine filaments which has a great affinity for uranyl salts (*42*). In less developed spermatogenic cells (*e.g.*, pachytene spermatocytes) the chromatoid body does not seem to be present. However, another structure very much related to the chromatoid body and, referred to as satellite, is distinguishable (*42*). Electron microscopy studies of the M-R fraction isolated from rat testis, revealed the presence of abundant smooth and rough ER, together with large masses of ribosome-like material.* This ultrastructural feature is similar to the chromatoid body as seen *in situ*, and, therefore, we would like to suggest that poly(ADP-ribose) synthetase is also associated to this structure.

At the present time we are far from understanding the biological role of the cytoplasmic poly(ADP-ribose) synthetase in germinal cells. However, future work, not only will answer this question, but also will shed some light on the important biological function that the intermitochondrial cement, and the chromatoid body, must be playing during gametogenesis.

ACKNOWLEDGMENTS

We wish to thank Mrs. M. Angélica Espinoza for her skilful typing and Mr. Risnel Bustos for the illustration. This work was supported by Grant S-79-20 and S-82-01 from the Research Fund of the Universidad Austral de Chile.

* J. Figueroa, I. Concha, and L. Burzio, unpublished results.

## REFERENCES


1. Hayaishi, O. and Ueda, K. Poly(ADP-ribose) and ADP-ribosylation of proteins. Annu. Rev. Biochem., *46*: 95–116, 1977.
2. Ueda, K., Ogata, N., Kawaichi, M., Inada, S., and Hayaishi, O. ADP-ribosylation reactions. Curr. Top. Cell. Result., *21*: 175–187, 1982.
3. Smulson, M. E. and Sugimura, T. (eds.) Novel ADP-Ribosylations of Regulatory Enzymes and Proteins, Elsevier/North-Holland Publ. Co., New York, 1980.
4. Miwa, M., Saikawa, N., Yamaizumi, Z., Nishimura, S., and Sugimura, T. Structure of poly(adenosine diphosphate ribose): Identification of 2′-[1″-ribosyl-2″-(or 3″-) (1‴-ribosyl)] adenosine-5′,5″,5‴ tris(phosphate) as a branch linkage. Proc. Natl. Acad. Sci. U.S., *76*: 595–599, 1979.
5. Miwa, M., Ishihara, M., Takishima, S., Takasuka, N., Maeda, M., Yamaizumi, Z., Sugimura, T., Yokoyama, S., and Miyazawa, T. The branching and linear portions of poly(ADP-ribose) have the same (1–2) ribose-ribose linkage. J. Biol. Chem., *256*: 2916–2921, 1981.
6. Yoshihara, K., Hashida, T., Tanaka, Y., Ohgushi, H., Yoshihara, H., and Kamiya, T. Bovine thymus poly(adenosine diphosphate ribose) polymerase. J. Biol. Chem., *253*: 6459–6466, 1978.
7. Kawaichi, M., Ueda, K., and Hayaishi, O. Initiation of poly(ADP-ribosyl)histone synthesis by poly(ADP-ribose) synthetase. J. Biol. Chem., *255*: 816–819, 1980.
8. Agemori, M., Kagamiyama, H., Nishikimi, M., and Shizuta, Y. Purification and properties of poly(ADP-ribose) synthetase from mouse testicle. Arch. Biochem. Biophys., *215*: 621–627, 1982.
9. Riquelme, P. T., Burzio, L. O., and Koide, S. S. ADP-ribosylation of rat liver lysine-rich histone *in vitro*. J. Biol. Chem., *254*: 3018–3028, 1979.
10. Burzio, L. O., Riquelme, P. T., and Koide, S. S. ADP-ribosylation of rat liver nucleosomal core histones. J. Biol. Chem., *254*: 3029–3037, 1979.
11. Ogata, N., Ueda, K., Kagamiyama, H., and Hayaishi, O. ADP-ribosylation of histone H1. J. Biol. Chem., *255*: 7616–7620, 1980.
12. Ogata, N., Ueda, K., and Hayaishi, O. ADP-ribosylation of histone H2B. J. Biol. Chem., *255*: 7610–7615, 1980.
13. Yoshihara, K., Hashida, T., Yoshihara, H., Tanaka, Y., and Ohgushi, H. Enzyme-bound early product of purified poly(ADP-ribose) polymerase. Biochem. Biophys. Res. Commun., *78*: 1281–1288, 1977.
14. Ogata, N., Ueda, K., Kawaichi, M., and Hayaishi, O. Poly(ADP-ribose) synthetase, a main acceptor of poly(ADP-ribose) in isolated nuclei. J. Biol. Chem., *256*: 4135–4137, 1981.
15. Nishikimi, M., Ogasawara, K., Kameshita, I., Taniguchi, T., and Shizuta, Y. Poly-(ADP-ribose) synthetase. The DNA binding domain and the automodification domain. J. Biol. Chem., *257*: 6102–6105, 1982.
16. Burzio, L. O., Koide, S. S., Puigdomenech, P., and Ruiz-Carrillo, A. ADP-ribosylation of rat liver H1 and nucleosomal core histones. *In;* M. E. Smulson and T. Sugimura (eds.), Novel ADP-Ribosylations of Regulatory Enzymes and Proteins, pp. 345–353, Elsevier/North-Holland Publ. Co., New York, 1980.
17. Poirier, G. G., de Murcia, G., Jongstra-Bilen, J., Niedergang, C., and Mandel, P. Poly(ADP-ribosylation) of polynucleosomes causes relaxation of chromatin structure. Proc. Natl. Acad. Sci. U.S., *79*: 3423–3427, 1982.
18. Butt, T. R., Jump, D. B., and Smulson, M. E. Nucleosome periodicity in HeLa cell

chromatin as probed by microsomal nuclease. Proc. Natl. Acad. Sci. U.S., *76*: 1628–1632, 1979.

19. Durkacz, B., Omidiji, O., Gray, D., and Shall, S. (ADP-ribose)$_n$ participates in DNA excision repair. Nature, *283*: 593–596, 1980.
20. Berger, N. A. and Sikorski, G. W. Poly(adenosine diphosphoribose) synthesis in ultraviolet-irradiated xeroderma pigmentosum cells reconstituted with *Micrococcus luteus* UV endonuclease. Biochemistry, *20*: 3610–3614, 1981.
21. Creissen, D. and Shall, S. Regulation of DNA ligase activity by poly(ADP-ribose). *296*: 271–272, 1982.
22. Whitby, A. J., Stone, P. R., and Whish, W. J. Effect of polyamines and $Mg^{2+}$ on poly(ADP-ribose) synthesis and ADP-ribosylation of histones in wheat. Biochem. Biophys. Res. Commun., *90*: 1295–1304, 1979.
23. Ikai, K., Ueda, K., Fukushima, M., Nakamura, T., and Hayaishi, O. Poly(ADP-ribose) synthesis, a marker of granulocyte differentiation. Proc. Natl. Acad. Sci. U.S., *77*: 3682–3685, 1980.
24. Kanai, Y., Tanuma, S., and Sugimura, T. Immunofluorescent staining of poly(ADP-ribose) *in situ* in HeLa cell chromosomes in the M phase. Proc. Natl. Acad. Sci. U.S., *78*: 2801–2804, 1981.
25. Burzio, L. O., Reich, L., and Koide, S. S. Poly(adenosine diphosphoribose) synthetase activity of isolated nuclei or normal and leukemic leukocytes. Proc. Soc. Exp. Biol. Med., *149*: 933–938, 1975.
26. Roberts, J. H., Stark, P., Giri, C., and Smulson, M. Cytoplasmic poly(ADP-ribose) polymerase during the HeLa cell cycle. Arch. Biochem. Biophys., *171*: 305–315, 1975.
27. Kun, E., Zimber, P. H., Chang, A. C., Puschendorf, B., and Grunicke, H. Macromolecular enzymatic product of $NAD^+$ in liver mitochondria. Proc. Natl. Acad. Sci. U.S., *72*: 1436–1440, 1975.
28. Burzio, L. O., Zuvić, T., Phillips, D. M., and Koide, S. S. Poly(adenosine diphosphate ribose) synthetase during oogenesis of the *Xenopus laevis*. *In;* M. A. Siddiqui, M. Krauskopf, and H. Weissbach (eds.), Molecular Approaches to Gene Expression and Protein Structure, pp. 149–171, Academic Press, New York, 1981.
29. Burzio, L. O. and Koide, S. S. Stimulation of poly(adenosine diphosphate ribose) synthetase activity of *Xenopus* germinal vesicle by progesterone. Ann. N. Y. Acad. Sci., *286*: 398–407, 1977.
30. Dawid, I. B. Evidence for the mitochondrial origin of frog egg cytoplasmic DNA. Proc. Natl. Acad. Sci. U.S., *56*: 269–275, 1966.
31. Smith, L. D. and Ecker, R. E. Role of the oocyte nucleus in physiological maturation in *Rana pipiens*. Dev. Biol., *19*: 281–309, 1969.
32. Gurdon, J. B. Nuclear transplantation and control of gene activity in animal development. Proc. Roy. Soc. B., *176*: 303–314, 1970.
33. Callan, H. G. and Lloyd, L. Lampbrush chromosomes of crested newts *Triturus cristatus*. Phil. Trans. Roy. Soc. B., *243*: 135–210, 1960.
34. Burzio, L. O., Sáez, L., and Cornejo, R. Poly(ADP-ribose) synthetase activity in rat testis mitochondria. Biochem. Biophys. Res. Commun., *103*: 369–375, 1981.
35. Clermont, Y. and Perey, B. Quantitative study of the cell population of the seminiferous tubules in immature rats. Am. J. Anat., *100*: 241–268, 1955.
36. Turkington, R. W. and Majumder, G. C. Gene Activation during Spermatogenesis. J. Cell. Physiol., *85*: 485–508, 1975.
37. Concha, I. I., Brito, M., and Burzio, L. O. Isolation and characterization of two rat testis specific histones H1. Arch. Biol. Med. Exp. (Chile), 1982, in press.
38. Meistrich, M. L., Bruce, W. R., and Clermont, Y. Cellular composition of fractions

of mouse testis cells following velocity sedimentation separation. Exp. Cell Res., *79*: 213–227, 1973.

39. Noll, H. *In;* P. N. Campbell and J. R. Sargent (eds.), Polysomes: Analysis of Structure and Function in Techniques in Protein Biosynthesis, pp. 101–179, Academic Press, New York, 1969.
40. Momii, A. and Koide, S. S. RNA synthesis and poly(ADP-ribose) synthetase activity in developing mouse testis. Arch. Androl.
41. Clérot, J. C. Mise en Evidence par Cytochimie Ultrastructurale De L'Emission de Protéines par le Noyau D'auxocytes de Batraciens. J. Microscopie, *7*: 973–992, 1968.
42. Fawcett, D. W., Eddy, E. M., and Phillips, D. M. Observations on the fine structure and relationships of the chromatoid body in mammalian spermatogenesis. Biol. Reprod., *2*: 129–153, 1970.

# ADP-RIBOSYLATION AND DNA REPAIR

ADP-RIBOSYLATION, DNA REPAIR AND CANCER, M. MIWA ET AL. (EDS.),
JAPAN SCI. SOC. PRESS, TOKYO/VNU SCIENCE PRESS, UTRECHT, PP. 155-163, 1983

# Functional Aspects of Mono- and Poly(ADP-ribosyl)ation: Subcellular Distribution and ADP-ribosyl Turnover under Conditions of Repair and 'Starvation'

H. Hilz, K. Wielckens, P. Adamietz, R. Bredehorst, and A. Kreymeier

*Institut für Physiologische Chemie der Universität Hamburg, D-2000 Hamburg 20, West Germany*

*Abstract:* Three types of ADP-ribosyl proteins (poly(ADP-ribose) conjugates, $NH_2OH$ sensitive and $NH_2OH$ resistant mono(ADPR) conjugates) could be found in all eukaryotic cells so far studied. They changed independently under various conditions and showed an uneven subcellular distribution suggesting independent functions.

Treatment of Ehrlich ascites tumor (EAT) cells with monofunctional or cross-linking alkylating agents led to rapid fragmentation of DNA and depletion of NAD while poly(ADPR) polymerase activity showed a retarded increase. Endogenous amounts of poly(ADPR) groups increased 4- to 30-fold, depending on dose, with the same initial kinetics as the loss of NAD and the appearance of DNA strand breaks. Turnover of poly(ADPR) was determined from the decay rate of the polymer after the addition of benzamide to alkylated cells. At peak elevation of poly(ADPR), an apparent half-life of about 1 min was obtained (control cells: $t/2 \gg 3$ hr). There was also an accumulation of nuclear mono(ADPR) conjugates with a half-life of about 10 min.

In contrast to *in vitro* experiments, histone H1 *in vivo* proved to be only a minor acceptor of ADPR groups in rat liver and in hepatoma cells. It carried less than 0.2% of total monomeric, and less than 2% of total polymeric ADPR residues. Alkylation of cells increased mono(ADP-ribosyl)ation of histone H1 to a much higher degree than poly(ADP-ribosyl)ation. Addition of benzamide to alkylated cells inhibited poly(ADPR) formation and NAD depletion, but interfered with neither DNA fragmentation nor with DNA resealing. Nevertheless, benzamide was a very effective co-cytostatic. "Starvation" of EAT cells also induced fragmentation of DNA and a concomitant increase in poly(ADPR) that approached the values found in dimethylsulfate-treated cells. Turnover of poly(ADPR) was also greatly stimulated. However, NAD was not depleted under these conditions.

"Starvation"-induced fragmentation of DNA could be suppressed by benzamide indicating that DNA fragmentation in this system was dependent on poly(ADPR) formation. These data suggest that poly(ADPR) synthesis is not, or at least not exclusively, triggered by DNA fragmentation but is induced by certain structural features of the chromatin.

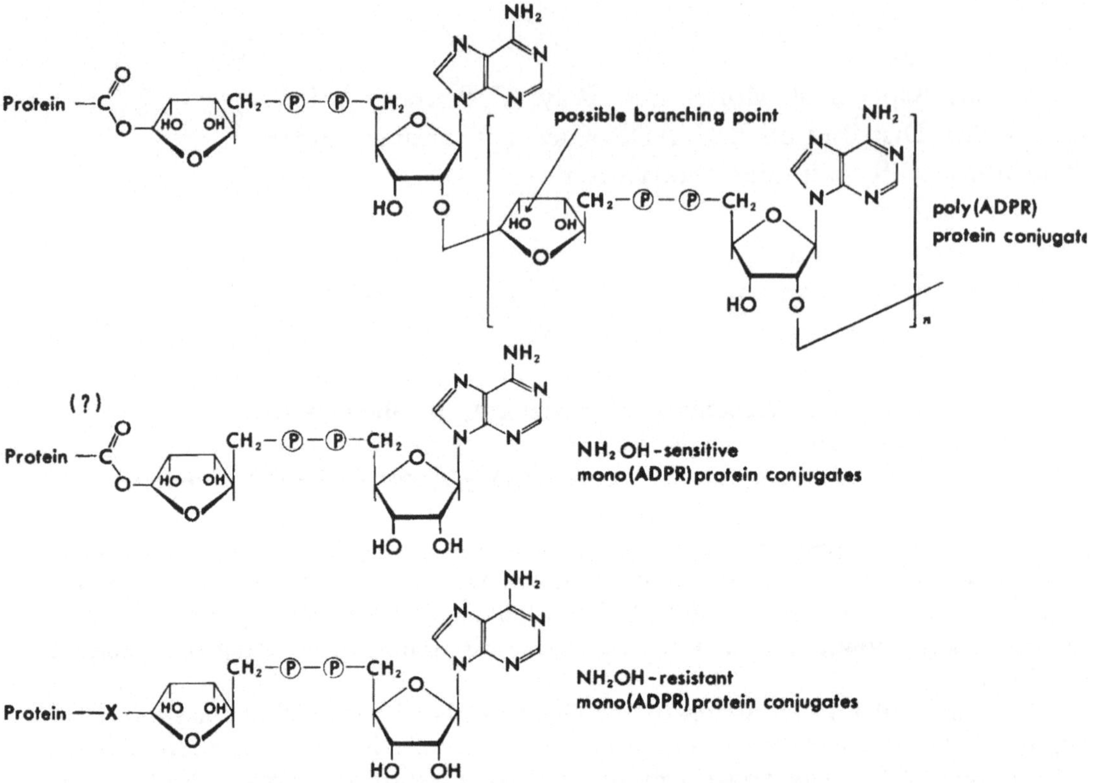


FIG. 1. Three types of poly(ADPR) protein conjugates. (From ref. *25*).

Transfer of ADP-ribose (ADPR) residues from NAD to proteins has been detected in eukaryotic cells and as a function of various bacterial toxins and phages (for a review see refs. *1–3*). Three types of (ADP-ribosyl) protein conjugates could be differentiated in eukaryotic cells: conjugates carrying single ADPR residues either in a hydroxylamine-sensitive or in a hydroxylamine-resistant linkage, and poly(ADPR) conjugates (Fig. 1). Procedures developed in this laboratory (*4*, *5*) have allowed quantitation of these conjugates in various tissues and under different conditions of growth and metabolism (*6–9*). These data provided suggestive evidence of the independent functions of these three types of conjugates as deduced from independent changes in developing rat liver (*8*), in normal *versus* leukemic lymphocytes (*7*), during progression through the cell cycle (*10*), in testosterone-dependent alterations of mouse kidney (*9*), or in differentiating *Dictyostelium discoideum* (*6*). Nevertheless, deductions as to the functions of ADP-ribosylation and poly(ADP-ribosyl)ation on a more detailed basis were not possible from these analyses.

### *Subcellular Distribution of Poly(ADPR) Protein Conjugates*

Since the three types of poly(ADPR) proteins serve independent functions, it was to be expected that their subcellular distribution was not identical.

When the distribution of total mono(ADPR) protein conjugates was analyzed in

TABLE 1. Total Protein-bound Mono(ADPR) Residues in Homogenates and in Nuclei from Different Tissues

| Preparation | Mono(ADPR) residues (pmol/mg DNA) | Mono(ADPR) residues (% of total) |
|---|---|---|
| Rat liver | | |
| homogenate | 7,830 ± 800 | 100 |
| nuclei (neutral sucrose) | 70 ± 15 | 1 |
| nuclei (citric acid) | 220 ± 25 | 3 |
| Mouse kidney | | |
| homogenate | 1,960 ± 110 | 100 |
| nuclei (citric acid) | 90 ± 15 | 4 |
| EAT cells | | |
| homogenate | 730 ± 35 | 100 |
| nuclei (citric acid) | 310 ± 30 | 42 |

From ref. *11*.

rat liver, it became apparent that by far the most conjugates were located outside the nucleus (Table 1). This was also true for rat kidney. The situation appeared to be different in Ehrlich ascites tumor (EAT) cells which showed a nearly equal distribution of mono(ADPR) conjugates between nucleus and cytoplasm. However, it should be pointed out that the absolute amounts of these conjugates located in EAT cell nuclei are in the same range as in the non-proliferating tissues, the difference being caused by the low cytoplasmic values in EAT cells (*11*).

Extranuclear mono(ADPR) protein conjugates again showed an uneven distribution. The bulk of the $NH_2OH$-sensitive conjugates was present in the microsomal fraction, while $NH_2OH$-resistant conjugates were mainly in the mitochondrial and plasma membrane fractions (*11*).

Poly(ADPR) protein conjugates, on the other hand, appeared to be located in the nucleus (*11*, *12*). This suggested that poly(ADP-ribosyl)ation is involved in chromatin structure and function.

### *Augmentation of Endogenous Poly- and Mono(ADPR) Residues by Alkylating Agents*

When EAT cells were treated with triaziquonum (a trifunctional aziridine) or with dimethylsulfate, poly(ADPR) residues rapidly increased to reach a peak 4–20 times normal between 30 and 90 min, depending on the dose applied. There was also a significant increase in protein-bound mono(ADPR) residues. This augmentation of monomeric ADPR residues, however, is a specific nuclear event. Separation into nuclear and cytoplasmic mono(ADPR) protein conjugates revealed that it was the nuclear fraction only that increased by a factor of 8 (*13*) with kinetics similar to the rise in the poly(ADPR) conjugates. On the other hand, activation of poly(ADPR) polymerase as measured in permeabilized cells reached maximal values only rather late (*13*).

In intact tissues, histone H1 appears to be a minor acceptor of mono- and poly-(ADPR) groups. In rat liver and in hepatoma cells (AH 7974) histone H1 contained less than 0.2% of total monomeric and less than 3% of total polymeric ADPR re-

TABLE 2. Dimethylsulfate-induced Changes in Total and in Histone Hl-associated ADPR Residues

| Fraction | ADPR residues | | |
|---|---|---|---|
| | Control (pmol/10⁸ cells) | Dimethylsulfate-treated (pmol/10⁸ cells) | Dimethylsulfate-treated (stimulation) |
| A. Total conjugates | | | |
| mono (ADPR) | 1,030 | 2,020 | 2× |
| poly (ADPR) | 7.1 | 210 | 30× |
| B. Histone Hl conjugates | | | |
| mono (ADPR) | 1.6 | 188 | 118× |
| poly (ADPR) | 0.2 | 4.7 | 24× |

From ref. *16*.

sidues. In alkylated cells both histone H1-associated poly(ADPR) residues and mono-(ADPR) residues increased considerably, but to a different extent (Table 2 and ref. *14*). The predominant change was a marked increase in $NH_2OH$-resistant mono-(ADPR) groups that exceeded by far the moderate augmentation of total protein-bound mono(ADPR) residues under these conditions. The data demonstrate that induction of DNA repair affects ADP-ribosylation of histone H1 in a special way.

NAD was lost in alkylated cells with initial kinetics that were comparable to the augmentation of protein-bound poly(ADPR) residues and therefore suggestive of a link between these two parameters. However, it should be kept in mind that the observed increase of poly(ADPR) residues cannot explain the depletion of NAD stores, which are present in much higher concentrations than protein-bound polymeric+monomeric ADPR residues (see below). Only if the augmentation of poly(ADPR) residues is accompanied by a high turnover of the ADPR groups would a significant drain of NAD be expected.

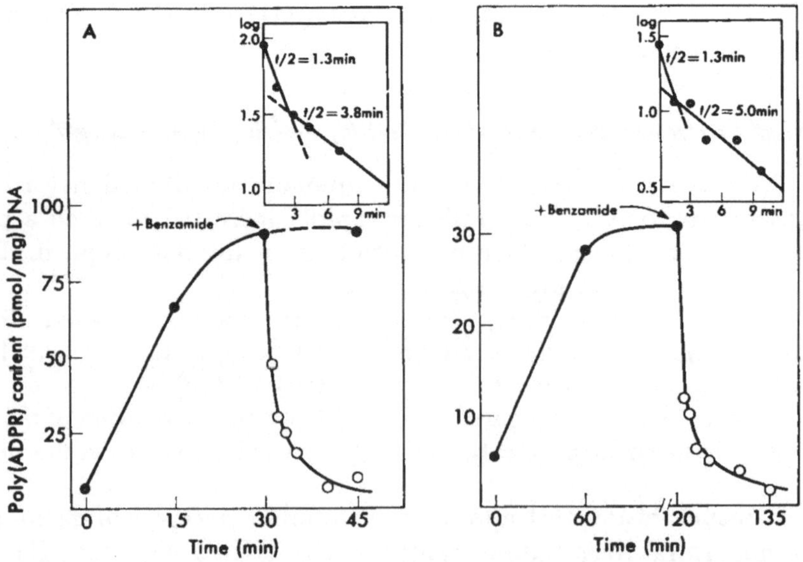


FIG. 2. Poly(ADPR) turnover in alkylated and in starved EAT cells. (From ref. *24*).

### *Dramatic Stimulation of Poly(ADPR) Turnover by Alkylation*

To analyze the turnover of poly(ADPR) groups *in vivo*, we used benzamide as the most potent inhibitor of poly(ADPR) polymerase activity (*28*). This compound is also of advantage because of the high lipophilia allowing rapid penetration into cells (*13*). When benzamide at saturating concentrations (10 $\mu$M) was added to dimethylsulfate-treated cultures at peak elevation of poly(ADPR) residues, an immediate drop of the polymer was observed. From the decay curve (Fig. 2A) it was concluded that the half-life of poly(ADPR) residues is about 1 min (*13*). This shows that the turnover of poly(ADPR) groups which in undisturbed cells exhibit a $t/2 \gg$ 3 hr, becomes dramatically stimulated in alkylated cells. There is also a small fraction of the polymer with a somewhat slower turnover. Nuclear mono(ADP-ribosyl) groups exhibit a half-life (8–10 min) that is definitely longer than the bulk of the polymeric ADPR residues. It is tempting to speculate that it is the removal of the "primary" ADPR groups at the acceptor sites which is the rate-limiting step in the overall poly(ADP-ribosyl) degradation thus leading to an accumulation of mono(ADPR) residues under conditions of increased poly(ADPR) turnover.

### *Poly(ADPR) Turnover and NAD Depletion in Alkylated vs. "Starved" EAT Cells*

In cells treated with alkylating agents or ionizing radiation, NAD rapidly declines to low values (*16–19*). This has been interpreted as a depletion of substrate by an accelerated poly(ADPR) formation (*20–22*). However, NAD is present in EAT cells at 80,000 pmol per $10^8$ cells, while nuclear ADPR residues at peak elevation amount to about 2,300, and polymeric ADPR groups to about 220 pmol/$10^8$ cells. The contribution of mono(ADPR) groups to the depletion of NAD is negligible because of the relatively slow half-life (requiring about 100 pmol/min). Polymeric ADPR residues induced by alkylation rise to about 200 pmol per $10^8$ cells while "active ADPR" (=NAD) is depleted under these conditions with a $t/2 \approx 15$ min. To explain this depletion as caused by ADP-ribosylation, a turnover of the poly(ADPR) groups has to be assumed that is much higher than that measured in our experiment (Fig. 2A), namely $t/2 < 0.1$ min instead of 1 min. This assumption is supported by recent experiments in which poly(ADPR) turnover could be stimulated to apparently the same extent ($t/2 \approx 1$ min), while NAD levels were not significantly affected.

When EAT cells were subjected to "starvation" conditions, they responded rapidly with an increasing fragmentation of DNA and an increasing accumulation of poly(ADPR) (*23*). When analyzed at 120 min poly(ADPR) had increased by 800% and exhibited a high turnover, the half-life being in the range of 1 min. This value was identical to that found with alkylating agents (Fig. 2B). In contrast to the alkylated cells, however, NAD did not decrease under "starvation" conditions (*23*). When calculated from the half-lives and the amounts of NAD *versus* poly(ADPR), it became clear that a half-life of poly(ADPR) of about 1 min could effect a decrease of endogenous NAD by not more than 5%.

The apparently identical half-lives of the polymer under both conditions in spite of a completely different stability of the substrate NAD suggests that in alkylated cells, but not in "starved" cells, the penetration of benzamide to the nucleus is the rate-limiting step in the determination of poly(ADPR) half-life. It is reasonable to assume, therefore, that the decrease of NAD induced by alkylating agents is caused by a stimulation of poly(ADP-ribosyl)ation to such an extent that $t/2$ values $<0.1$ min are reached. The enzymic capacity of the EAT cells would allow such extreme turnover rates. It should be pointed out, however, that other causes of NAD depletion cannot be completely ruled out at present.

### *Poly(ADPR) and DNA Repair*

In the preceding section, it was shown that fragmentation of DNA by alkylating agents was associated with a marked stimulation of poly(ADPR) turnover. In order to analyze a possible involvement of poly(ADP-ribosyl)ation in DNA repair processes, we have also determined the influence of the poly(ADPR) polymerase inhibitor benzamide on dimethylsulfate-induced fragmentation of DNA and resealing. Alkylation of EAT cells with dimethylsulfate led to the rapid appearance of DNA strand breaks which were rejoined within 3 to 4 hr. Benzamide at a concentration leading to a complete inhibition of poly(ADPR) formation neither blocked DNA fragmentation nor had a significant effect on resealing when the values with benzamide alone were subtracted (nucleoid procedure). Failure of benzamide to inhibit DNA repair was also seen in human lymphocytes after chloroambucil treatment and in HeLa cells after dimethylsulfate (manuscript in preparation). It thus appears that poly(ADP-ribosyl)ation is not an essential step in DNA excision repair as analyzed by fragmentation and rejoining of DNA. This, however, does not exclude that poly(ADPR) formation is involved in the prevention or error-prone repair, or of error-prone transcription, or in the reconstitution of an intact chromatin structure. Such an interpretation is supported by the effects of poly(ADPR) polymerase inhibitors on the plating efficiency of alkylated cells: when combined with certain cytostatic agents, benzamide is indeed a very efficient co-cytostatic and cocytotoxic compound, acting synergistically with various agents on the proliferation of different cell lines (*cf.* refs. *24*, *25*).

We also observed that benzamide was able to relieve the inhibition of RNA synthesis induced by dimethylsulfate treatment, similar to the work of Taniguchi *et al.* (*26*). This effect of benzamide on RNA synthesis was restricted to a relatively narrow concentration range and it constitutes one of the most sensitive parameters influenced by benzamide *in vivo* so far observed.

### *Fragmentation of DNA: No Prerequisite of Poly(ADP-ribosyl)ation*

"Starvation" of EAT cells induced a progressive fragmentation of DNA concomitant with the increase in poly(ADPR) and poly(ADPR) turnover (*23*). Nevertheless, fragmentation of DNA is not a prerequisite for the stimulation of poly(ADPR) synthesis in this system. Contrary to expectation, suppression of poly(ADP-ribosyl)ation by benzamide prevented fragmentation of DNA in "starved" cells (*23*).

These data demonstrate that poly(ADP-ribosyl)ation reactions can also be induced very efficiently by processes other than those which operate in cells after treatment with DNA fragmenting agents or after induction of differentiation (*27*).

## CONCLUSIONS

Alkylation of cells induces an accumulation of both poly(ADPR) and (nuclear) mono(ADPR) protein conjugates. Histone H1 serves only as a minor acceptor of ADPR groups in unstimulated as well as in dimethylsulfate-treated hepatoma cells. Upon alkylation, histone H1 is modified mainly by mono(ADPR) residues. Depletion of NAD by treatment of cells with alkylating agents appears to be caused by a dramatic stimulation of poly(ADPR) turnover ($t/2 < 0.1$ min), and less so by an increased turnover of nuclear mono(ADPR) groups ($t/2 \approx 10$ min).

In alkylated cells, poly(ADP-ribosyl)ation is not an essential step in DNA fragmentation, repair replication or resealing of naked DNA, yet benzamide is an efficient co-cytostatic agent and it relieves dimethylsulfate-induced inhibition of RNA synthesis at low concentrations. It is proposed, therefore, that the stimulation of nuclear ADPR turnover by repair-inducing agents either serves to prevent an error-prone repair of DNA or an erroneous transcription, or it is involved in the reconstitution of the correct chromatin structure.

Accumulation of poly(ADPR) and acceleration of polymer turnover can also be induced by "starvation". Under these conditions DNA fragmentation is a consequence rather than a prerequisite of increased poly(ADP-ribosyl)ation.

These data suggest to us that it is a certain structural state of the chromatin and not necessarily DNA fragmentation that induces high rates of ADP-ribosylation.

## REFERENCES


1. Hilz, H. and Stone, P. R. Poly(ADP-ribose) and ADP-ribosylation of proteins. *In;* F. Lynen (ed.), Reviews of Physiology, Biochemistry and Pharmacology, Vol. 76, pp. 1–58, Springer-Verlag, Berlin, 1976.
2. Hayaishi, O. and Ueda, K. Poly(ADP-ribose) and ADP-ribosylation of proteins. Annu. Rev. Biochem., *46*: 95–116, 1977.
3. Purnell, M. R., Stone, P. R., and Whish, W.J.D. ADP-ribosylation of nuclear proteins. Biochem. Soc. Trans., *8*: 215–227, 1980.
4. Bredehorst, R., Ferro, A. M., and Hilz, H. Determination of ADP-ribose and poly (ADP-ribose) by a new radioimmunoassay. Eur. J. Biochem., *82*: 115–121, 1978.
5. Wielckens, K., Bredehorst, R., Adamietz, P., and Hilz, H. Protein-bound polymeric and monomeric ADPR residues in hepatic tissues: Comparative analysis using a new procedure for the quantitation of poly(ADPR). Eur. J. Biochem., *117*: 69–74, 1981.
6. Bredehorst, R., Klapproth, K., Hilz, H., Scheidegger, C., and Gerisch, G. Protein-bound mono(ADP-ribose) residues in differentiating cells of *Dictyostelium discoideum.* Cell Differ., *9*: 95–103, 1980.
7. Wielckens, K., Garbrecht, M., Kittler, M., and Hilz, H. ADP-ribosylation of proteins in normal lymphocytes and in low-grade malignant non-Hodgkin lymphoma cells. Eur. J. Biochem., *104*: 279–287, 1980.
8. Bredehorst, R., Wielckens K., Adamietz, P., Steinhagen-Thiessen, E., and Hilz, H. Mono(ADP-ribosyl)ation and poly(ADP-ribosyl)ation of proteins in developing liver and in hepatomas. Eur. J. Biochem., *120*: 267–274, 1981.
9. Gartemann, A., Bredehorst, R., Strätling, W. H., Wielckens, K., and Hilz, H. Mono- and poly-ADP-ribosylation of proteins in mouse kidney following castration and testosterone treatment. Biochem. J., *198*: 37–44, 1981.

10. Wielckens, K., Sachsenmaier, W., and Hilz, H. Protein-bound mono(adenosine diphosphate-ribose) levels during the cell cycle of the slime mold physarum polycephalum. Hoppe-Seyler's Z. Physiol. Chem., *360*: 39–43, 1979.
11. Adamietz, P., Wielckens, K., Bredehorst, R., Lengyel, H., and Hilz, H. Subcellular distribution of mono(ADP-ribose) protein conjugates in rat liver. Biochem. Biophys. Res. Commun., *101*: 96–103, 1981.
12. Hilz, H., Wielckens, K., and Bredehorst, R. Quantitation of mono- and poly(ADPR) protein conjugates. *In;* O. Hayaishi and K. Ueda (eds.), ADP-Ribosylation Reactions, pp. 305–321, Academic Press, London, New York, 1982.
13. Wielckens, K., Schmidt, A., George, E., Bredehorst, R., and Hilz, H. DNA fragmentation and NAD depletion: Their relation to the turnover of endogenous mono(ADP-ribosyl) and poly(ADP-ribosyl) proteins. J. Biol. Chem., in press.
14. Kreimeyer, A., Wielckens, K., and Hilz, H. ADP-ribosylation *in vivo* of histone H1 in alkylated hepatoma cells. In preparation.
15. Roitt, J. M. The inhibition of carbohydrate metabolism in ascites-tumor cells by ethyleneimines. Biochem. J., *63*: 300–307, 1956.
16. Holzer, H., Glogner, P., and Sedlmayr, O. Zum Mechanismus der Glykolysehemmung durch carcinostatisch wirkende Äthyleniminverbindungen. Biochem. Z., *330*: 59–72, 1958.
17. Hilz, H., v. Gossler, M., Oldekop, M., and Scholz, M. Effects of X-Ray irradiation in ascites tumor cells: Partial restoration of DPN content and DNA synthesis by nicotinamide. Biochem. Biophys. Res. Commun., *6*: 379–384, 1961.
18. Hilz, H., Hlavica, P., and Bertram, B. Die entscheidende Bedeutung der DPNase-Aktivierung für den DPN-Abfall in bestrahlten Ascitestumorzellen. Biochem. Z., *338*: 283–299, 1963.
19. Maas, H., Rathgen, G. H., Künkel, H. A., and Schubert, G. Versuche zur Deutung der strahleninduzierten Glykolysehemmung in Yoshida-Ascites-Tumorzellen. Z. Naturforsch., *13b*: 735–741, 1958.
20. Whish, W.J.D., Davies, M. I., Shall, S. Stimulation of poly(ADP-ribose) polymerase activity by the anti-tumor antibiotic, streptozotocin. Biochem. Biophys. Res. Commun., *65*: 722–730, 1975.
21. Berger, N. A., Sikorski, G. W., and Petzold, S. J. Association of poly(ADP-ribose) synthesis with DNA strand breaks in replication and repair. *In;* M. Smulson and T. Sugimura (eds.), Novel ADP-ribosylations of Regulatory Enzymes and Proteins, pp. 185–194, Elsevier/North Holland, New York, 1980.
22. Juarez-Salines, H., Sims, J. L., and Jacobson, M. K. Poly(ADP-ribose) levels in carcinogen-treated cells. Nature, *282*: 740–741, 1979.
23. Wielckens, K., George, E., Pless, T., and Hilz, H. Stimulation of poly(ADP-ribosyl) ation during Ehrlich ascites tumor cell "starvation" and suppression of concomitant DNA fragmentation by benzamide. J. Biol. Chem., *258*: 4098-4104, 1983.
24. Durkacz, B. W., Omidiji, O., Gray, D. A., and Shall, S. $(ADP\text{-ribose})_n$ participates in DNA excision repair. Nature, *283*: 593–596, 1980.
25. Hilz, H. ADP-Ribosylation—a multifunctional process. Hoppe-Seyler's Z. Physiol. Chem., *362*: 1415–1425, 1981.
26. Taniguchi, T., Agemori, M., Kameshita, I., Nishikimi, M., and Shizuta, Y. Participation of poly(ADP-ribosyl)ation in the depression of RNA synthesis caused by treatment of mouse lymphoma cells with methylnitrosourea. J. Biol. Chem., *257*: 4027–4030, 1982.
27. Kanai, M., Miwa, M., Kondo, T., Tanaka, Y., Nakayasu, M., and Sugimura, T.

Involvement of poly(ADP-ribose) metabolism in induction of differentiation of HL-60 promyelocytic leukemia cells. Biochem. Biophys. Res. Commun., *105*: 404–411.
28. Shall, S. Experimental manipulation of the specific activity of poly(ADP-ribose) polymerase. J. Biochem., *77*: 2 p, 1975.

ADP-RIBOSYLATION, DNA REPAIR AND CANCER, M. MIWA ET AL. (EDS.),
JAPAN SCI. SOC. PRESS, TOKYO/VNU SCIENCE PRESS, UTRECHT, PP. 165-174, 1983

# Mono- and Poly(ADP-ribose) Metabolism Following DNA Damage

Myron K. JACOBSON,*[1] Janice Yoder SMITH,*[2] Mingkwan MINGMUANG,*[2] D. Michael PAYNE,*[1] and Elaine L. JACOBSON*[2]

*Departments of Chemistry and Biochemistry, North Texas State University/Texas College of Osteopathic Medicine, Denton, Texas 76203*[1] and Department of Biology, Texas Woman's University, Denton, Texas 76204,*[2] U.S.A.*

*Abstract:* Newly developed chemical methods have been applied to study ADP-ribosyl transferase reactions in intact cells following DNA damage. The intracellular levels of NAD and protein-bound monomeric and polymeric ADP-ribose residues were measured in cultured human cells following UV irradiation and in cultured mouse cells treated with N-methyl-N′-nitro-N-nitrosoguanidine (MNNG). UV irradiation of cells caused a rapid increase in the levels of poly(ADP-ribose). Individual molecules of the polymer are present only transiently and the overall rate of conversion of NAD to poly(ADP-ribose) is proportional to the cellular content of DNA strand breaks. Treatment of cells with MNNG also causes a rapid increase in the levels of both monomers and polymers of ADP-ribose. Non-toxic levels of members of two different classes of ADP-ribosyl transferase inhibitors prevent recovery of cell division following treatment of C3H10T1/2 cells with MNNG, while closely related compounds that are not inhibitory have no effect. These studies demonstrate that DNA damage results in a rapid perturbation of ADP-ribose metabolism and suggest that ADP-ribosyl transferase activity is necessary for cellular recovery from DNA damage.

We have previously studied the mechanism by which N-methyl-N′-nitro-N-nitrosoguanidine (MNNG) causes a rapid depletion of the cellular NAD pools in 3T3 cells (*1–3*). These studies have concluded that (i) the effect is specific for the NAD pool as other pools such as NADP and ATP are unaffected, (ii) MNNG does not affect the biosynthesis of NAD, (iii) MNNG treatment of cells does not affect the activity of NAD glycohydrolase or NAD pyrophosphorylase *in vitro*, (iv) MNNG treatment of cells results in an increased activity of poly(ADP-ribose) polymerase activity *in vitro* and a rapid increase in the levels of poly(ADP-ribose) *in vivo*, and (v) that the depletion of the NAD pool can be blocked by ADP-ribosyl transferase inhibitors. Further, our studies have shown that lowering of NAD pools is observed following treatments with several different classes of DNA damaging agents (*4*).

We describe here studies of the metabolism of monomeric and polymeric ADP-ribose following DNA damage and the effect of inhibitors of ADP-ribosyl transferases on cellular recovery from DNA damage.

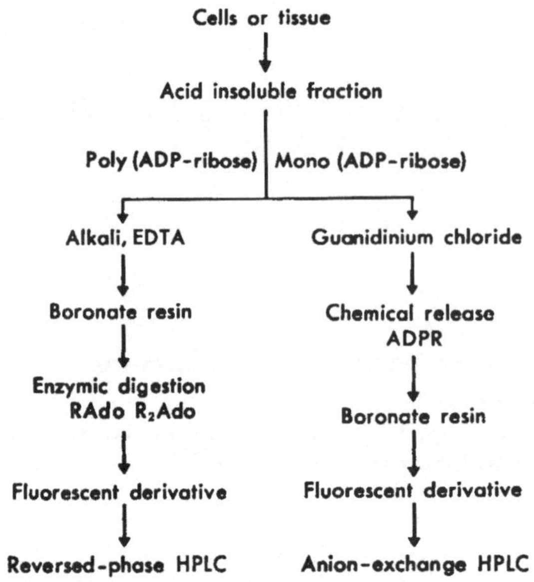


FIG. 1. Analytical scheme for the measurement of monomeric and polymeric ADP-ribosyl residues from intact cells. RAdo, ribosyladenosine; $R_2Ado$, diribosyladenosine; HPLC, high performance liquid chromatography; ADPR, adenosine diphosphoribose.

*Measurement of Monomeric and Polymeric ADP-ribosyl Residues in Intact Cells*

The study of mammalian ADP-ribose metabolism *in vivo* requires methodology capable of measuring the amount of protein-bound monomers and polymers of ADP-ribose present in a limited quantity of cultured cells or animal tissues. The methods for making such measurements must be both highly selective and highly sensitive, since the cellular content of ADP-ribosyl residues is very low relative to chemically similar compounds (*5*). Figure 1 shows the overall approach developed in our laboratory for the measurement of both monomeric and polymeric ADP-ribosyl residues in intact cells. These complementary methods have two common features that provide the needed selectivity and sensitivity. The first is the use of an immobilized boronate resin which allows the quantitative separation of small amounts of ADP-ribosyl residues from the bulk of interfering substances present in crude extracts of acid-insoluble material. The boronate affinity resin functions by the reversible formation of a complex between boronate and molecules that contain 1,2 *cis*-diol groups (*6*). The second key feature of both methods involves the quantitative conversion of adenine-containing compounds to highly fluorescent 1, $N^6$-etheno derivatives (*7*), greatly enhancing the sensitivity of detection of the desired compounds.

*Poly(ADP-ribose) Metabolism in UV Irradiated Human Fibroblasts*

Figure 2 shows the total cellular content of poly(ADP-ribose) (A) and its precursor, NAD, (B) in human diploid fibroblasts following treatment with 5 J/m$^2$ of UV light. These measurements were made in CF-3 cells which were seeded in 100

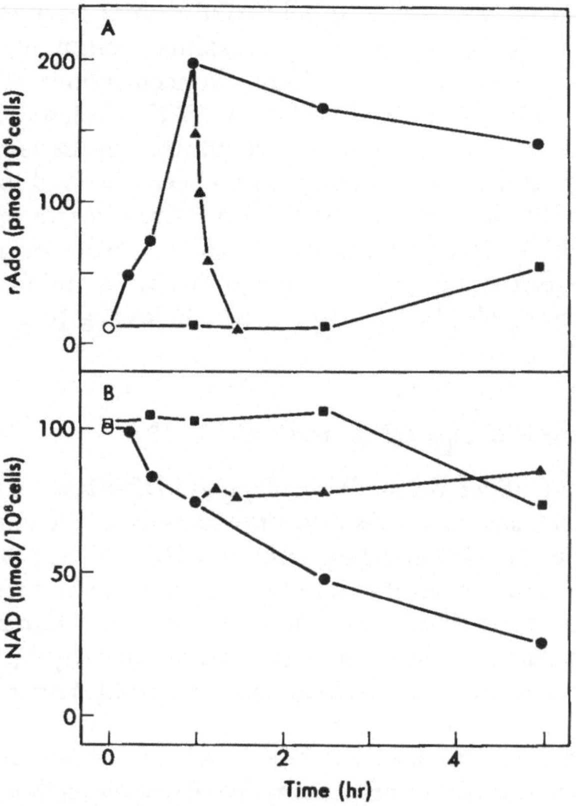


FIG. 2. Total intracellular content of poly(ADP-ribose) and NAD following treatment of human diploid fibroblasts with 5 J/m² of UV irradiation. The final concentration of 3-aminobenzamide (ABA) added at 0 hr was 3 mM and at 1 hr was 10 mM. rAdo, ribosyladenosine. A, poly(ADP-ribose). B, NAD. ● UV only; ▲ ABA at 1 hr; ■ ABA at 0 hr.

mm dishes and allowed to grow to confluence in medium containing 10% fetal calf serum. The cells were then mitotically arrested by replacing the medium with medium containing 0.5% fetal calf serum for 5–7 days. The total poly(ADP-ribose) in these non-dividing cells prior to UV treatment was 12 pmol/$10^8$ cells of ribosyladenosine derived from poly(ADP-ribose). By one hour following UV treatment the level of poly(ADP-ribose) increased to approximately 200 pmol/$10^8$ cells and remained elevated for 5 hr. The increase over basal level was approximately 100% at 15 min, 1,300% at 1 hr and 1,100% by 5 hr following UV treatment. The initial content of NAD was approximately 100 nmol/$10^8$ cells and no change in NAD content was detected at 15 min following UV treatment after which a progressive decrease in the cellular NAD pool was observed. Between 1 and 5 hr the NAD pool size was decreasing at an approximate rate of 200 pmol/min/$10^8$ cells. The effects of the poly(ADP-ribose) polymerase inhibitor, 3-aminobenzamide, when added to the culture medium are also shown. When added prior to UV treatment, no detectable increase in poly(ADP-ribose) content or decrease in NAD was observed for 2.5 hr. By 5 hr, however, both an increase of poly(ADP-ribose) and decrease in NAD were observed. When 3-aminobenzamide was added to the culture medium at 1 hr following UV treatment, no further loss of NAD occurred and poly(ADP-ribose) disappeared with a half life of approximately 6 min. This represents a maximum estimate of half life since it cannot be determined how rapidly inhibitory concentrations of 3-aminobenzamide reach the nucleus. If it is assumed that all the NAD is being converted to poly(ADP-ribose), the half life of poly(ADP-ribose) would actually be less than one min. A rapid turnover of poly(ADP-ribose) *in vivo* has also been observed in 3T3 cells following MNNG treatment (*2*). In general, these observations *in vivo* are in contrast to the nature of poly(ADP-ribose) synthesis in isolated nuclei or permeable cells in which the polymer is quite stable (*8*, *9*), although Benjamin and Gill (*10*, *11*) have observed rapid turnover of poly(ADP-ribose) *in vitro* following treatment with X-rays.

### *Regulation of ADP-ribose Synthesis Following UV Irradiation*

Studies of the factors that affect the activity of poly(ADP-ribose) polymerase *in vitro* have indicated that DNA strand breaks affect the activity of the enzyme. Activity is stimulated by nucleases (*10–13*) and Benjamin and Gill (*10*, *11*) have quantitatively related polymerase activity to the number of nuclease induced strand breaks. Cohen and Berger (*14*) have shown that the presence of UV damage in the SV40 minichromosome is not sufficient to support polymerase activity but that the enzyme is active following treatment of the minichromosome with *Micrococcus luteus* UV endonuclease.

We describe here two lines of evidence that following UV irradiation the rate of synthesis of poly(ADP-ribose) in intact cells is regulated by the cellular content of DNA strand breaks. Many studies have shown that following UV irradiation of human fibroblasts there is a transient increase in the number of DNA strand breaks which are associated with DNA excision repair (*15–18*). Cells from patients with the genetic disease xeroderma pigmentosum have greatly reduced rates of removal of UV-induced pyrimidine dimers and subsequent repair replication relative to normal cells (*19*). The excision repair of UV light-induced pyrimidine dimers requires an endonucleolytic incision into DNA near the dimer. Xeroderma pigmentosum cells are known to be deficient in the introduction of endonucleolytic incisions into DNA (*20*) although the exact nature of the defect is poorly understood.

Table 1 shows levels of NAD following treatment with either UV or MNNG in a control cell line of human fibroblasts, and three cell lines carrying the xeroderma pigmentosum genotype including xeroderma pigmentosum complementation groups

TABLE 1. NAD Levels in Normal and Xeroderma Pigmentosum Cells

| Cell line | Control | Treatment UV<br>NAD, % zero time | MNNG |
|---|---|---|---|
| IMR90 | 94.5±0.7 (2) | 17.4±2.8 (5) | 6.5±1.3 (4) |
| XP12BE (Group A) | 94.8±5.0 (10) | 89.0±4.7 (13) | 15.7±0.5 (6) |
| XP2BE (Group C) | 98.5±5.7 (4) | 73.0±9.9 (8) | 23.0±3.5 (4) |
| XP6BE (Group D) | 95.4±2.0 (5) | 89.8±4.4 (5) | 8.5±2.0 (2) |

(*n*) = number of determinations.

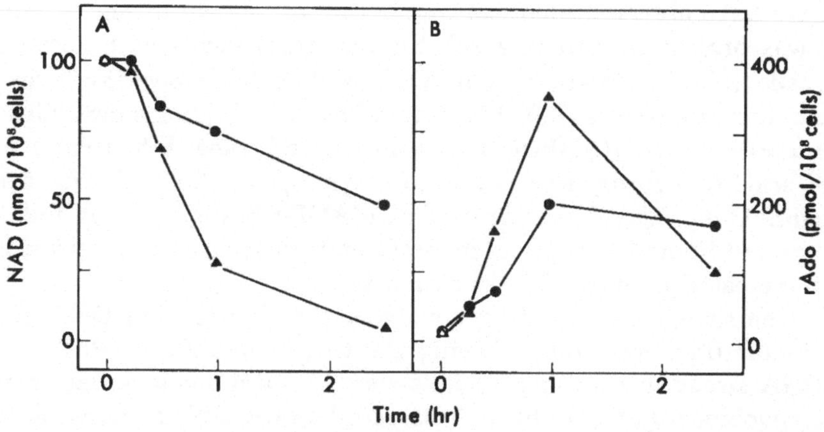


FIG. 3. Effect of 0.1 mM araC on the total intracellular content of NAD and poly(ADP-ribose) following 5 J/m$^2$ of UV irradiation of human diploid fibroblasts. A, NAD; B, poly(ADP-ribose). ▲ UV+araC; ● UV.

A, C, and D. The data are expressed as the percentage of toal cellular NAD content present at time zero. The absolute levels of NAD were similar in all of the cell lines, ranging from 2.7 to 3.5 pmol/$\mu$g of protein. Doses of MNNG or UV light were chosen that cause a large amount of reduction of the NAD content of the control cell line. Xeroderma pigmentosum cells are proficient in the excision of damage caused by MNNG (*21*) and all of the xeroderma pigmentosum cell lines depleted their NAD pools following MNNG treatment. Xeroderma pigmentosum cells of complementation groups A and D did not show a significant change in NAD levels following UV treatment, and xeroderma pigmentosum cells of complementation group C showed a significant but greatly reduced amount of lowering of NAD relative to normal cells. These data provide a highly specific correlation between the ability of the cells to convert NAD to poly(ADP-ribose) and the known ability of these cell lines to introduce DNA strand breaks following UV irradiation.

The formation of DNA strand breaks following UV damage is dependent on endonuclease activity that appears to be the rate limiting step in repair of pyrimidine dimers (*17*). The presence of either hydroxyurea or 1-$\beta$-D-arabinofuranosylcytosine (araC) results in an increase in the cellular content of DNA strand breaks following UV treatment by decreasing the rate of the polymerization step at sites under repair without altering the rate of incision (*17*). Figure 3 shows the effect of 0.1 mM araC on cellular NAD and poly(ADP-ribose) content following UV treatment. The rate of depletion of the NAD pool was stimulated by the presence of araC (A). In the period from 15 min to 1 hr following UV treatment, the amount of NAD depletion was increased by 150% by the presence of araC. Figure 3 B shows that the content of poly(ADP-ribose) at 15 min following UV irradiation was similar in the presence or absence of araC but that in the presence of araC the content was increased 120% at 30 min and 80% at 1 hr, relative to cells treated with UV alone. By 2.5 hr, the intracellular content of poly(ADP-ribose) was lower in the presence of araC, presumably due to turnover of poly(ADP-ribose) and substrate depletion in the araC treated cells which had only 5% of their initial NAD content remaining at that time. We have observed identical results to those shown in Fig. 3 if 10 mM hydroxyurea was present instead of araC. Similar experiments with xeroderma pigmentosum cells (complementation group A) showed no NAD lowering in the presence or absence of these inhibitors (*24*). These data have ruled out the possibility that the observed increase in poly(ADP-ribose) synthesis following UV treatment in the presence of araC or hydroxyurea is due to a synergistic effect of these compounds and UV light. Thus, the enhancement of poly(ADP-ribose) synthesis and NAD depletion in normal human fibroblasts by araC or hydroxyurea can be attributed directly to an increased number of DNA strand breaks.

It has been shown that the drug novobiocin, an inhibitor of DNA gyrase (*22*) inhibits endonucleolytic incisions following UV treatment. As a result the number of transient DNA strand breaks formed following UV treatment is decreased in the presence of novobiocin (*22*). We have measured the rate of NAD conversion to poly-(ADP-ribose) following UV treatment in the presence of this drug. Both the initial rate and total amount of NAD were decreased by 40% in the presence of novobiocin (*23*) where DNA strand breaks are decreased relative to controls.

In summary, the observations on the relationship of poly(ADP-ribose) synthesis to DNA strand breaks in intact cells presented here argue that earlier, *in vitro* studies of the mechanism of regulation of poly(ADP-ribose) polymerase activity by DNA strand breaks are physiologically relevant.

### *Protein-bound Mono(ADP-ribose) Following MNNG Treatment in C3H10T1/2 Cells*

Table 2 shows the total cellular content of protein-bound mono(ADP-ribose) in asynchronously dividing C3H10T1/2 cells. Following treatment with MNNG, the monomeric ADP-ribose residues increased from 37 pmol/$10^8$ cells to 209 pmol/$10^8$ cells in 20 min. The corresponding increase in polymeric residues in this experiment are also shown in Table 2. The total number of polymeric residues was 51 pmol/$10^8$ cells prior to treatment. Twenty min following treatment with MNNG the number of polymeric residues increased to 2,760 pmol/$10^8$ cells, twelve times the increase observed in protein-bound mono(ADP-ribose). The increases in both types of residues were blocked if 3-methoxybenzamide (MBA) was added to the culture medium prior to MNNG treatment. Since the polymeric residues of ADP-ribose are rapidly turned over under these conditions (*2*), it is possible that the increase in monomeric residues resulted from polymer turnover. Alternatively, these residues may be the result of mono(ADP-ribosyl) transferase activity.

TABLE 2. Monomeric and Polymeric ADP-ribosyl Residues in C3H10T1/2 Cells

| Treatment | ADP-ribosyl residues (pmol/$10^8$ cells) | |
|---|---|---|
| | Monomeric | Polymeric |
| None | 37 | 51 |
| MNNG | 209 | 2,760 |
| MNNG+MBA | 38 | 65 |

50 μg/ml MNNG for 20 min, MBA at 3 mM.

*Effect of ADP-ribosyl Transferase Inhibitors on Recovery from DNA Damage*

Inhibitors of ADP-ribosyl transferase have been shown to be synergistic with DNA damaging agents in inhibiting colony formation in mammalian cells (*25–28*). Two separate classes of inhibitors are the methylated xanthines such as theophylline and benzamide and its derivatives. The corresponding benzoates are not inhibitors. Table 3 shows that 1 mM MBA greatly enhanced the cytotoxicity of MNNG in C3H10T1/2 cells. In an attempt to develop an experimental system in which ADP-ribosyl transferase reactions and other specific biochemical events can be studied under the same conditions employed to study biological consequences of DNA damage such as recovery of cell division, malignant transformation, and mutagenesis, we have examined the effect of ADP-ribosyl transferase inhibitors on asynchronously dividing monolayer cultures of C3H10T1/2 cells following MNNG treatment.

The C3H10T1/2 cell line has an optimal generation time of 12–14 hr as shown in Fig. 4. When these cells were treated with MNNG (5 μg/ml), no increase in cell

TABLE 3. Effect of MBA and MNNG on Colony Formation in C3H10T1/2 Cells

| Treatment | % survival |
|---|---|
| None | 100 |
| MBA | 105 |
| MNNG | 42 |
| MNNG+MBA | 0.13 |

MBA was present at 1 mM and MNNG treatment was 5 μg/ml for 20 min.

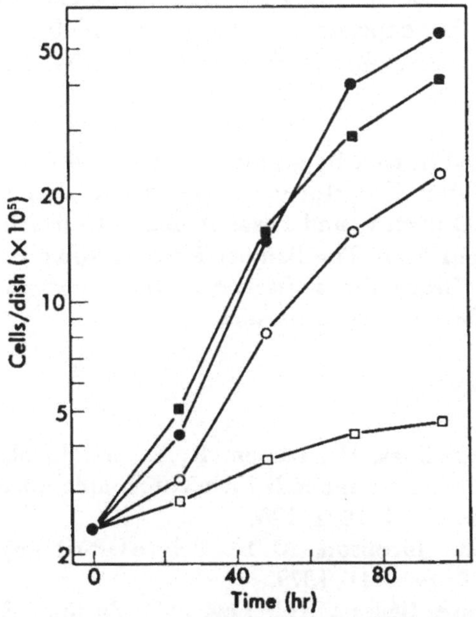


FIG. 4. Effect of 1 mM MBA on growth of C3H10T1/2 cells following treatment with MNNG. The cells were treated with 5 μg/ml MNNG for 20 min. ● control; ■ MBA; ○ MNNG; □ MNNG+MBA.

TABLE 4. Effect of ADP-ribosyl Transferase Inhibitors on Growth in C3H10T1/2 Cells Following Treatment with MNNG

| Addition | Control | | MNNG | |
|---|---|---|---|---|
| | Population doublings | Ratio | Population doublings | Ratio |
| None | 4.26±0.56 | — | 3.04±0.62 | — |
| MBA | 3.81±0.50 | 0.89 | 0.53±0.23 | 0.17 |
| None | 4.70±0.64 | — | 3.18±0.04 | — |
| Theophylline | 4.03±1.3 | 0.86 | 0.75±0.48 | 0.23 |

number was observed for approximately 15 hr at which time growth resumed with a generation time of approximately 22 hr (Fig. 4). The addition of MBA, an ADP-ribosyl transferase inhibitor, immediately following MNNG treatment inhibited recovery of cell division. Addition of MBA to untreated cultures had little or no effect on growth. Table 4 shows the number of population doublings which occurred in 96 hr in MNNG treated or untreated cells in the presence or absence of MBA or theophylline. In 6 separate experiments the mean number of population doublings in MNNG treated cells at 96 hr was 3.04, while addition of MBA to the cultures limited growth to 0.53 populations doubling. Theophylline showed a similar effect. We have observed results similar to those shown in Fig. 4 and Table 4 with benzamide and 3-aminobenzamide. The corresponding benzoates were completely inactive. These data confirm earlier reports that ADP-ribosyl transferase inhibitors enhance the cytotoxicity of DNA damaging agents and suggest that ADP-ribosyl transferase activity is essential during cellular recovery from DNA damage. Since the C3H10T1/2 cell line has been widely used to study malignant transformation and mutagenesis, it should provide a valuable experimental model in which to study the function of ADP-ribosyl transferase reactions during recovery from DNA damage.

## ACKNOWLEDGMENTS

The research described was supported in part by grants CA23994 and CA29357 from the National Institutes of Health, B-633 from The Robert A. Welch Foundation, and institutional grants from North Texas State University and Texas Woman's University. D.M.P. was supported by a predoctoral fellowship from The Samuel Roberts Noble Foundation. We thank Toni Benjamin and Lynne Gracy for assistance in the preparation of the manuscript and Gilberto Duran-Torres for technical assistance.

## REFERENCES

1. Jacobson, M. K., Levi, V., Juarez-Salinas, H., Barton, R. A., and Jacobson, E. L. Effect of carcinogenic N-alkyl-N-nitroso compounds on nicotinamide adenine dinucleotide metabolism. Cancer Res., *40*: 1797–1802, 1980.
2. Juarez-Salinas, H., Sims, J. L., and Jacobson, M. K. Poly(ADP-ribose) levels in carcinogen-treated cells. Nature, *282*: 740–741, 1979.
3. Jacobson, M. K., Sims, J. L., Juarez-Salinas, H., Levi, V., Barton, R. A., and Jacobson, E. L. Mechanism of carcinogen-induced alteration of NAD metabolism. *In;* M. Smulson and T. Sugimura (eds.), Novel ADP-ribosylations of Regulatory Enzymes and Proteins, pp. 239–250, Elsevier/North-Holland, New York, 1980.

4. Rankin, P. W., Jacobson, M. K., Mitchell, V. R., and Busbee, D. L. Reduction of nicotinamide adenine dinucleotide levels by ultimate carcinogens in human lymphocytes. Cancer Res., *40*: 1803–1807, 1980.
5. Jacobson, M. K., Payne, C. M., and Jacobson, E. L. Mono- and poly(ADP-ribosylation) in intact cells. Oklahoma Medical Research Foundation Conference on Post-translational Modification of Proteins for Function, Oklahoma City, OK, 1982.
6. Rosenberg, M., Wieber, J., and Gilham, P. Studies on the interactions of nucleotides, polynucleotides and nucleic acids with dihydroxyboryl-substituted celluloses. Biochemistry, *11*: 3623–3628, 1972.
7. Barrio, J., Secrist, J., III, and Leonard, N. Fluorescent adenosine and cytidine derivatives. Biochem. Biophys. Res. Commun., *46*: 597–604, 1972.
8. Berger, N. A., Petzold, S. J., and Berger, S. J. Association of poly(ADP-ribose) synthesis with cessation of DNA synthesis and fragmentation. Biochim. Biophys. Acta, *564*: 90–104, 1979.
9. Levi, V., Juarez-Salinas, H., and Jacobson, M. K. Increased poly(ADP-ribose) synthesis in quiescent cells not associated with DNA fragmentation. Fed. Proc., *40*: 171, 1981.
10. Benjamin, R. C. and Gill, D. M. ADP-ribosylation in mammalian cell ghosts. J. Biol. Chem., *255*: 10493–10501, 1980.
11. Benjamin, R. C. and Gill, D. M. Poly(ADP-ribose) synthesis *in vitro* programmed by damaged DNA. J. Biol. Chem., *255*: 10502–10508, 1980.
12. Miller, E. G. Effect of deoxyribonuclease I on the number and length of chains of poly(ADP-ribose) synthesized *in vitro*. Biochim. Biophys. Acta, *395*: 191–200, 1975.
13. Miller, E. G. Stimulation of nuclear poly(adenosine diphosphate-ribose) polymerase activity from HeLa cells by endonucleases. Biochem. Biophys. Res. Commun., *66*: 280–286, 1975.
14. Cohen, J. J. and Berger, N. A. Activation of poly(adenosine diphosphate ribose) polymerase with UV irradiated and UV endonuclease treated SV 40 minichromosome. Biochem. Biophys. Res. Commun., *98*: 268–274, 1981.
15. Collins, A.R.S., Schor, S. L., and Johnson, R. T. The inhibition of repair in UV irradiated human cells. Mutat. Res., *42*: 413–432, 1977.
16. Hiss, E. A. and Preston, R. J. The effect of cytosine arabinoside on the frequency of single strand breaks in DNA of mamalian cells following irradiation or chemical treatment. Biochim. Biophys. Acta, *478*: 1–8, 1977.
17. Erixon, K. and Ahnström, G. Single-strand breaks in DNA during repair of UV-induced damage in normal human and xeroderma pigmentosum cells as determined by alkaline DNA unwinding and hydroxylapatite chromatography: effects of hydroxyurea, 5-fluorodeoxyuridine and 1-$\beta$-D-arabinofuranosylcytosine on the kinetics of repair. Mutat. Res., *59*: 257–271, 1979.
18. Collins, A.R.S., Downes, C. S., and Johnson, R. T. Cell cycle-related variations in UV damage and repair capacity in Chinese hamster (CHO-K1) cells. J. Cell. Physiol., *103*: 179–191, 1980.
19. Cleaver, J. E. Xeroderma pigmentosum. *In;* J. B. Stanbury, J. B. Wyngaarden, and D. S. Frederickson (eds.), The Metabolic Basis of Inherited Disease, pp. 1072–1095, McGraw-Hill Book Co., New York, 1978.
20. Hashem, N., Bootsma, D., Keijzer, W., Greene, A., Coriell, L., Thomas, G., and Cleaver, J. E. Clinical characteristics, DNA repair, and complementation groups in xeroderma pigmentosum patients from Egypt. Cancer Res., *40*: 13–18, 1980.
21. Cleaver, J. E. DNA repair with purines and pyrimidines in radiation- and carcinogen-

damaged normal and xeroderma pigmentosum human cells. Cancer Res., *33*: 362–369, 1973.

22. Collins, A.R.S. and Johnson, R. T. Novobiocin: an inhibitor of the repair of UV induced damage in mammalian cells. Nucleic Acids Res., *7*: 1311–1320, 1979.
23. Antol, K. M. Poly(ADP-ribose) metabolism following DNA damage. Ph. D. Dissertation, 1982.
24. Jacobson, E. L., Antol, K. M., Juarez-Salinas, H., and Jacobson, M. K. Poly(ADP-ribose) metabolism in UV irradiated human fibroblasts. J. Biol. Chem., *258*: 103-107, 1983.
25. Durkacz, B. W., Omidiji, O., Gray, D. A., and Shall, S. (ADP-ribose)$_n$ participates in DNA excision repair. Nature, *283*: 593–596, 1980.
26. Nduka, N., Skidmore, C. J., and Shall, S. The enhancement of cytotoxicity of N-methyl-N-nitrosourea and of $\gamma$-radiation by inhibitors of poly(ADP-ribose) polymerase. Eur. J. Biochem. *105*: 525–530, 1980.
27. Durrant, L. G., Margison, G. P., and Boyle, J. M. Effect of 5-methylnicotinamide on mouse L1210 cells exposed to N-methyl-N-nitrosourea: mutation induction, formation and removal of methylation products in DNA, and unscheduled DNA synthesis. Carcinogenesis, *2*: 1013–1017, 1981.
28. Durrant, L. G. and Boyle, J. M. Potentiation of cell killing by inhibitors of poly(ADP-ribose) polymerase in four rodent cell lines exposed to N-methyl-N-nitrosourea or UV light. Chem.-Biol. Interact., *38*: 325–338, 1982.

ADP-RIBOSYLATION, DNA REPAIR AND CANCER, M. MIWA ET AL. (EDS.), JAPAN SCI. SOC. PRESS, TOKYO/VNU SCIENCE PRESS, UTRECHT, PP. 175-182, 1983

# Inhibition of DNA Ligase Activity by Histones and Its Reversal by Poly(ADP-ribose)

Kunihiro UEDA, Yasuhiro OHASHI, Kazuyuki HATAKEYAMA, and Osamu HAYAISHI

*Department of Medical Chemistry, Kyoto University Faculty of Medicine, Kyoto 606, Japan*

*Abstract:* The molecular mechanism by which poly(ADP-ribose) participates in DNA repair was investigated using purified DNA ligase in DNA-histone systems. The ligase activity was markedly inhibited by histones; the inhibition was >80% with histone H1 at concentrations equal to DNA. This inhibition was reversed efficiently by poly(ADP-ribose), either added exogenously or synthesized *in situ* with poly(ADP-ribose) synthetase. The reversal effect was specific for poly(ADP-ribose); other polyanions such as mRNA, rRNA's, tRNA, and synthetic poly(A) were less effective or totally ineffective. The poly(ADP-ribose) effect appeared to be caused by binding to histones and decreasing DNA-histone interactions. Poly(ADP-ribose) also had high affinity for DNA ligase. These observations, together with the findings of absolute dependence of poly(ADP-ribose) synthetase activity on DNA strand ends and extensive automodification of the synthetase in DNA-damaged cells, suggested a possible mechanism of poly(ADP-ribose) action in DNA repair, in which automodified poly(ADP-ribose) synthetase serves as a link between DNA damage and activation of DNA ligase.

Involvement of poly(ADP-ribose) in DNA excision repair has been suggested by various lines of evidence (*1*). For the molecular mechanism of this involvement, Durkacz *et al.* (*2*) first reported that inhibition of poly(ADP-ribose) synthesis retarded the rejoining of damaged DNA. Soon afterwards, we made a similar observation with our different system (*3*) (Fig. 1); when mouse lymphoblasts, L1210 cells, were treated with N-methyl-N′-nitro-N-nitrosoguanidine (MNNG), the DNA underwent rapid fragmentation and subsequent rejoining, as judged by alkaline sucrose gradient centrifugation. However, when 3-aminobenzamide, a potent inhibitor of poly(ADP-ribose) synthetase, was included in the post-treatment incubation, essentially no rejoining was observed after 4 hr, but the majority of DNA fragments became even smaller. Because 3-aminobenzamide did not inhibit repair DNA synthesis (*4*), these results suggested that poly(ADP-ribose) synthesis was required for the step of rejoining damaged DNA. This notion coincided with the view of Shall

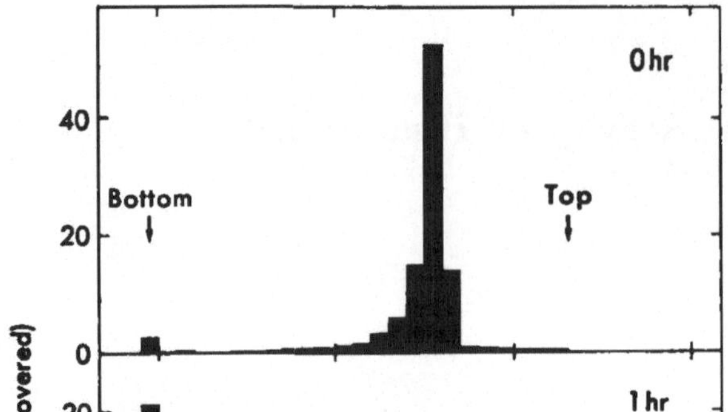


FIG. 1. DNA repair in MNNG-treated lymphocytes and its inhibition by 3-aminobenzamide. Mouse lymphoblasts, L1210 cells, prelabeled with [$^3$H]thymidine, were treated with MNNG. The cells were washed and incubated in a fresh medium. At intervals, the cells were lysed, and the DNA analyzed by alkaline sucrose gradient centrifugation. Where indicated, 3-aminobenzamide (3AB) was added in the post-treatment incubation. From Hayaishi *et al.* (*3*).

(*5*). The present study was performed to elucidate the mechanism of poly(ADP-ribose) action in DNA rejoining.

### *Possible Mechanism of Poly(ADP-ribose) Function in DNA Rejoining*

The following four possibilities have been considered in our laboratory. (i) Activation of DNA ligase. The activation may be elicited by modification of the ligase itself (*6*), or indirectly in some other way. (ii) Localization of DNA ligase at strand breaks. This mechanism may be important for effective repair of damages that take place at random in the nucleus. (iii) Modification of DNA-histone interactions. This mechanism may involve condensation (*7*) or decondensation (*8*) of polynucleosomal solenoids, or relaxation of nucleosome cores. (iv) Inhibition of protease(s) and/or DNase(s). This possibility has been suggested by the findings of inhibition of a chromatin protease by poly(ADP-ribose) (*9*), and inactivation of $Mg^{2+}$, $Ca^{2+}$-

dependent endonuclease by oligo(ADP-ribosyl)ation (*10*). This mechanism may serve for protection of damaged nucleosomes from degradation. The results to be presented suggest the first possibility as the main mechanism, and the second and third possibilities as those supporting this mechanism.

*Inhibition of DNA Ligase Activity by Histones*

At the beginning of the attempt to reconstitute *in vitro* a DNA rejoining system, the effect of histones on DNA ligase activity was examined, because DNA of eukaryotes exists in a complex with histones. DNA ligase was purified about 500-fold from calf thymus. As shown in Fig. 2, the ligase activity was markedly inhibited by histones. The inhibition was dependent on the subclasses of histones and the ratio of histone to DNA. Histone H1 alone was more effective than whole histone mixture, and the inhibition by H1 was >80% at concentrations equal to or higher than that of DNA. Histone H1 carrying, on the average, 1.5 ADP-ribose residues was equally effective as unmodified H1. In view of the fact that histones are modified *in vivo* mostly by monomers and short oligomers (*11*), this result suggested that ADP-ribosylation of histones hardly affects the inhibitory effect. Similar inhibition by histones was found by Zimmerman and Levin (*12*) for rat liver DNA ligase.

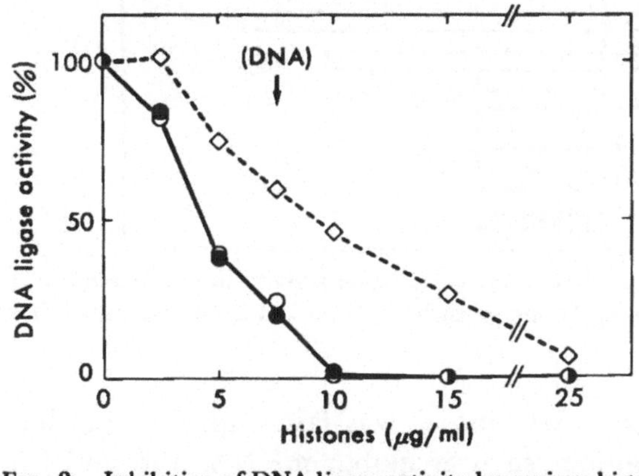


FIG. 2. Inhibition of DNA ligase activity by various histones. The ligase activity was assayed by conversion of [5'-$^{32}$P]phosphate in nicked DNA into the alkaline phosphatase-resistant form. Arrow indicates the concentration of nicked DNA employed. □ whole histone; ○ (ADP-ribose)$_{1.5}$-H1; ● H1.

*Reversal of Histone Effect by Poly(ADP-ribose)*

Using this system of DNA and histone, the effect of poly(ADP-ribose) on DNA ligase activity was examined (Fig. 3). It is evident that poly(ADP-ribose) reversed the histone inhibition efficiently; the activity of DNA ligase, which was inhibited to about 15% of the original activity by histone H1, was restored almost completely by the addition of increasing amounts of poly(ADP-ribose). At the concentration of 7.5 μg/ml, that was equal to DNA and histone, poly(ADP-ribose) stimulated DNA ligase activity >4-fold, and recovered 65% of the original activity. In the absence of histone H1, no significant activation was observed.

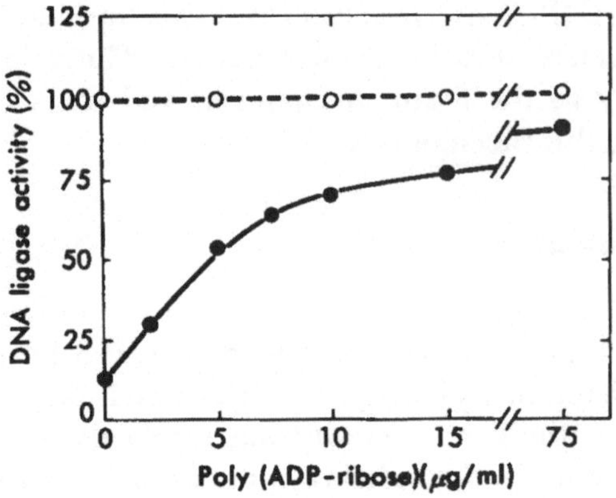


FIG. 3. Reversal of histone inhibition of DNA ligase activity by poly(ADP-ribose). The ligase activity was assayed using $[^{32}P]$nicked DNA, with or without histone H1 (7.5 μg/ml each), and various concentrations of poly(ADP-ribose). ○ −histone; ● +histone.

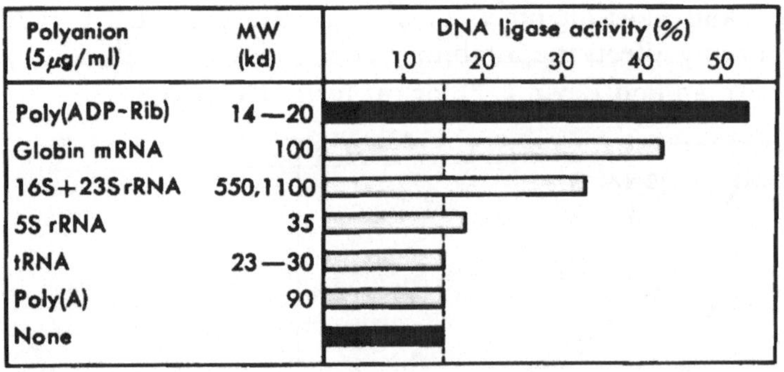


FIG. 4. Effects of various polyanions on DNA ligase activity in the presence of histone. The ligase activity was assayed using 7.5 μg/ml each of $[^{32}P]$nicked DNA and histone H1, and various polyanions as indicated.

The activating effect of poly(ADP-ribose) was fairly specific for this polymer. As shown in Fig. 4, other polyanions such as globin mRNA, 16S and 23S rRNA's were less effective, and 5S rRNA, tRNA and a synthetic polymer, poly(A), were almost inert. These data appeared to support the view that poly(ADP-ribose) is the most potent and physiological activator, because the syntheses *in vivo* of all kinds of RNA after DNA damage have been reported to decrease, rather than increase (*13*), whereas the synthesis of poly(ADP-ribose) increases 10- to 100-fold (*1*, *14*, *15*).

Poly(ADP-ribose) synthesized *in situ* on protein acceptors also effectively activated the DNA ligase activity inhibited by histones. The experiment shown in Fig. 5 consisted of two stages. The first stage was the synthesis of poly(ADP-ribose) from NAD with varying amounts of poly(ADP-ribose) synthetase (left panel). The reaction mixture also contained nicked DNA and histone H1. After poly(ADP-ribose) synthesis was terminated by the addition of 3-aminobenzamide, DNA ligase was added, and the second stage (the ligase reaction) was carried out (right panel). The activity of DNA ligase, which was inhibited by histone up to 12%, increased with increasing amounts of the synthetase, almost in parallel to the increase in poly(ADP-

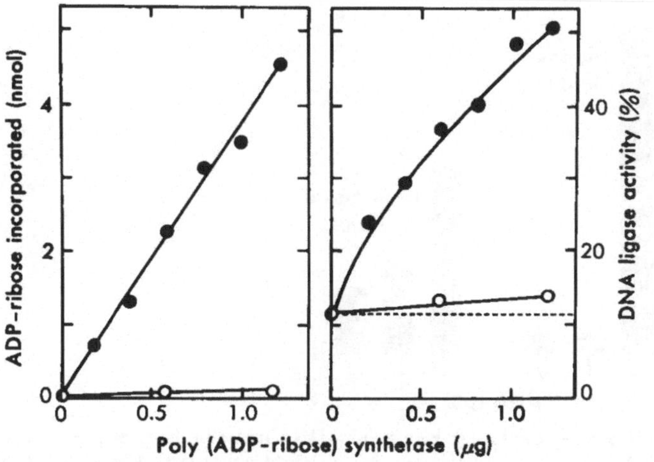


FIG. 5. Effect of poly(ADP-ribose) synthesis on DNA ligase activity in the presence of histone. After poly(ADP-ribose) synthesis for 1 hr with [$^{14}$C]NAD and varying amounts of poly(ADP-ribose) synthetase in the presence of [$^{32}$P]nicked DNA and histone H1 (7.5 μg/ml each) (left), 3-aminobenzamide and DNA ligase were added, and the ligase reaction carried out for 30 min (right). ● −3AB; ○ +3AB. + or −3AB indicates the addition or no addition of 3-aminobenzamide in the stage of poly(ADP-ribose) synthesis (3-aminobenzamide was present in all incubations at the stage of DNA ligase reaction).

ribose) synthesis. When 3-aminobenzamide was included in the first stage ("+3AB"), no significant activation was observed, indicating that the activation was not caused by the synthetase protein but its product, poly(ADP-ribose). Thus, the *in vivo* effect of 3-aminobenzamide on DNA rejoining (Fig. 1) was reproduced *in vitro*.

Parallel experiments showed that poly(ADP-ribose) synthesized under these conditions was attached almost exclusively to the synthetase and not histone H1, and that the effect of the synthetase-bound poly(ADP-ribose) was nearly identical with that of free polymer.

Creissen and Shall (*6*) recently reported that DNA ligase II, an isozyme, might be ADP-ribosylated and thereby activated in DNA-damaged cells. In our experiment (Fig. 5), the possibility of ADP-ribosylatoin of DNA ligase did not exist, because 3-aminobenzamide was included in the second incubation with DNA ligase. Furthermore, we have not detected ADP-ribosylation of our DNA ligase preparation (*16*), nor a significant activity suggestive of DNA ligase II during our purification.

### *Alteration in DNA-histone Interaction by Poly(ADP-ribose)*

The reversal of histone effect on DNA ligase activity by poly(ADP-ribose) was probably caused by a modification of DNA-histone interactions. Poly(ADP-ribose) had a capacity to bind very efficiently to histones (Fig. 6). This binding appeared to be ionic and nonspecific, because the binding of poly(ADP-ribose) was mostly replaced by DNA of lesser amounts than histone. Additional evidence for the change in DNA-histone interactions induced by poly(ADP-ribose) was obtained by analysis of nuclease susceptibility of DNA; the DNA in a complex with histone H1 became

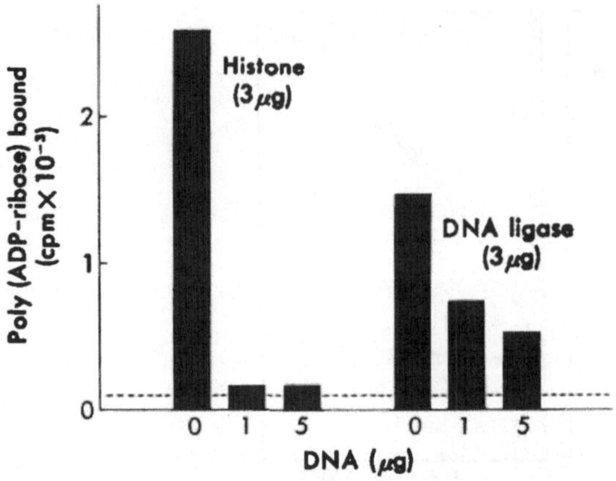


FIG. 6. Poly(ADP-ribose) binding to histone and DNA ligase in the presence of varying amounts of DNA. The binding was assayed by retention of [$^{14}$C]poly(ADP-ribose) on a Millipore filter. The dotted line indicates the background level observed with bovine serum albumin.

more susceptible to micrococcal nuclease upon poly(ADP-ribose) synthesis (K. Hatakeyama, unpublished results).

### *Poly(ADP-ribose) Binding to DNA Ligase*

Although poly(ADP-ribose) had no direct influence on DNA ligase activity (Fig. 3), there appeared to be a direct interaction between poly(ADP-ribose) and DNA ligase. As shown in Fig. 6, DNA ligase had a capacity to bind a considerable amount of poly(ADP-ribose) (*16*). The amount of bound poly(ADP-ribose) in the absence of DNA was less than that to histone, but about half of the poly(ADP-ribose) remained bound in the presence of excess DNA. The DNA ligase preparation used in this experiment was the gift of Drs. H. Teraoka and K. Tsukada (Tokyo Medical and Dental University), and was purified about 4,000-fold from calf thymus (*17*). Referring to the findings that DNA ligase is not a basic protein like histones (*17*), and that poly(ADP-ribose) did not affect DNA ligase activity at all (Fig. 3), the binding in the presence of excess DNA appeared to represent a specific binding to the ligase at sites distinct from those for DNA binding. The dissociation constant of the complex, as estimated from Scatchard plots, was about $2 \times 10^{-8}$ M. This means that the binding was fairly tight. Furthermore, the binding reaction proceeded very rapidly; it was completed within 30 sec at 37°C.

### *Poly(ADP-ribose) Synthetase as a Link between DNA Damage and DNA Ligase*

These combined observations led to the following working hypothesis about the poly(ADP-ribose) action in DNA repair (Fig. 7). When a DNA strand is broken by various damaging agents and/or endonuclease(s) and exonuclease(s), poly(ADP-ribose) synthetase binds to the broken ends and synthesizes many poly(ADP-ribose) molecules on itself. In the meantime, DNA polymerase fills in the gap. Poly(ADP-ribose) molecules concentrated on the synthetase attract and bind DNA ligase, and

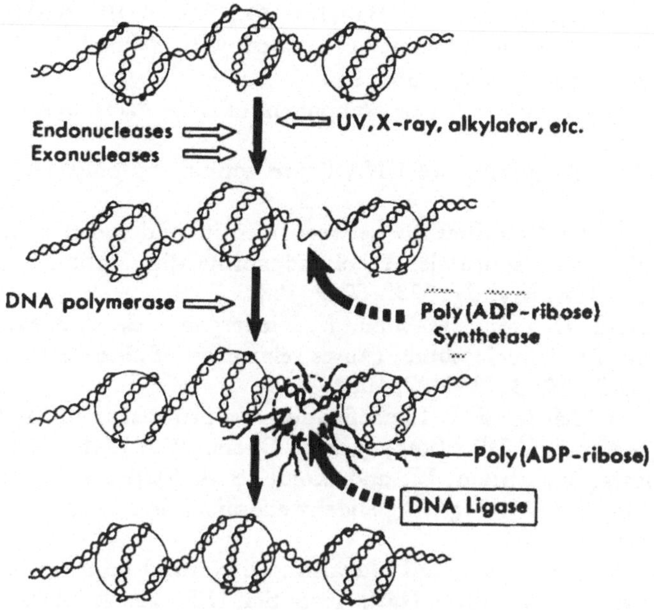


FIG. 7. A possible mechanism of poly(ADP-ribose) action in DNA repair.

support the ligase action by counteracting histones. Poly(ADP-ribose) synthetase thus serves as a link between DNA damage and DNA ligase. The facts that poly-(ADP-ribose) synthetase requires DNA strand ends for its activity (*18, 19*), and that the synthetase automodifies itself with about 15 molecules of long poly(ADP-ribose) (*20*) support this hypothesis. Although this series of events is yet to be consolidated by future experiments, this model appears to explain the significance of automodification of poly(ADP-ribose) synthetase which predominates in DNA-damaged cells (*4, 21*). This mechanism also has the potential to allow the repair of DNA not only in linker regions but in core regions of polynucleosomes.

ACKNOWLEDGMENT

This work was supported in part by grants-in-aid for Scientific and Cancer Research from the Ministry of Education, Science and Culture, Japan, and by a research grant from the Princess Takamatsu Cancer Research Fund.

## REFERENCES

1. Shall, S. ADP-ribose and DNA repair. *In;* O. Hayaishi and K. Ueda (eds.), ADP-ribosylation Reactions: Biology and Medicine, pp. 477–520, Academic Press, New York, 1982.
2. Durkacz, B. W., Nduka, N., Omidiji, O., Shall, S., and Zia'ee, A. ADP-ribose and DNA repair. *In;* M. E. Smulson and T. Sugimura (eds.), Developments in Cell Biology, Vol. 6, Novel ADP-ribosylations of Regulatory Enzymes and Proteins, pp. 207–216, Elsevier/North-Holland Publ. Co., New York, 1980.
3. Hayaishi, O., Kawaichi, M., Ogata, N., and Ueda, K. Automodification of poly(ADP-ribose) synthetase and DNA repair. *In;* H. Holzer (ed.), Metabolic Interconversion of Enzymes 1980, pp. 4–9, Springer-Verlag, Berlin-Heidelberg-New York, 1981.

4. Ogata, N., Kawaichi, M., Ueda, K., and Hayaishi, O., Poly(ADP-ribosyl)ation of 110,000 dalton protein in human lymphocytes treated with *N*-methyl-*N'*-nitro-*N*-nitrosoguanidine. Biochem. Int., *1*: 229–236, 1980.
5. Shall, S. ADP-ribose in DNA repair: A new component of DNA excision repair. Adv. Rad. Biol., 1982, in press.
6. Creissen, D. and Shall, S. Regulation of DNA ligase activity by poly(ADP-ribose). Nature, *296*: 271–272.
7. Butt, T. R. and Smulson, M. Relationship between nicotinamide adenine dinucleotide concentration and *in vitro* synthesis of poly(adenosine diphosphate ribose) on purified nucleosomes. Biochemistry, *19*: 5235–5242, 1980.
8. Poirier, G. G., de Murcia, G., Jongstra-Bilen, J., Niedergang, C., and Mandel, P. Poly(ADP-ribosyl)ation of polynucleosomes causes relaxation of chromatin structure. Proc. Natl. Acad. Sci. U.S., *79*: 3423–3427, 1982.
9. Inagaki, T., Miura, K., and Murachi, T. Identification of a protease inhibitor from rat peritoneal macrophages as poly(ADP-ribose). J. Biol. Chem., *255*: 7746–7750, 1980.
10. Yoshihara, K., Tanigawa, Y., Burzio, L., and Koide, S. S. Evidence for adenosine diphosphate ribosylation of $Ca^{2+}$, $Mg^{2+}$-dependent endonuclease. Proc. Natl. Acad. Sci. U.S., *72*: 289–293, 1975.
11. Ueda, K., Omachi, A., Kawaichi, M., and Hayaishi, O. Natural occurrence of poly-(ADP-ribosyl) histones in rat liver. Proc. Natl. Acad. Sci. U.S., *72*: 205–209, 1975.
12. Zimmerman, S. B. and Levin, C. J. DNA ligase activity on chromatin and its analogs. Rejoining of DNA strands in polylysine-DNA complexes and in reconstituted chromatins. Biochemistry, *14*: 1671–1677, 1975.
13. Taniguchi, T., Agemori, M., Kameshita, I., Nishikimi, M., and Shizuta, Y. Participation of poly(ADP-ribosyl)ation in the depression of RNA synthesis caused by treatment of mouse lymphoma cells with methylnitrosourea. J. Biol. Chem., *257*: 4027–4030, 1982.
14. Juarez-Salinas, H., Sims, J. L., and Jacobson, M. K. Poly(ADP-ribose) levels in carcinogen-treated cells. Nature, *282*: 740–741, 1979.
15. Berger, N. A., Sikorski, G. W., Petzold, S. J., and Kurohara, K. K. Association of poly(adenosine diphosphoribose) synthesis with DNA damage and repair in normal human lymphocytes. J. Clin. Invest., *63*: 1164–1171, 1979.
16. Ueda, K., Kawaichi, M., Ogata, N., and Hayaishi, O. Poly(ADP-ribosyl)ation of nuclear proteins. *In;* K. Mizobuchi *et al.* (eds.), Nucleic Acid Research: Future development, Academic Press, New York, 1983, in press.
17. Teraoka, H. and Tsukada, K. Eukaryotic DNA ligase. Purification and properties of the enzyme from bovine thymus, and immunochemical studies of the enzyme from animal tissues. J. Biol. Chem., *257*: 4758–4763, 1982.
18. Ohgushi, H., Yoshihara, K., and Kamiya, T. Bovine thymus poly(adenosine diphosphate ribose) polymerase. Physical properties and binding to DNA. J. Biol. Chem., *255*: 6205–6211, 1980.
19. Benjamin, R. C. and Gill, D. M. Poly(ADP-ribose) synthesis *in vitro* programmed by damaged DNA. A comparison of DNA molecules containing different types of strand breaks. J. Biol. Chem., *255*: 10502–10508, 1980.
20. Kawaichi, M., Ueda, K., and Hayaishi, O. Multiple autopoly(ADP-ribosyl)ation of rat liver poly(ADP-ribose) synthetase. Mode of modification and properties of automodified synthetase. J. Biol. Chem., *256*: 9483–9489, 1981.
21. Ueda, K., Ogata, N., Kawaichi, M., Inada, S., and Hayaishi, O. ADP-ribosylation reactions. *In;* E. Stadtman and B. Horrecker (eds.), Current Topics in Cellular Regulation, Vol. 21, pp. 175–187, Academic Press, New York, 1982.

ADP-RIBOSYLATION, DNA REPAIR AND CANCER, M. MIWA ET AL. (EDS.),
JAPAN SCI. SOC. PRESS, TOKYO/VNU SCIENCE PRESS, UTRECHT, PP. 183-193, 1983

# $Ca^{2+}$, $Mg^{2+}$-dependent Endonuclease and ADP-ribosylation

Koichiro YOSHIHARA, Yasuharu TANAKA, and Tomoya KAMIYA
*Department of Biochemistry, Nara Medical University, Kashihara, Nara 634, Japan*

*Abstract:* The molecular mechanism of the inhibition of $Ca^{2+}$, $Mg^{2+}$-dependent endonuclease by ADP-ribosylation was studied by using purified bull seminal plasma $Ca^{2+}$, $Mg^{2+}$-dependent endonuclease, endonuclease-stimulating proteins, and poly(ADP-ribose) polymerase. The activity of an essentially homogeneous preparation of the endonuclease was markedly suppressed by its preincubation with $NAD^+$, poly(ADP-ribose) polymerase, DNA, and $Mg^{2+}$. These four components of the incubation mixture were all essential for the suppression of the activity. Analyses of the initial and the chased reaction product by Sephadex G-100 column chromatography and sodium dodecyl sulfate (SDS)-polyacrylamide gel electrophoresis revealed that $Ca^{2+}$, $Mg^{2+}$-dependent endonuclease was ADP-ribosylated during the incubation and its activity was markedly inhibited by the elongation of the ADP-ribose polymer covalently attached to the endonuclease. When the suppressed enzymes were mildly treated with an alkaline pH of 10.0, the activity was restored almost to the level of the unmodified control sample. These facts indicate that the linkage between the enzyme and poly(ADP-ribose) is hydrolyzed at this pH, and that the liberated polymer itself does not appreciably affect the endonuclease activity. These results also suggest that an electric repulsion between negative charges on DNA and poly(ADP-ribose) attached to $Ca^{2+}$, $Mg^{2+}$-dependent endonuclease is the basis for the observed suppression of the enzyme by ADP-ribosylation.

Though histone H2B and H1 are shown to be as good endonuclease-stimulators (*1*) as they are good acceptors of ADP-ribose in poly(ADP-ribose) polymerase reaction (*2*), ADP-ribosylation of these two proteins did not affect their endonuclease-stimulating ability appreciably, at least under the conditions used.

In 1970, Burzio and Koide (*3*) reported an interesting finding that ADP-ribosylation in rat liver nuclei markedly suppressed their DNA-synthesizing activity. Later, the observed suppression of DNA synthesis was shown to be caused by two mechanisms: the one is the release of a DNA polymerase from chromatin (*4*) and the other is the suppression of a template (primer)-activating process occurring in rat liver nuclei in the presence of both $Ca^{2+}$ and $Mg^{2+}$ ions (*5*). Further studies on the suppression of the template(primer)-activating process by ADP-ribosylation revealed that a chromatin-bound $Ca^{2+}$, $Mg^{2+}$-dependent endonuclease was directly inhibited by ADP-ribosylation of the enzyme molecule itself (*6*).

Contrary to these findings, Roberts *et al.* (*7*) have reported that the DNA synthesis of HeLa cell nuclei was stimulated by ADP-ribosylation. Tanigawa *et al.* (*8*) have reported that, upon preincubation with $NAD^+$, the liver nuclei from adult hen and chick embryo showed quite contrasting responses, suppression in the former tissue and stimulation of DNA synthesis in the latter. Recently they showed that the observed stimulation of DNA synthesis by ADP-ribosylation in chick embryo liver nuclei was apparently due to an increase in the sensitivity of ADP-ribosylated chromatin to the attack of a nuclear endonuclease (*9*), while the suppression observed with adult hen liver nuclei was due to a strong inhibitory effect of free polymer of ADP-ribose on a $Mg^{2+}$-dependent endonuclease localized in the liver cell nuclei (*10*). Yamada *et al.* (*11*) also have reported that an exonuclease localized in rat liver nuclei is suppressed by free poly(ADP-ribose).

These results indicate that ADP-ribosylation occurring in various cell nuclei has diverse indirect effects on the activity of various nuclear DNases. Since in our previous study we used rather crude preparations of $Ca^{2+}$, $Mg^{2+}$-dependent endonuclease and poly(ADP-ribose) polymerase, and thus could not separately evaluate the different effects of ADP-ribosylation on the endonuclease, we attempted to demonstrate directly the ADP-ribosylation of $Ca^{2+}$, $Mg^{2+}$-dependent endonuclease by using purified enzymes. As described elsewhere (*12*), recently we succeeded in purifying $Ca^{2+}$, $Mg^{2+}$-dependent endonuclease from bull seminal plasma to an essentially homogeneous preparation. During this study, we found that several DNA-binding proteins with a high endonuclease-stimulating activity were separated from the enzyme during the purification process. Comparisons of the endonuclease-stimulating activities of these endogenous proteins with other basic proteins including histones, protamines, and basic homopolyamino acids revealed that this endonuclease is markedly stimulated by rather broad categories of basic proteins with DNA-binding affinity (*1*). Since it is well known that histones are good acceptors of ADP-ribose (*13*) and, in addition, three endogenous endonuclease-stimulative proteins seemed to be functioning as ADP-ribose acceptors in poly(ADP-ribose) polymerase reaction as will be described later, we reexamined the mechanism of the inhibition of $Ca^{2+}$, $Mg^{2+}$-dependent endonuclease using the purified enzyme and its stimulator proteins.

### *Bull Seminal $Ca^{2+}$, $Mg^{2+}$-dependent Endonuclease and Its Stimulation by Endogenous DNA-binding Proteins and Histones*

$Ca^{2+}$, $Mg^{2+}$-dependent endonuclease was purified to near homogeneity from lyophilized powder of bull seminal plasma by a successive purification through carboxymethylcellulose, phosphocellulose, and DNA-cellulose column chromatography, with a recovery of 33%. A summary of a typical purification is shown in Table 1. The detail of the purification will appear elsewhere (*12*). Since we found, upon purification of the enzyme, that three endogenous protein fractions with a high endonuclease-stimulating activity were separated from it, we separately purified these proteins (*1*).

TABLE 1. Purification of Bull Seminal $Ca^{2+}$, $Mg^{2+}$-dependent Endonuclease

| Purification step | Protein (mg) | Endonuclease activity (unit × $10^{-5}$) | Yields (%) | Specific activity (unit/mg × $10^{-3}$) | Purification (fold) |
|---|---|---|---|---|---|
| 1) Lyophilized powder | 2,970 | 16.8 | 100 | 0.57 | 1 |
| 2) CM-cellulose[a)] | 178 | 14.6 | 87 | 8.2 | 14 |
| 3) P-Cellulose[b)] | 63.3 | 9.9 | 59 | 15.6 | 27 |
| 4) DNA-cellulose | 1.04 | 5.6 | 33 | 539.5 | 950 |

[a)] Carboxymethylcellulose. [b)] Phosphocellulose. (From Hashida *et al.* (*12*)).

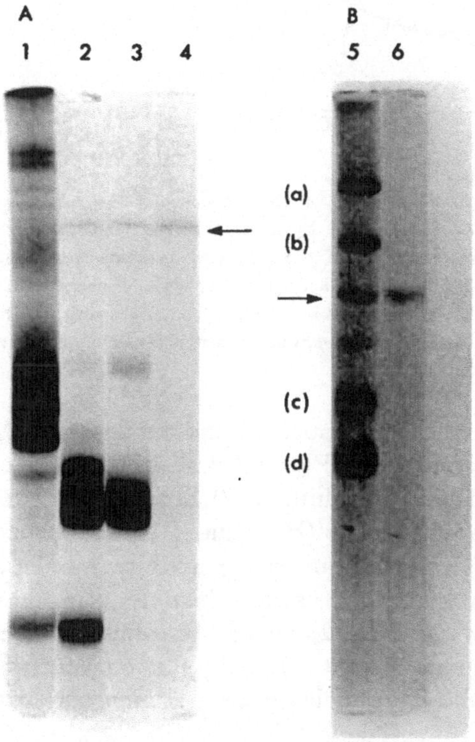


FIG. 1. Acid-urea and SDS-polyacrylamide gel electrophoresis of purified $Ca^{2+}$, $Mg^{2+}$-dependent endonuclease. A. Acid-urea polyacrylamide gel electrophoresis of enzyme fractions from various purification steps. The amount of enzyme applied is: (1) 200 units of crude enzyme (368 $\mu$g); (2) 1,000 units of carboxymethylcellulose enzyme (122 $\mu$g); (3) 1,000 units of phosphocellulose-step enzyme (64 $\mu$g); and (4) 1,000 units of DNA-cellulose step enzyme (1.9 $\mu$g), respectively. B. SDS-polyacrylamide gel electrophoresis of purified enzyme from DNA-cellulose step; (5) 10 $\mu$g each of molecular marker proteins, bovine serum albumin (a), ovalbumin (b), trypsin inhibitor (c), cytochrome *c* (d), and purified enzyme (2,000 units, 3.8 $\mu$g); (6) purified enzyme (2,000 units, 3.8 $\mu$g). Arrows indicate the position of purified enzyme. (From Hashida *et al.* (*12*)).

These endogenous activator proteins are seen as major protein bands on the analysis of crude endonuclease preparations at carboxymethylcellulose and phosphocellulose column chromatography steps by acid-urea polyacrylamide gel electro-

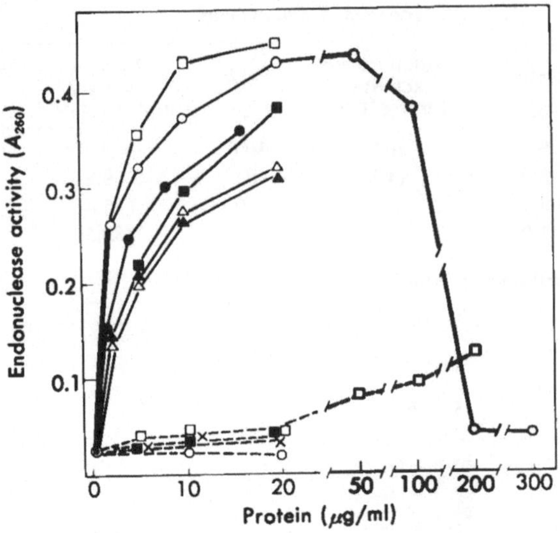


FIG. 2. Effect of various DNA-binding proteins on the activity of $Ca^{2+}$, $Mg^{2+}$-dependent endonuclease. Endonuclease assay was carried out by measuring the formation of acid-soluble DNA fragments. The indicated concentration of the following proteins were added into the reaction mixture. O—O histone H2B; ● ● activator PI; Δ—Δ activator PII′; ▲—▲ activator PIII; □—□ poly-L-lysine; ■ ■ salmine; □---□ bovine serum albumin; ■---■ ovalbumin; ×---× γ-globulin; ●---● poly-L-glutamic acid; ⊙ enzyme activity without activator protein. (From Yoshihara *et al.* (*1*)).

phoresis (Fig. 1A) and are eliminated from the purified endonuclease (Fig. 1 A and B). These proteins were designated as activators PI, PII′, and PIII, according to the order of their elution from the phosphocellulose column (*12*). Since these endonuclease-stimulating proteins showed DNA-binding ability (Hashida, T., unpublished results), we examined whether a similar endonuclease-stimulating activity was exhibited by other known DNA-binding proteins including various histones, protamines, and basic homopolyamino acids. As shown in Fig. 2, all of these basic proteins tested markedly stimulated the endonuclease activity at relatively low concentrations (2 to 20 µg/ml, the protein/DNA ratio in the endonuclease assay corresponded to 0.02 to 0.2 on a weight basis).

### *ADP-ribose Accepting Ability of Endogenous Activator Proteins and Purified $Ca^{2+}$, $Mg^{2+}$-dependent Endonuclease*

The finding that the activity of $Ca^{2+}$, $Mg^{2+}$-dependent endonuclease was markedly influenced by several basic proteins such as histones, suggests a possibility that ADP-ribosylation of such an endonuclease-stimulating protein might cause an apparent suppression of the enzyme as we observed previously (*6*), since the partially purified endonuclease from bull seminal plasma (carboxymethyl (CM)-cellulose step) is shown to be contaminated by large amounts of endogenous activator proteins (Fig. 1A). Thus, we attempted to estimate the ADP-ribose accepting ability of purified $Ca^{2+}$, $Mg^{2+}$-dependent endonuclease and endogenous activator proteins. As we reported earlier (*15*), when $Mg^{2+}$ was omitted from the reaction mixture and a min-

TABLE 2. Stimulation of Poly(ADP-ribose) Polymerase Reaction by Various Proteins Including $Ca^{2+}$, $Mg^{2+}$-dependent Endonuclease

| Protein added | Amount ($\mu g$) | ADP-ribose incorporated (pmol) |
|---|---|---|
| 1) None | — | 28 |
| | — | |
| 2) Endogenous activator protein PIII′ | 2.5 | 213 |
| | 5.0 | 306 |
| 3) $Ca^{2+}$, $Mg^{2+}$-dependent endonuclease | 1.5 | 144 |
| | 3.0 | 263 |
| | 4.5 | 363 |
| 4) Histone H2B | 2.5 | 238 |
| | 5.0 | 403 |
| 5) Histone H1 | 1 | 253 |
| | 2 | 538 |
| 6) Bovine serum albumin | 4 | 25 |
| 7) DNase I | 5 | 21 |
| 8) DNase II | 5 | 14 |

imum amount of an appropriate DNA (active DNA (*14*)) was used to support poly-(ADP-ribose) polymerase activity, the ADP-ribose-transferring activity of this enzyme became almost completely dependent on exogenously added histone. Under this reaction condition, the automodification reaction of the enzyme was limited to a minimum and the main product was ADP-ribosylated histone (*15*). Thus, we can estimate the ADP-ribose accepting ability of exogenously added proteins by measuring the stimulation of this reaction by these proteins.

As shown in Table 2, both purified $Ca^{2+}$, $Mg^{2+}$-dependent endonuclease and activator PII′ were good stimulators of the reaction as histones H2B and H1. The stimulation of the ADP-ribose transfer by $Ca^{2+}$, $Mg^{2+}$-dependent endonuclease should be ascribed to its acceptor function since the main product in this reaction was found to be associated with the endonuclease as will be shown later. Activators PI and PIII also showed similar stimulation (data not shown).

### *Effect of ADP-ribosylation of Histones H2B and H1 on Their Endonuclease Stimulating Activity*

To examine whether ADP-ribosylation of various basic proteins affects their endonuclease-stimulating activity, we ADP-ribosylated histones H2B and H1 by poly-(ADP-ribose) polymerase under the histone-dependent reaction condition (*15*) and compared the endonuclease-stimulating activities of the modified and unmodified histones. As shown in Fig. 3, the modified histones H2B (0.5 ADP-ribose residue per molecule on an average) and H1 (7 ADP-ribose / molecule) showed essentially the same activities as those of their unmodified control samples. Although the final conclusion should be reserved because the synthesis of a larger molecule of poly(ADP-ribose) bound to histone may decrease their binding affinity for DNA, and thus may affect their endonuclease-stimulating activity, the results indicate that the observed

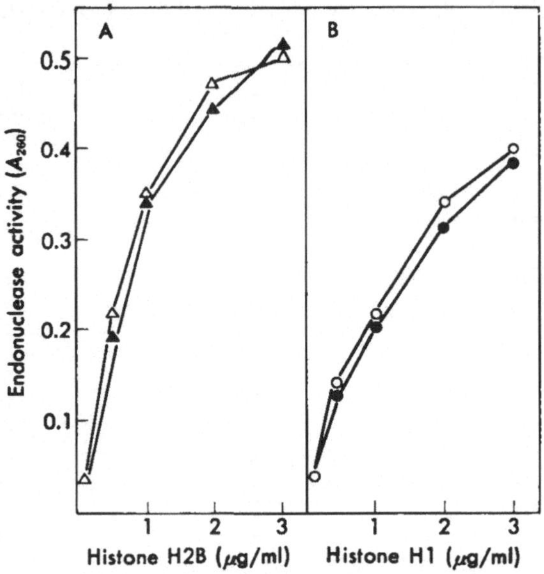


FIG. 3. Effect of ADP-ribosylation of histones H2B and H1 on their endonuclease-stimulating activity. Histones were ADP-ribosylated as described in a previous report (*15*). The indicated concentrations of modified histone H2B (▲ containing 0.5 ADP-ribose residue per molecule), histone H1 (● containing 7 ADP-ribose residue per molecule), and unmodified control histones (△, ○) were added to the reaction mixture for endonuclease. The endonuclease assay was carried out using DNA-cellulose as substrate as will be described elsewhere (*12*).

amounts of ADP-ribose associated with histones do not affect their endonuclease-stimulating activity.

### *Suppression of Purified $Ca^{2+}$, $Mg^{2+}$-dependent Endonuclease by ADP-ribosylation*

When purified $Ca^{2+}$, $Mg^{2+}$-dependent endonuclease was incubated with $NAD^+$ and poly(ADP-ribose) polymerase in the presence of $Mg^{2+}$ and DNA, the activity of the endonuclease was markedly and rapidly suppressed in a time-dependent manner. When any one of $NAD^+$, poly(ADP-ribose) polymerase, or $Mg^{2+}$ was omitted from the reaction mixture, this rapid decrease in endonuclease activity was suppressed and only a gradual decrease, probably due to spontaneous denaturation of the enzyme, was observed. When *Neurospora* NADase was used in place of poly(ADP-ribose) polymerase in the reaction, no suppression was seen. The presence of the endonuclease in the reaction mixture during the synthesis of poly(ADP-ribose) was essential for the suppression of endonuclease activity. When poly(ADP-ribose) was synthesized previous to the incubation of the endonuclease and then the endonuclease was incubated in the mixture, no significant suppression of the enzyme by the product was observed. Since our $Ca^{2+}$, $Mg^{2+}$-dependent endonuclease preparation was still contaminated by a relatively small amount of an endogenous activator protein PII′ (10 to 15%), we further examined whether the contaminating protein was involved in the mechanism of the suppression. The experiment gave a rather negative result because the addition of further amounts of activator PII′ into the preincubation mixture rather

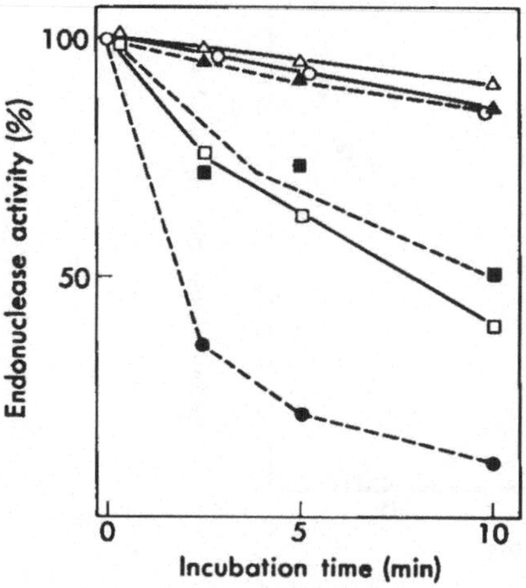


FIG. 4. Effect of ADP-ribosylation on various endonuclease. $Ca^{2+}$, $Mg^{2+}$-dependent endonuclease (●---●; 0.45 μg), DNase I (▲---▲; 0.6 μg), and micrococcal nuclease (■---■; 0.6 μg) were preincubated in the reaction mixture for poly(ADP-ribose) synthesis containing 0.2 mM $NAD^+$, 10 mM $MgCl_2$, 0.5 μg each of active DNA and poly(ADP-ribose) polymerase, and other components for the reaction (*14*). At the indicated time, an aliquot of the sample was taken out and the activity of the endonuclease was assayed using DNA-cellulose as substrate as will be described elsewhere (*12*). Respective open symbols with solid lines indicate control experiments, in which the endonuclease were incubated as above except that poly(ADP-ribose) polymerase was omitted from the mixture.

abolished the suppression of the endonuclease activity. This effect of activator PII′ may be due either to its interaction with the endonuclease (*12*) or to a competition between these two ADP-ribose accepting proteins.

Thus, the results also suggest that the endonuclease is suppressed by the ADP-ribosylation of the enzyme itself. This suppression of $Ca^{2+}$, $Mg^{2+}$-dependent endonuclease by ADP-ribosylation seems to be rather specific for this type endonuclease, since no significant suppression in others tested (DNase I and micrococcal nuclease) was observed (Fig. 4).

### *Analysis of ADP-ribosylated $Ca^{2+}$, $Mg^{2+}$-dependent Endonuclease*

As we mentioned in the previous section of this article, $Ca^{2+}$, $Mg^{2+}$-dependent endonuclease is inhibited in an ADP-ribosylating system containing $Mg^{2+}$. Thus, we first attempted to see whether $Ca^{2+}$, $Mg^{2+}$-dependent endonuclease was ADP-ribosylated under the reaction condition by analyzing the product by SDS-polyacrylamide gel electrophoresis. This attempt, however, was unsuccessful because the main product formed under the reaction condition was recovered at the top of the gel (data not shown), suggesting that the main product is ADP-ribosylated poly(ADP-ribose) polymerase formed as the result of automodification reaction of this enzyme, which is shown to occur preferentially in the presence of $Mg^{2+}$ (*2*). Therefore, we tried to demonstrate ADP-ribosylation of $Ca^{2+}$, $Mg^{2+}$-dependent endonuclease using other reaction conditions of poly(ADP-ribose) polymerase, where the incorporation of ADP-ribose into acid-insoluble material become essentially proportional to the amount of

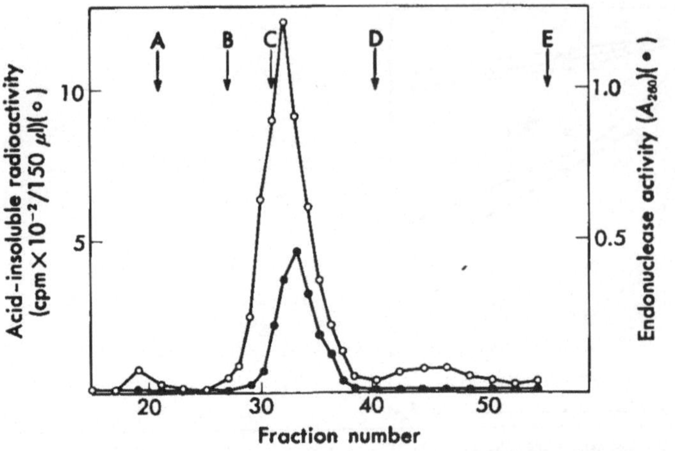


FIG. 5. Sephadex G-100 column chromatography of ADP-ribosylated $Ca^{2+}$, $Mg^{2+}$-dependent endonuclease. The reaction product was synthesized under the histone-dependent reaction condition of poly(ADP-ribose) polymerase using [$^{3}$H]NAD as substrate as described previously (*15*), except that histones were replaced by purified $Ca^{2+}$, $Mg^{2+}$-dependent endonuclease. All reaction products were analyzed by Sephadex G-100 column chromatography.

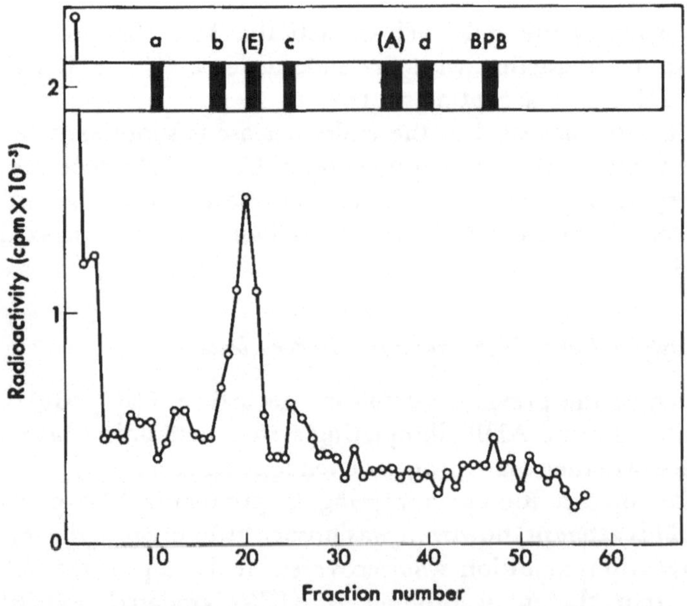


FIG. 6. SDS-polyacrylamide gel electrophoresis of ADP-ribosylated $Ca^{2+}$, $Mg^{2+}$-dependent endonuclease. The same product as shown in Fig. 5 was analyzed by SDS-polyacrylamide gel electrophoresis, with molecular markers of bovine serum albumin (a), ovalbumin (b), carbonic anhydrase (c), cytochrome *c* (d), and bromophenol blue (BPB). The position of $Ca^{2+}$, $Mg^{2+}$-dependent endonuclease (E) and an endogenous endonuclease-stimulating protein PII′ (A) are indicated on a schematic illustration of the stained gel.

exogenous acceptor added (Table 2). As shown in Fig. 5, when the reaction product was analyzed by Sephadex G-100 column chromatography, the major product was eluted at a position very near to the endonuclease, indicating that the main product of this reaction is the ADP-ribosylated endonuclease. A slightly faster elution of the radioactivity peak of the modified endonuclease than the authentically added endonuclease may be due to the change in either the enzyme molecular size, or molecular conformation, or both by ADP-ribosylation. The results also indicate that the automodification of poly(ADP-ribose) polymerase is not appreciably occurring under the reaction conditions. When the product was analyzed by SDS-polyacrylamide gel electrophoresis (Fig. 6), a peak of radioactivity (23%) was found to be associated with the endonuclease, indicating that the linkage between poly(ADP-ribose) and the endonuclease is of a covalent type.

These results clearly indicate that $Ca^{2+}$, $Mg^{2+}$-dependent endonuclease can be ADP-ribosylated under the acceptor-dependent reaction condition of poly(ADP-ribose) polymerase, even in the absence of $Mg^{2+}$ in the reaction mixture. The endonuclease activity, however, was not suppressed under the reaction condition, that is, in the absence of $Mg^{2+}$. In order to elucidate the role of $Mg^{2+}$ in the endonuclease suppression, we carried out a chase experiment in which [$^3$H]ADP-ribosyl-$Ca^{2+}$, $Mg^{2+}$-dependent endonuclease, formed in the reaction without $Mg^{2+}$, was further incubated with a high concentration of cold $NAD^+$ in the presence of $Mg^{2+}$. After the chase, the endonuclease activity decreased by approximately 80%. The analysis of the chased product by Sephadex G-100 column chromatography clearly indicated that the molecular size of the endonuclease increased enormously during the chase incubation in the presence of $Mg^{2+}$, since approximately 80% of the radioactivity of the initial product, which had originally been found near the position of the endonuclease (Fig. 5), was eluted between the void volume and the position of bovine serum albumin (*16*).

Similar results were obtained in an analysis of the same products by SDS-polyacrylamide gel electrophoresis (*16*). These results suggest that the suppression of $Ca^{2+}$, $Mg^{2+}$-dependent endonuclease is caused by a highly polymerized poly(ADP-ribose), which is covalently bound to the endonuclease.

### *Restoration of the Activity of Poly(ADP-ribosyl)ated $Ca^{2+}$, $Mg^{2+}$-dependent Endonuclease by Treatment with Alkaline pH*

Since it is well known that poly(ADP-ribose) polymerase catalyzes a transfer reaction of ADP-ribose to a carboxyl residue of a protein forming an ester linkage, which is extremely labile to an alkaline pH (*13*), we attempted to hydrolyze this linkage and to see whether the suppressed endonuclease activity can be reversed by treating the modified enzyme with an alkaline pH 10. As shown in Table 3, a mild treatment of the modified enzyme with pH 10 reversed the suppressed endonuclease activity to almost the same level as that of the unmodified control enzyme, while the activity of the control enzyme remained unchanged during the treatment. The results clearly indicated that a covalent binding of a large molecule or a large number of poly(ADP-ribose) to the endonuclease is the basis for the suppression of the enzyme

TABLE 3. Treatment of ADP-ribosylated $Ca^{2+}$, $Mg^{2+}$-dependent Endonuclease with Alkaline pH

| Endonuclease samples | Treatment[b)] | Endonuclease activity[c)] | |
|---|---|---|---|
| | | Before treatment | After treatment |
| 1) Control I (-NAD)[a)] | pH 10.0 | 100 | 94 |
| 2) Control II (-ADPR polymerase)[a)] | pH 10.0 | 95 | 89 |
| 3) ADP-ribosylated $Ca^{2+}$, $Mg^{2+}$-endonuclease | pH 10.0 | 13 | 79 |
| 4) ADP-ribosylated $Ca^{2+}$, $Mg^{2+}$-endonuclease | pH 7.4 | 13 | 22 |

[a)] The two control endonuclease samples were obtained by treating the endonuclease in reaction mixture for ADP-ribosylation as samples 3) and 4) except that either $NAD^+$ (control I) or poly(ADP-ribose) polymerase (control II) was omitted from the reaction mixture. [b)] Endonuclease samples were treated in a dilution medium for the endonuclease containing 0.1 mg/ml of histone H2B to keep the enzyme activity stable (*12*). The pH was changed as indicated. Samples were kept on ice for 9 hr. [c)] The initial activity of the control sample (I) was set at 100%.

and that poly(ADP-ribose) itself liberated from the enzyme was unable to suppress the enzyme.

## DISCUSSION

All of the results described above clearly indicate that $Ca^{2+}$, $Mg^{2+}$-dependent endonuclease itself is ADP-ribosylated by poly(ADP-ribose) polymerase and suppressed as a result of an extensive synthesis of polymer covalently associated with the endonuclease.

Taking into consideration that the suppression occurred only after an extensive polymerization of the product and that the polymer liberated from the enzyme was unable to suppress the enzyme, the molecular basis for the suppression of the modified endonuclease may be an electric repulsion between the negative charged groups of the substrate (DNA) and the modified enzyme. At present, we are trying to confirm this point experimentally.

In a previous report, in which we studied the mechanism of the inhibition of $Ca^{2+}$, $Mg^{2+}$-dependent endonuclease using a rather crude enzyme preparation, we suggested that this enzyme may be inhibited by an attachment of a single residue of ADP-ribose (*6*). The present results studied with the purified enzyme, however, indicated that a relatively small amount of ADP-ribose bound to the enzyme does not appreciably affect the endonuclease activity. Since our recent study (*12*) has shown that a crude endonuclease preparation contains a very large amount of proteins with ADP-ribose accepting activity, and that one of these proteins, activator PII′, is located very near bull seminal $Ca^{2+}$, $Mg^{2+}$-dependent endonuclease on Sephadex G-100 column chromatography (data not shown), the previous incorrect estimation of the size or the amount of the poly(ADP-ribose) causing the suppression of the endonuclease is probably due to the analysis of a wrong product of ADP-ribosylated contaminants in the enzyme preparation.

## REFERENCES


1. Yoshihara, K., Tanaka, Y., Hashida, T., Ura, H., Kamiya, T., Tanigawa, Y., and Koide, S. S. Stimulation of bull seminal $Ca^{2+}$, $Mg^{2+}$-dependent endonuclease by various DNA-binding proteins. Biochem. Biophys. Res. Commun., *106*: 803–810, 1982.
2. Yoshihara, K., Hashida, T., Tanaka, Y., Matsunami, N., Yamaguchi, A., and Kamiya, T. Mode of enzyme-bound poly(ADP-ribose) synthesis and histone modification by reconstituted poly(ADP-ribose) polymerase-DNA-cellulose complex. J. Biol. Chem., *256*: 3471–3478, 1981.
3. Burzio, L. and Koide, S. S. A functional role of poly(ADPR) in DNA synthesis. Biochem. Biophys. Res. Commun., *40*: 1013–1020, 1970.
4. Yoshihara, K. and Koide, S. S. Release of DNA polymerase from rat liver chromatin on incubation with NAD. FEBS Lett., *35*: 262–264, 1973.
5. Burzio, L. and Koide, S. S. Activation of the template activity of isolated rat liver nuclei for DNA synthesis and its inhibition by NAD. Biochem. Biophys. Res. Commun., *53*: 572–579, 1973.
6. Yoshihara, K., Tanigawa, Y., Burzio, L., and Koide, S. S. Evidence for adenosine diphosphate ribosylation of $Ca^{2+}$, $Mg^{2+}$-dependent endonuclease. Proc. Natl. Acad. Sci. U.S., *72*: 289–293, 1975.
7. Roberts, J. H., Stark, P., and Smulson, M. E. Poly(ADP-ribose): Release of template restriction in HeLa cells. Proc. Natl. Acad. Sci. U.S., *71*: 3212–3216, 1974.
8. Tanigawa, Y., Kawamura, M., Kitamura, A., and Shimoyama, M. Suppression and stimulation of DNA synthesis by ADP-ribosylation of nuclear proteins from adult hen and chick embryo liver. Biochem. Biophys. Res. Commun., *81*: 1278–1285, 1978.
9. Tanigawa, Y., Kitamura, A., Kawamura, M., and Shimoyama, M. Effect of poly-(ADP-ribose) formation on DNA synthesis in chick-embryo-liver nuclei. Eur. J. Biochem., *92*: 261–269, 1978.
10. Tanigawa, Y., Sato, Y., and Shimoyama, M. Poly(ADP-ribose)-sensitive $Mg^{2+}$-dependent endonuclease. Seikagaku, *54*: 689, 1982 (in Japanese).
11. Yamada, M., Nagao, M., Miwa, M., and Sugimura, T. Inhibition of deoxyribonuclease in an extract of rat liver nuclei by poly(ADP-ribose). Biochem. Biophys. Res. Commun., *56*: 1093–1099, 1974.
12. Hashida, T., Tanaka, Y., Matsunami, N., Yoshihara, K., Kamiya, T., Tanigawa, Y., and Koide, S. S. Purification and properties of bull seminal plasma $Ca^{2+}$, $Mg^{2+}$-dependent endonuclease. J. Biol. Chem., *257*, 13114–13119, 1982.
13. Nishizuka, Y., Ueda, K., Yoshihara, K., Yamamura, H., Takeda, M., and Hayaishi, O. Enzymic adenosine diphosphoribosylation of nuclear proteins. Cold Spring Harbor Symp. Quant. Biol., *34*: 781–786, 1969.
14. Yoshihara, K., Hashida, T., Tanaka, Y., Ohgushi, H., Yoshihara, H., and Kamiya, T. Bovine thymus poly(adenosine diphosphate ribose) polymerase. J. Biol. Chem., *253*: 6459–6466, 1978.
15. Tanaka, Y., Hashida, T., Yoshihara, H., and Yoshihara, K. Bovine thymus poly-(ADP-ribose) polymerase histone-dependent and $Mg^{2+}$-dependent reaction. J. Biol. Chem., *254*: 12433–12438, 1979.
16. Tanaka, Y. and Yoshihara, K. Molecular basis for the suppression of $Ca^{2+}$, $Mg^{2+}$-dependent endonuclease by ADP-ribosylation, unpublished.

ADP-RIBOSYLATION, DNA REPAIR AND CANCER, M. MIWA ET AL. (EDS.), JAPAN SCI. SOC. PRESS, TOKYO/VNU SCIENCE PRESS, UTRECHT, PP. 195-207, 1983

# Poly(ADP-ribose): Spectator or Participant in Excision Repair of DNA Damage

J. E. CLEAVER, W. J. BODELL,*[1] C. BOREK,*[2] W. F. MORGAN, and J. L. SCHWARTZ*[3]

*Laboratory of Radiobiology and Environmental Health, University of California, San Francisco, California 94143, U.S.A.*

*Abstract:* Inhibition of poly(ADP-ribose) synthesis by 3-aminobenzamide in various human and hamster cell types influenced the responses to DNA damage from methyl methanesulfonate (MMS), but not from UV light. Excision of the major alkylation products and pyrimidine dimers was unaffected by 3-aminobenzamide. After exposure of cells to methyl methanesulfonate, 3-aminobenzamide increased the strand break frequency in all cell types studied, but stimulated repair replication only in lymphoid and HeLa cells, suggesting these are independent effects. 3-Aminobenzamide also inhibited the pathway for *de novo* synthesis of DNA purines, suggesting that some of its effects, particularly on repair replication, may be due to disturbance of precursor pathways. 3-Aminobenzamide stimulated sister chromatid exchange formation and mutagenesis but inhibited transformation, suggesting that some of these endpoints involve ADP-ribosylation by ways other than repair. Poly(ADP-ribose) synthesis appears to regulate the ligation stage of repair of alkylation damage by modulating a dynamic balance between incision and ligation, so as to minimize the frequency of DNA breaks.

A variety of DNA-damaging agents, including radiations and carcinogenic chemicals, decrease cellular NAD content and stimulate poly(ADP-ribose) synthesis. Single-strand breaks appear to act as stimuli for poly(ADP-ribose) polymerase activity, which modifies both the polymerase itself and chromatin and nonhistone proteins by ADP-ribosylation (*1–5*). Inhibitors such as 3-aminobenzamide and 5-methylnicotinamide prevent the decrease in cellular NAD content in intact cells (*6*) and prevent the synthesis of poly(ADP-ribose) in permeable cells (*6*).

There are, however, several puzzling or inconsistent features concerning the role

Present address:

*[1] Brain Tumor Research Center, University of California, San Francisco, California 94143, U.S.A.

*[2] Department of Radiology, Columbia University College of Physicians and Surgeons, 630 W. 168th Street, New York, New York 10032, U.S.A.

*[3] Laboratory of Radiobiology, Harvard School of Public Health, 665 Huntington Avenue, Boston, Massachusetts 02115, U.S.A.

of poly(ADP-ribose) synthesis in DNA repair. Polymerization seems especially prominent in cells damaged by alkylating agents that induce many DNA breaks (*5–12*). X rays, however, also make numerous strand breaks, but inhibitors of poly-(ADP-ribose) synthesis do not affect cell survival (*12*) or unscheduled synthesis (*9*) in X-irradiated cells. Repair of UV damage, on the other hand, involves few single-strand breaks in DNA (*13, 14*), but there are many reports that inhibitors of poly-(ADP-ribose) polymerase stimulate unscheduled DNA synthesis in UV-irradiated lymphoid cells (*8, 11, 15–17*). All of these reports, however, measured repair after irradiation by the amount of [$^3$H]thymidine incorporated into DNA in the presence of 10–20 mM hydroxyurea. Because hydroxyurea retards the completion of repair and increases the frequency of single-strand breaks in DNA (*14, 18–21*), the use of excessive concentrations of hydroxyurea may create a role for poly(ADP-ribose) synthesis that it does not normally have during repair of UV damage. Another problem in current studies is that most investigations into possible relationships of poly(ADP-ribose) to DNA repair have used lymphocytes, lymphoid, or lymphoma cell lines (*6, 8, 9, 12, 17*), and some responses may be unique to these cell types.

We have therefore examined the synthesis of poly(ADP-ribose) in human fibroblasts and investigated the consequence of its inhibition by 3-aminobenzamide in a variety of cell types damaged by methyl methanesulfonate (MMS) or UV light (*22*). Specific endpoints included excision, strand breaks, repair replication, sister chromatid exchanges, mutagenesis, and transformation.

*Procedures*

Cultures of normal human fibroblasts from a variety of donors were established from skin biopsies. An SV40-transformed human cell line (GM637) was obtained from the Human Genetic Mutant Cell Repository, Camden, N. J. A Chinese hamster ovary (CHO-AA8) cell line was obtained from L. Thompson, Lawrence Livermore Laboratories. The HeLa S3 cell line (CCL2.2) originated from the American Type Culture Collection. Cultures were maintained in Eagle's minimal essential medium, 10% fetal calf serum, and antibiotics. The WIL-2 normal human lymphoid line was obtained from W. Thilly and maintained in RPMI medium, 10% fetal calf serum, and antibiotics. The HeLa S3 line was maintained in suspension culture but transferred 2 to 4 times in monolayer before experiments. Peripheral lymphocytes were isolated from 25 ml of whole blood using Ficoll-Hypaque, centrifuged, and resuspended in RPMI medium plus 10% serum. When uniformly labeled DNA was required for alkaline sucrose gradient analysis or DNA repair assays, cells were grown for 4 or more days in 0.01 $\mu$Ci [$^{14}$C]thymidine per ml (specific activity 56 Ci/mol) followed by at least one day in unlabeled medium. Cells were exposed to UV light (254 nm; 1.3 J/m$^2$/sec), X rays, or various concentrations of MMS for 30 min in serum-free medium.

Poly(ADP-ribose) synthesis was measured in permeabilized cells after irradiation or exposure to MMS using the method of Berger *et al.* (*4*), and cellular NAD concentrations were measured chemically by the method of Skidmore *et al.* (*23*). The frequency of single-strand breaks in DNA was measured by alkaline elution (*24*). Repair replication was measured in isopycnic gradients (*13*).

The excision of alkylation products from CHO and HeLa cells exposed to [$^{14}$C]-MMS (25 μCi/ml; 5 Ci/mol) was determined by Sephadex G-10 column chromatography of DNA isolated from pairs of cultures grown with or without 3-aminobenzamide (5 mM) and harvested 6–7 or 26 hr after exposure (*22*). DNA was digested by HCl and analyzed by Sephadex G-10 column chromatography (*25*). The excision of UV photoproducts was determined by a radioimmune assay (*26*).

Sister chromatid exchanges were measured in CHO cells by the method of Wolff (27). MMS-induced mutations to 6-thioguanine and ouabain resistance were determined by the methods described by Cleaver (*28*). Morphologic transformants in mouse 10T1/2 cells were scored 14 days after exposure to ionizing radiation as described by Borek *et al.* (*29*).

### *3-Aminobenzamide and Poly(ADP-ribose) Synthesis*

Synthesis of poly(ADP-ribose) from [$^3$H]NAD in permeable human cells depended markedly on the treatment cells received before permeabilization. MMS-damaged cells showed a strong stimulation of up to 5-fold above control that was dose dependent between 2 and 8 mM. After damage from UV light, however, the stimulation

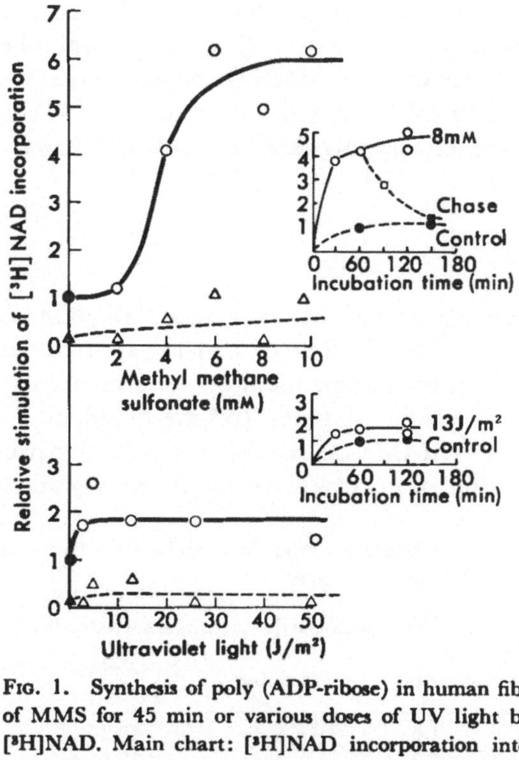


FIG. 1. Synthesis of poly (ADP-ribose) in human fibroblasts exposed to various concentrations of MMS for 45 min or various doses of UV light before permeabilization and incubation in [$^3$H]NAD. Main chart: [$^3$H]NAD incorporation into acid-insoluble material during 60 min incubation with (△) or without (○) 2 mM 3-aminobenzamide. Inset charts: incorporation of [$^3$H]NAD into MMS- or UV- damaged cells as a function of incubation time with (●) or without (○) 2 mM 3-aminobenzamide. □ $^3$H label in acid-insoluble material of permeable cells incubated for 60 min in [$^3$H]NAD and then incubated in nonradioactive NAD. (Reprinted from Cleaver *et al.* (*22*) by permission of *J. Biol. Chem.*).

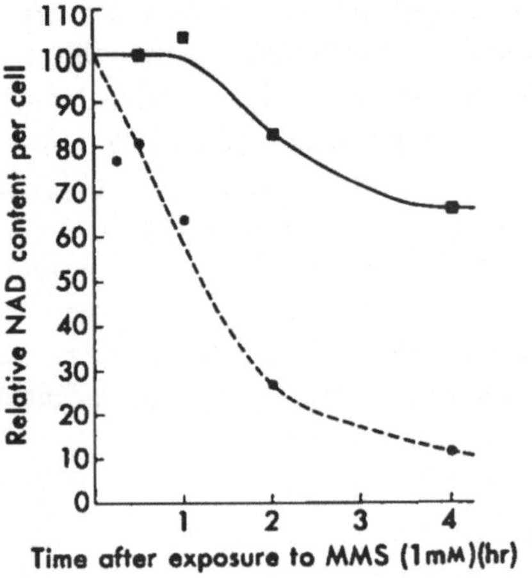


FIG. 2. Relative NAD content of intact human cells exposed to 1 mM MMS for 30 min and grown in fresh medium (●) or medium containing 5 mM 3-aminobenzamide (■) for various times before assaying NAD by the method of Skidmore *et al.* (*23*).

of poly(ADP-ribose) synthesis was less than 2-fold (Fig. 1). When [$^3$H]NAD was replaced by nonradioactive NAD at the same concentration (0.3 mM), label in the polymer was lost within 1–2 hr because of turnover (Fig. 1). 3-Aminobenzamide inhibited synthesis of poly(ADP-ribose) in control, MMS-damaged, and UV-damaged cells (Fig. 1). Exposure of intact cells to MMS caused a decrease in NAD concentration, owing to its consumption in synthesis of poly(ADP-ribose), which was also prevented by 3-aminobenzamide (Fig. 2).

### *Excision of Damaged Bases*

Excision of UV-induced photoproducts and MMS-induced alkylated bases was unaffected by 5 mM 3-aminobenzamide. Within 2 hr of 13 J/m$^2$ approximately 80% of the UV photoproducts had been excised, independent of the presence of 3-aminobenzamide. The specific activities of alkylated bases (N$^3$-methyladenine and N$^7$-methylguanine) of cells treated with [$^{14}$C]MMS decreased over a 26 hr period (Table 1). The rate of excision was faster and more extensive for N$^3$-methyladenine than

TABLE 1. Specific Activity of DNA Purines and of Alkylation Products in HeLa S3 Cells Exposed to 5 mM [$^{14}$C]MMS and Incubated with or without 5 mM 3-Aminobenzamide

| Treatment | Incubation (hr) | Specific activity (cpm/absorbance unit) | | | |
|---|---|---|---|---|---|
| | | Adenine | Guanine | N$^3$-Methyl-adenine | N$^7$-Methyl-guanine |
| [$^{14}$C]MMS | 0 | 1.6 | 1.1 | 40.2 | 416.2 |
| [$^{14}$C]MMS | 7 | 2.9 | 6.5 | 17.1 | 426.1 |
| [$^{14}$C]MMS+3AB | 7 | 2.8 | 2.2 | 18.5 | 442 |
| [$^{14}$C]MMS | 26 | 295 | 113 | 15.9 | 353 |
| [$^{14}$C]MMS+3AB | 26 | 2.5 | 1.9 | 12.1 | 410 |

3AB, 3-aminobenzamide.

for $N^7$-methylguanine, and after 26 hr 40% of $N^3$-methyladenine and 85% of $N^7$-methylguanine remained in the DNA of HeLa cells. Similar results were obtained with CHO cells.

After treatment of cells with [$^{14}$C]MMS, $^{14}$C radioactivity that originated from the incorporation of $^{14}$C-labeled methyl groups into the carbon-1 pool (*30*) was also detected in DNA purines (Table 1). This metabolic labeling was inhibited by 3-aminobenzamide (Table 1).

## *Strand Breakage during Repair*

After exposure to UV light, fibroblasts and lymphoid cells initially contained strand breaks during repair, but after 2–4 hr the elution rate had decreased, indicating a slower rate of repair (Fig. 3). This corresponds to the known kinetics of excision repair based on unscheduled synthesis, dimer excision, and other methods of determining strand breaks (*13, 31*). Elution was maintained at a rapid rate when cells were incubated in 2–10 mM hydroxyurea, but not when grown in 3-aminobenzamide. Similar studies using nucleoid sedimentation and alkaline sucrose gradients confirmed that 3-aminobenzamide had no effect on strand break formation and sealing during UV repair in human fibroblasts (*32*).

After exposure to MMS, 3-aminobenzamide maintained rapid elution rates, indicating a higher frequency of strand breaks under these conditions (Fig. 3). This increased frequency of strand breaks could result either from long-lived blocks to ligation, such as occur with the DNA polymerase inhibitor aphidicolin (*33*), or from

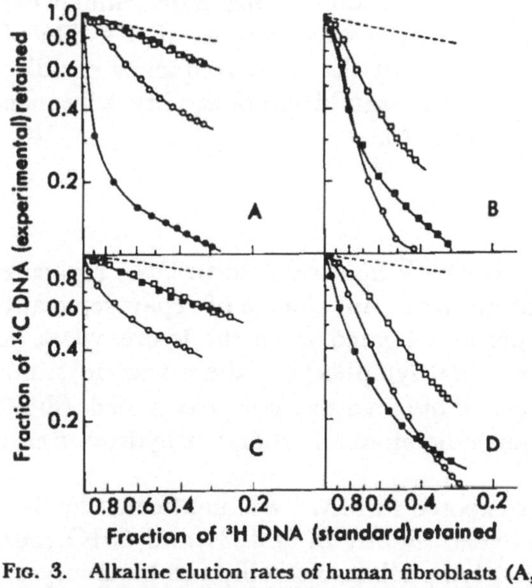


FIG. 3. Alkaline elution rates of human fibroblasts (A, B) or WIL-2 lymphoid cells (C, D) exposed to 13 J/m² UV light (A, C) or 2 mM MMS (B, D) and grown for 2 hr with or without 10 mM hydroxyurea or 5 mM 3-aminobenzamide. --- control; ○ harvested immediately after 13 J/m² UV or 2 mM MMS; harvested 2 hr after 13 J/m² UV or 2 mM MMS and grown without inhibitor (□), in 10 mM hydroxyurea (●), or in 5 mM 3-aminobenzamide (■). (Reprinted from Cleaver *et al.* (*22*) by permission of *J. Biol. Chem.*).

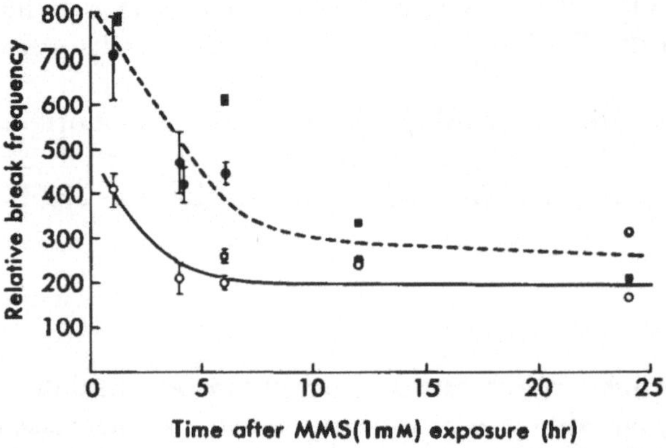


FIG. 4. Frequency of DNA strand breaks in human fibroblasts exposed to MMS and grown for various periods of time in fresh medium (○), in medium containing 5 mM 3-aminobenzamide (●), in medium containing 5 mM 3-aminobenzamide followed by 30 min in fresh medium immediately before harvesting (□), or in fresh medium followed by 30 min in 3-aminobenzamide immediately before harvesting (■). The frequency of strand breaks was determined from alkaline elution profiles (Fig. 3) that had been calibrated by comparison with ionizing radiation and is expressed in rads.

a reduction in the rate of ligation. The latter would shift the dynamic balance of strand breakage and ligation in the direction of a higher break frequency. These alternatives were investigated by exposing cells to 3-aminobenzamide for short intervals at 4–6 hr after exposure to MMS and comparing the break frequencies with those observed after continuous exposure to 3-aminobenzamide. Similar experiments were done by removing 3-aminobenzamide for various times after 4–6 hr of continuous exposure. When these changes were made the break frequency rapidly changed, indicating that brief exposures to 3-aminobenzamide produced break frequencies similar to those seen after long exposures (Fig. 4).

### *Repair Replication*

Exposure to 5 mM MMS or 13 J/m$^2$ UV light resulted in DNA repair replication in all cell types (Fig. 5). There was no significant change of repair replication of UV damage in any of the five cell types investigated when the hydroxyurea concentration was 0–2 mM (Fig. 5). In peripheral lymphocytes, there was no stimulation of repair by 3-aminobenzamide unless 10 mM hydroxyurea was added (Fig. 5). Such stimulation of UV repair replication is therefore an artifact of hydroxyurea unique to lymphocytes.

The amount of repair in cells exposed to MMS was unaffected by 3-aminobenzamide in primary fibroblasts, increased slightly in GM637 and CHO cells, and increased 3- to 5-fold in HeLa, WIL-2 lymphoblastoid cells, and lymphocytes (Fig. 5). The presence of 10 mM hydroxyurea increased the apparent stimulation of repair by 3-aminobenzamide in WIL-2 cells and lymphocytes.

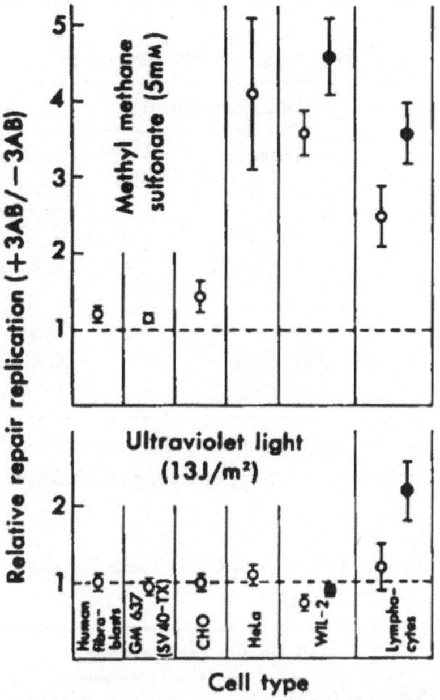


FIG. 5. Relative amounts of repair replication (ratios of amounts of cells grown in 5 mM 3-aminobenzamide to the amount in cells without 3-aminobenzamide in various mammalian cells. [$^{14}$C]thymidine-labeled cells or unlabeled lymphocytes were grown for 1 hr in bromodeoxyuridine (BrdUrd) ($10^{-5}$ M) and fluorodeoxyuridine (FdUrd) ($2\times10^{-6}$ M), then exposed to UV (13 J/m$^2$) or to 5 mM MMS for 45 min. Cells were then incubated for 4 hr in 10 $\mu$Ci/ml [$^3$H]-thymidine (specific activity 60 Ci/mmol), BrdUrd ($10^{-5}$ M), and FdUrd ($2\times10^{-6}$ M), and the DNA was isolated and analyzed on alkaline cesium chloride-cesium sulfate isopycnic gradients. In some experiments 10 mM hydroxyurea was included in the labeling mixture. Repair replication was calculated from the $^3$H/$^{14}$C ratios or from the $^3$H specific activity per $\mu$g DNA. 3-Aminobenzamide (5 mM) was added at the beginning of growth in BrdUrd. ○ no hydroxyurea; ● 10 mM hydroxyurea; 3AB, 3-aminobenzamide.

TABLE 2. SCE in Chinese Hamster and Human Cells Exposed to MMS and 3-Aminobenzamide

| Cell type | MMS (M) | SCE frequency per chromosome[a)] | | |
|---|---|---|---|---|
| | | 3-Aminobenzamide | | |
| | | 0 | 1 mM | 5 mM |
| CHO | 0 | 0.46 | 1.15 | 2.15 |
| | $5\times10^{-6}$ | 0.74 | 1.97 (1.43) | 3.60 (2.43) |
| | $1\times10^{-5}$ | 0.98 | 2.34 (1.67) | 3.76 (2.67) |
| | $5\times10^{-5}$ | 2.72 | 5.62 (3.41) | —[b)] |
| GM637 | 0 | 0.29 | 0.29 | 0.51 |
| | $5\times10^{-6}$ | 0.39 | 0.36 (0.38) | 0.72 (0.61) |
| | $1\times10^{-5}$ | 0.62 | 0.58 (0.62) | 0.97 (0.84) |
| | $5\times10^{-5}$ | 0.70 | 0.94 (0.70) | 1.13 (0.92) |

a) Each value is the mean of 2 experiments. Numbers in parentheses are the expected value if the increase above background for MMS and 3-aminobenzamide were independent and additive. b) No second-division cells.

TABLE 3. SCE in Chinese Hamster and Human Cells Exposed to UV Light and 3-Aminobenzamide

| Cell type | UV fluence ($J/m^2/sec$) | SCE frequency per chromosome[a)] | | |
|---|---|---|---|---|
| | | | 3-Aminobenzamide | |
| | | 0 | 1 mM | 5 mM |
| CHO | 0 | 0.51 | 1.10 | 2.19 |
| | 0.7 | 0.52 | 1.14 (1.11) | 2.33 (2.20) |
| | 1.3 | 0.67 | 1.33 (1.26) | 2.55 (2.35) |
| GM637 | 0 | 0.29 | 0.29 | 0.51 |
| | 0.7 | 0.33 | 0.37 (0.32) | 0.44 (0.55) |
| | 1.3 | 0.47 | 0.66 (0.47) | 0.78 (0.69) |

a) Each value is the mean of 2 experiments. The numbers in parentheses are the expected value if the increase above background from UV light and 3-aminobenzamide were independent and additive.

TABLE 4. Frequency of 6-Thioguanine- and Ouabain-resistant Mutants of CHO Cells Exposed to EMS and Grown for 7 Days with or without 3-Aminobenzamide

| EMS (mM) | 3-Aminobenzamide (mM) | Mutation frequencies | |
|---|---|---|---|
| | | 6-Thioguanine[a)] ($\times 10^6$) | Ouabain[b)] ($\times 10^6$) |
| 1 | 0 | 44.1± 1.7 | 29.5± 6.8 |
| | 10 | 82.4±17.7 | 102.2±64.1 |
| 2 | 0 | 123.0±31.5 | — |
| | 10 | 232.6± 5.9 | — |

a) Resistance was measured to a concentration of $6\times 10^{-5}$ M. b) Resistance was measured to a concentration of $3\times 10^{-3}$ M.

### *Sister Chromatid Exchanges (SCEs), Mutation, and Transformation*

Several long-term cytogenetic and genetic consequences of DNA damage and repair changed in distinctive ways when cells were grown in 3-aminobenzamide after exposure to MMS or UV (*34–36*; Tables 2–4).

Treatment for two complete cell cycles with 1 or 5 mM 3-aminobenzamide increased baseline SCE frequency in CHO and GM637 cells (Tables 2, 3). Exposure to MMS (Table 2) or UV light (Table 3) alone also increased the SCE frequency in both cell types. However, the combination of MMS or UV and 3-aminobenzamide resulted in greater increases in SCEs than expected if the results from both agents were additive (Tables 2, 3). Some synergism between the two treatments therefore stimulated SCE production.

MMS induced mutations to both 6-thioguanine and ouabain resistance, but 3-aminobenzamide alone was not mutagenic. Growth in 3-aminobenzamide after exposure to ethyl methanesulfonate (EMS), however, caused a stimulation in the induced mutation frequency at both gene loci (Table 4).

Morphologic transformation assayed 14 days after exposure of mouse 10T1/2 cells to ionizing radiation indicated a high transformation rate of $9.5\times 10^{-4}$ per survivor. This rate was reduced to zero by exposure to either benzamide or 3-aminobenzamide. Morphologic transformation therefore responds to inhibition of poly(ADP-ribose) synthesis in a way opposite to that of SCEs and mutagenesis.

## DISCUSSION

A role for poly(ADP-ribose) synthesis in cells exposed to DNA-damaging agents has been ascribed on the basis of a variety of evidence (*1–4, 6–12, 15–17, 37, 38*), obtained largely by measurements of repair and poly(ADP-ribose) synthesis in lymphocytes, lymphoid, and lymphoma cell types (*3, 4, 6, 8–10, 12, 16, 17, 39*). We therefore carried out the present studies to extend the experimental base of evidence with which to interpret the role of poly(ADP-ribose) in damaged cells.

Poly(ADP-ribose) polymerase is a tightly bound nuclear enzyme that is catalyzed by DNA strand breaks. It adds single ADP-ribose residues to proteins and forms a branched polymer with 1′–2′ glycoside links (*1, 5*). The polymer turns over through the action of a specific glycohydrolase and phosphodiesterase (*1, 5, 37*). This cycle of synthesis and degradation was observed in our experiments as the incorporation of [$^{3}$H]NAD into an acid-precipitable polymer by a 3-aminobenzamide-sensitive pathway, and its loss during incubation in nonradioactive NAD (Fig. 1). The extent of stimulation in permeable cells was dependent on the breaks introduced into DNA before permeabilization by exposure to MMS or UV light (Fig. 1). The stimulation was very small after UV irradiation, either because of the low frequency of single-strand breaks or because of their precise chemical structure.

The consequences of inhibiting poly(ADP-ribose) polymerase depended on the cell type investigated, the kinds of DNA damage involved, and the particular biological endpoint in question. We observed that 3-aminobenzamide produced pleoitropic effects after MMS exposure but none after UV exposure, except for an artifactual effect when high concentrations of hydroxyurea were also added to lymphocytes (Fig. 5). Therefore, although poly(ADP-ribose) synthesis is induced by exposure to MMS and UV light, there are important qualitative and quantitative differences. ADP-ribosylation has a major role in regulating the steps of repair after alkylation damage but an insignificant role after UV damage. Peripheral lymphocytes appear to be a special case, because the use of high concentrations of hydroxyurea with these cells may create a role for poly(ADP-ribose) synthesis after UV damage that is irrelevant to the normal operation of excision repair. High concentrations of hydroxyurea generate excess single-strand breaks because of an inhibition of repair polymerization and ligation (*14, 18, 19–21*; Figs. 3, 5). These excess strand breaks will stimulate far more poly(ADP-ribose) polymerization than normally occurs during UV repair. In general, therefore, it seems that the base excision repair pathway, which involves specific glycosylases and apurinic endonucleases for removal of damaged (alkylated) sites, involves poly(ADP-ribosyl)ation as an important regulatory mechanism. The nucleotide excision repair pathway, however, which involves endonucleolytic removal of large DNA adducts, does not appear to involve ribosylation to a major extent, except under exceptional and perturbed conditions.

Three aspects of excision repair have been investigated in detail in various cell types: excision, breakage and reunion, and repair replication. 3-Aminobenzamide had no effect on the excision of N$^{3}$-methyladenine, N$^{7}$-methylguanine, or pyrimidine dimers from DNA, but did appear to influence the incorporation of labeled methyl groups from the carbon-1 pool into purines by *de novo* synthesis. This may be either a direct effect of 3-aminobenzamide on pool sizes and the *de novo* synthesis pathways, or an indirect consequence of inhibition of DNA synthesis. These alternatives are relevant to the stimulation of repair replication that will be discussed below.

The single-strand break frequency during repair represents a dynamic balance among incision, polymerization, and ligation (*40*), and the rate of repair of various kinds of lesions may be limited at various sides of this balance. During repair of UV damage, incision is rate limiting (*13, 14, 31, 40*). Very few breaks are detected because ligation is rapid, despite the high frequencies of pyrimidine dimers produced by low to moderate biological doses (*13, 14, 31, 40*). During repair of alkylation damage the excision of alkylated sites ($N^7$-methylguanine, $N^3$-methyladenine) is rapid and ligation becomes rate limiting with the resulting increases in break frequencies. One of the effects of ADP-ribosylation appears to be an increase in the efficiency of DNA ligase (*39*), and therefore a shift in the dynamic balance of incision and ligation in the direction of fewer breaks. This was observed clearly in our experiments in which addition or removal of 3-aminobenzamide for short periods resulted in rapid shifts in the break frequency as ligation was retarded or accelerated, respectively (Fig. 4).

We observed large stimulations of repair replication after MMS exposure in CHO, HeLa, and WIL-2 cells and in lymphocytes but not in human fibroblasts (Fig. 5), even though delayed sealing of strand breaks occurs in all cell types investigated (Fig. 3) (*22*). Repair replication is therefore regulated by poly(ADP-ribose) differently from ligation, and varies independently of ligation when 3-aminobenzamide is present. Possible causes for increases in repair replication are an increased number of repaired sites, increased patch size, or increased specific activity of the patch. An increased number of patches is unlikely because 3-aminobenzamide had no influence on the excision of any alkylated base. An increased patch size could be a factor because this might occur as a natural consequence of the delayed ligation. But then it is difficult to explain why delayed ligation had no effect on repair replication in primary fibroblasts. It remains a possibility that the specific activity of the repair patch could increase, and be dependent on the initial contribution of the *de novo* pool in competition with the exogenous $^3H$.

Cellular consequences of delaying ligation and increasing the net frequency of single-strand breaks by inhibiting poly(ADP-ribose) synthesis include increased cell killing (*6, 7*), SCE formation (*34–36*; Tables 2, 3), and mutagenesis (*10*; Table 4), but decreased transformation. It is interesting that an inhibitor that has no effect on the actual removal of damaged sites from DNA (*10*; Table 1), but only alters the rate at which the patches are completed, should have such dramatic effects. Clearly, the integrity of DNA during excision has an important bearing on these endpoints. Some effects, however, may be a consequence of the extensive ADP-ribosylation that occurs throughout the cell nucleus and cytoplasm on proteins unrelated directly to the repair of DNA or of other effects of 3-aminobenzamide on the carbon-1 pool. It remains unexplained why a cellular response to DNA breakage involves widespread alterations in NAD and poly(ADP-ribosyl)ation throughout the nucleus and cytoplasm when the actual regulation of ligation is such a quantitatively insignificant part of this ADP-ribosylation scheme. The stimulation of SCE formation and mutagenesis and inhibition of transformation may perhaps reflect other consequences of ADP-ribosylation unrelated to the efficiency of ligation of DNA strand breaks.

ACKNOWLEDGMENTS

We are grateful to S. Shall and A. Lehmann, University of Sussex, England, for providing the initial samples of 3-aminobenzamide and stimulating our investigations, to J. Clarkson and D. L. Mitchell for measuring pyrimidine dimers by a radioimmune assay method, and to M. Teng, W. C. Charles, and A. Ong for technical assistance.

This work was supported by the U.S. Department of Energy and by grant TC32-CA 09215-03 from the National Institutes of Health (W.J.B.) and grant CA 12536-11 from the National Cancer Institute (C.B.).

## REFERENCES

1. Benjamin, R. C. and Gill, D. M. ADP-ribosylation in mammalian cell ghosts. Dependence of poly(ADP-ribose) synthesis on strand breakage in DNA. J. Biol. Chem., *255*: 10493–10501, 1980.
2. Benjamin, R. C. and Gill, D. M. Poly(ADP-ribose) synthesis *in vitro* programmed by damaged DNA. A comparison of DNA molecules containing different types of strand breaks. J. Biol. Chem., *255*: 10502–10508, 1980.
3. Berger, N. A., Sikorski, G. W., Petzold, S. J., and Kurohara, K. K. Association of poly(adenosine diphosphoribose) synthesis with DNA damage and repair in normal human lymphocytes. J. Clin. Invest., *63*: 1164–1171, 1979.
4. Berger, N. A., Sikorski, G. W., Petzold, S. J., and Kurohara, K. K. Defective poly-(adenosine diphosphoribose) synthesis in xeroderma pigmentosum. Biochemistry, *19*: 289–293, 1980.
5. Hayaishi, O. and Ueda, K. Poly(ADP-ribose) and ADP-ribosylation of proteins. Annu. Rev. Biochem., *46*: 95–116, 1977.
6. Durkacz, B. W., Omidiji, O., Gray, D. A., and Shall, S. (ADP-ribose)$_n$ participates in DNA excision repair. Nature, *283*: 593–596, 1980.
7. James, M. R. and Lehmann, A. R. Role of poly(adenosine diphosphate ribose) in deoxyribonucleic acid repair in human fibroblasts. Biochemistry, *21*: 4007–4013, 1982.
8. Bohr, V. and Klenow, H. 3-Aminobenzamide stimulates unscheduled DNA synthesis and rejoining of strand breaks in human lymphocytes. Biochem. Biophys. Res. Commun., *102*: 1254–1261, 1981.
9. Durkacz, B. W., Irwin, J., and Shall, S. Inhibition of (ADP-ribose)$_n$ biosynthesis retards DNA repair but does not inhibit DNA repair synthesis. Biochem. Biophys. Res. Commun., *101*: 1433–1441, 1981.
10. Durrant, L. G., Margison, G. P., and Boyle, J. M. Effects of 5-methylnicotinamide on mouse L1210 cells exposed to N-methyl-N-nitrosourea: mutation induction, formation and removal of methylation products in DNA, and unscheduled DNA synthesis. Carcinogenesis, *2*: 1013–1017, 1981.
11. Miwa, M., Kanai, M., Kondo, T., Hoshino, H., Ishihara, K., and Sugimura, T. Inhibitors of poly(ADP-ribose) polymerase enhance unscheduled DNA synthesis in human peripheral lymphocytes. Biochem. Biophys. Res. Commun., *100*: 463–470, 1981.
12. Nduka, N., Skidmore, C. J., and Shall, S. The enhancement of cytotoxicity of N-methyl-N-nitrosourea and of $\gamma$-radiation by inhibitors of poly(ADP-ribose) polymerase. Eur. J. Biochem., *105*: 525–530, 1980.

13. Cleaver, J. E. Methods for studying repair of DNA damaged by physical and chemical carcinogens. *In;* H. Busch (ed.), Methods in Cancer Research, Vol. 11, pp. 123–165, Academic Press, New York, 1975.
14. Erixon, K. and Ahnstrom, G. Single-strand breaks in DNA during repair of UV-induced damage in normal human and xeroderma pigmentosum cells as determined by alkaline DNA unwinding and hydroxylapatite chromatography. Effects of hydroxyurea, 5-fluorodeoxyuridine and 1-β-D-arabinofuranosylcytosine on the kinetics of repair. Mutat. Res., *59*: 257–271, 1979.
15. Althaus, F. R., Lawrence, S. D., Sattler, G. L., and Pitot, H. C. ADP-ribosyltransferase activity in cultured hepatocytes. Interactions with DNA repair. J. Biol. Chem., *257*: 5528–5535, 1982.
16. Berger, N. A. and Sikorski, G. W. Nicotinamide stimulates repair of DNA damage in human lymphocytes. Biochem. Biophys. Res. Commun., *95*: 67–72, 1980.
17. Sims, J. L., Sikorski, G. W., Catino, D. M., Berger, S. J., and Berger, N. A. Poly-(adenosine diphosphoribose) polymerase inhibitors stimulate unscheduled deoxyribonucleic acid synthesis in normal human lymphocytes. Biochemistry, *21*: 1813–1821, 1982.
18. Ben-Hur, E. and Ben-Ishai, R. DNA repair in ultraviolet light irradiated HeLa cells and its reversible inhibition by hydroxyurea. Photochem. Photobiol., *13*: 337–345, 1971.
19. Collins, A.R.S., Schor, S. L., and Johnson, R. T. The inhibition of repair in UV irradiated human cells. Mutat. Res., *42*: 413–432, 1977.
20. Cornelis, J. J. The influence of inhibitors on dimer removal and repair of single-strand breaks in normal and bromodeoxyuridine substituted DNA of HeLa cells. Biochim. Biophys. Acta, *521*: 134–143, 1978.
21. Francis, A A., Blevins, R. D., Carrier, W. L., Smith, D. P., and Regan, J. D. Inhibition of DNA repair in ultraviolet-irradiated human cells by hydroxyurea. Biochim. Biophys. Acta, *563*: 385–392, 1979.
22. Cleaver, J. E., Bodell, W. J., Morgan, W. F., and Zelle, B. Differences in the regulation by poly(ADP-ribose) of repair of DNA damage from alkylating agents and ultraviolet light according to cell type. J. Biol. Chem., in press.
23. Skidmore, C. J., Davies, M. I., Goodwin, P. M., Halldorsson, H., Lewis, P. J., Shall, S., and Zia'ee, A. The involvement of poly(ADP-ribose) polymerase in the degradation of NAD caused by γ-radiation and N-methyl-N-nitrosourea. Eur. J. Biochem., *101*: 135–142, 1979.
24. Kohn, K. W., Ewig, R.A.G., Erickson, L. C., and Zwelling, L. A. Measurement of strand breaks and cross-links by alkaline elution. *In;* E. C. Friedberg and P. C. Hanawalt (eds.), DNA Repair: A Laboratory Manual of Research Procedures, Vol. 1B, pp. 379–401, Marcel Dekker, New York, 1981.
25. Bodell, W. J., Singer, B., Thomas, G. H., and Cleaver, J. E. Evidence for removal at different rates of *O*-ethyl pyrimidines and ethylphosphotriesters in two human fibroblast cell lines. Nucleic Acids Res., *6*: 2819–2829, 1979.
26. Mitchell, D. L., Nairn, R. S., Alvillar, J. A., and Clarkson, J. M. Loss of thymine dimers from mammalian cell DNA. The kinetics for antibody-binding sites are not the same as that for T4 endonuclease V sites. Biochim. Biophys. Acta, *697*: 270–277, 1982.
27. Wolff, S. Measurement of sister chromatid exchange in mammalian cells. *In;* E. C. Friedberg and P. C. Hanawalt (eds.), DNA Repair: A Laboratory Manual of Research Procedures, Vol. 1B, pp. 575–586, Marcel Dekker, New York, 1981.
28. Cleaver, J. E. Induction of thioguanine- and ouabain-resistant mutants and single-

strand breaks in the DNA of Chinese hamster ovary cells by $^3$H-thymidine. Genetics, *87*: 129–138, 1977.

29. Borek, C., Miller, R., Pain, C., and Troll, W. Conditions for inhibiting and enhancing effects of the protease inhibitor antipain on X-ray-induced neoplastic transformation in hamster and mouse cells. Proc. Natl. Acad. Sci. U.S., *76*: 1800–1803, 1979.
30. Whittle, E. D. Methylation of rat-liver RNA *in vivo* by methyl methanesulphonate. Biochim. Biophys. Acta, *195*: 381–388, 1969.
31. Cleaver, J. E. DNA repair and its coupling to DNA replication in eukaryotic cells. Biochim. Biophys. Acta, *516*: 489–516, 1978.
32. Charles, W. C. and Cleaver, J. E. Comparison of nucleoid and alkaline sucrose gradients in the analysis of inhibitors of DNA repair in human fibroblasts. Biochem. Biophys. Res. Commun., *107*: 250–257, 1982.
33. Snyder, R. D. and Regan, J. D. Aphidicolin inhibits repair of DNA in UV-irradiated human fibroblasts. Biochem. Biophys. Res. Commun., *99*: 1088–1094, 1981.
34. Morgan, W. F. and Cleaver, J. E. 3-Aminobenzamide synergistically increases sister-chromatid exchanges in cells exposed to methyl methanesulfonate but not to ultraviolet light. Mutat. Res., *104*: 361–366, 1982.
35. Oikawa, A., Tohda, H., Kanai, M., Miwa, M., and Sugimura, T. Inhibitors of poly-(adenosine diphosphate ribose) polymerase induce sister chromatid exchanges. Biochem. Biophys. Res. Commun., *97*: 1311–1316, 1980.
36. Hori, T.-A. High incidence of sister chromatid exchanges and chromatid interchanges in the conditions of lowered activity of poly(ADP-ribose) polymerase. Biochem. Biophys. Res. Commun., *102*: 38–45, 1981.
37. Sugimura, T., Miwa, M., Saito, H., Kanai, Y., Ikejima, M., Terada, M., Yamada, M., and Utakoji, T. Studies of nuclear ADP-ribosylation. Adv. Enzyme Regul., *18*: 195–220, 1980.
38. McCurry, L. S. and Jacobson, M. K. Unscheduled synthesis of DNA and poly(ADP-ribose) in human fibroblasts following DNA damage. J. Supramol. Struct. Cell. Biochem., *17*: 87–90, 1981.
39. Creissen, D. and Shall, S. Regulation of DNA ligase activity by poly(ADP-ribose). Nature, *296*: 271–272, 1982.
40. Cleaver, J. E., Thomas, G. H., Trosko, J. E., and Lett, J. T. Excision repair (dimer excision, strand breakage and repair replication) in primary cultures of eukaryotic (bovine) cells. Exp. Cell Res., *74*: 67–80, 1972.

[illegible]

ADP-RIBOSYLATION, DNA REPAIR AND CANCER, M. MIWA ET AL. (EDS.),
JAPAN SCI. SOC. PRESS, TOKYO/VNU SCIENCE PRESS, UTRECHT, PP. 209-218, 1983

# Roles of Poly(ADP-ribose) Synthesis in Repair and Replication in Normal Human, Cockayne Syndrome, and Xeroderma Pigmentosum Fibroblasts after UV Irradiation

Yoshisada FUJIWARA,*[1] Kaoru GOTO,*[1],*[3] Kazuo YAMAMOTO,*[1] and Masamitsu ICHIHASHI*[2]

*Department of Radiation Biophysics*[1] and Department of Dermatology,*[2] Kobe University School of Medicine, Kobe 650, Japan*

*Abstract:* Roles of poly(ADP-ribose) in repair and replication were studied in UV-irradiated normal, xeroderma pigmentosum (XP), and Cockayne syndrome (CS) fibroblasts. UV radiation reduced cellular NAD with concurrent synthesis of poly(ADP-ribose) in a dose-dependent manner in normal and CS cells, but not in XP cells. Enzymatic incision of DNA of UV-irradiated XP cells by T4-endonuclease V activated poly(ADP-ribose) polymerase. Methyl methanesulfonate (MMS) also reduced the cellular NAD in all the above strains. However, inhibitors of poly(ADP-ribose) synthesis did not affect UV-induced unscheduled DNA synthesis (UDS) and UV and MMS survivals. Thus, poly(ADP-ribose) synthesis may have a chromatin-stabilizing effect, but not the key role in excision repair of UV damage, unlike in the repair of dimethyl sulfate alkylation. CS cells were deficient in the NAD pool and in the recovery of post-UV DNA synthesis, which were rescued by an exogenous supply of NAD. Such normalization in CS cells was inhibited by excess nicotinamide as in normal cells, suggesting that replicon reinitiation may require specific poly(ADP-ribosyl)ation in UV-irradiated human cells.

Poly(ADP-ribose) is synthesized from cellular NAD by tightly chromatin-bound poly(ADP-ribose) polymerase, and ADP-ribosylation of nuclear proteins results in diverse biological functions (*1*, *2*). Poly(ADP-ribose) may possibly participate in some aspects of DNA replication, damage, and repair in mammalian cells (*1–20*). Various DNA-damaging agents (many alkylating agents, bleomycin, X($\gamma$)-rays, UV, and DNases) decrease the cellular NAD level *in vivo* and stimulate poly(ADP-ribose) synthesis to various extents in permeabilized cells following treatments (*3–7*). Poly(ADP-ribose) synthesis increases as a function of DNA breakage *in vitro* and *in vivo* (*8*, *9*). Specific inhibitors of poly(ADP-ribose) polymerase such as 3-aminobenzamide (3-ABA) and 5-methylnicotinamide prevent rejoining of single strand DNA breaks caused by dimethyl sulfate and kill synergistically L1210 lymphoma (*4*) and human fibroblasts (*10*). Nutritional depletion of cellular NAD by nicotinamide starvation

*[3] Present address: Osaka City Institute of Public Health and Environmental Science, Osaka 543, Japan.

also prevents strand-break rejoining (*4, 11*). Theophylline, a known inhibitor of poly-(ADP-ribose) polymerase, inhibits both NAD reduction and repair synthesis after treatments with various reactive carcinogens of non-PHA-stimulated human lymphocytes (*12*). However, there have been inconsistent results showing either stimulation or no effect of inhibitors of poly(ADP-ribose) polymerase on unscheduled DNA synthesis (UDS) caused by alkylating agents or UV. Stimulation occurs mostly in lymphocytes and lymphoma cells in the presence of excessive hydroxyurea after treatment (*9, 13–15*), while there is no effect in lymphocytes and human fibroblasts in the absence or presence of less than 2 mM hydroxyurea (*9, 10*), despite greater stimulation after a high concentration of alkylating agents (*9, 10, 13–15*). In this regard, poly(ADP-ribose) may have a role in regulating the final ligation step of repair of alkylation damage to L1210 (*16*) and human fibroblasts (*9, 10*).

Excision-defective xeroderma pigmentosum (XP) cells of various complementation groups sustain no or minimal *in vivo* NAD decrease and poly(ADP-ribose) synthesis, despite a normal level of intrinsic enzyme activity, in response to large doses of 254 nm UV (10–60 J/m$^2$) (*17–20*), although proficient XP variant cells exhibit both phenomena (*18, 19*). These results suggest that incision breaks are most responsible for the activation of poly(ADP-ribose) synthesis in UV-irradiated cells. However, it is still uncertain whether poly(ADP-ribose) is actively involved in excision repair within physiological dose range. We have studied the roles of poly(ADP-ribose) in the repair and replication in UV-irradiated normal, XP and Cockayne syndrome (CS) cells, since the latter two are UV-sensitive, but XP and CS cells show a clear contrast in the repair capacity.

*DNA Repair Characteristics in XP and CS Cells*

Briefly, the XP6KO group-A and XP2BI group-G (courtesy of Dr. D. G. Harnden) cells used are most sensitive to UV killing, since they completely lack the initial incision of pyrimidine dimers (*22*). On the other hand, 5-fold more UV-sensitive CS cells possess no apparently detectable defect in any repair step, as evidenced by the various repair-detecting methods (*20–22*). However, these XP and CS cells show normal resistance to and repair of alkylation and ionizing radiation lesions (*21*). Excision-proficient CS cells manifest the defective recovery of DNA synthesis after such low doses that allow recovery in normal cells; this is different from the defective recovery observed in defective XP cells (*20–22*).

*Repair and Poly(ADP-ribose) Synthesis in UV-irradiated Cells*

Measurement of cellular NAD level after UV irradiation provides a sensitive method of monitoring the conversion of NAD to poly(ADP-ribose) (*19*). A dose of 5 J/m$^2$ 254 nm UV caused a maximum lowering of the NAD level to approximately 70% of control after 2 hr, followed by the subsequent increment to the control level in 6 hr after irradiation of NHSF6 (normal) (Fig. 1 A) and CSH4MI (CS4MI's mother, heterozygous) cells (Fig. 1 B). CS4MI (homozygous) cells, that have a 45%-reduced NAD level as in the other CS strains compared to NHSF6, CSH4MI, XP6KO group-A and XP2BI group-G cells (*20*) (Fig. 1), also exhibited a similiar

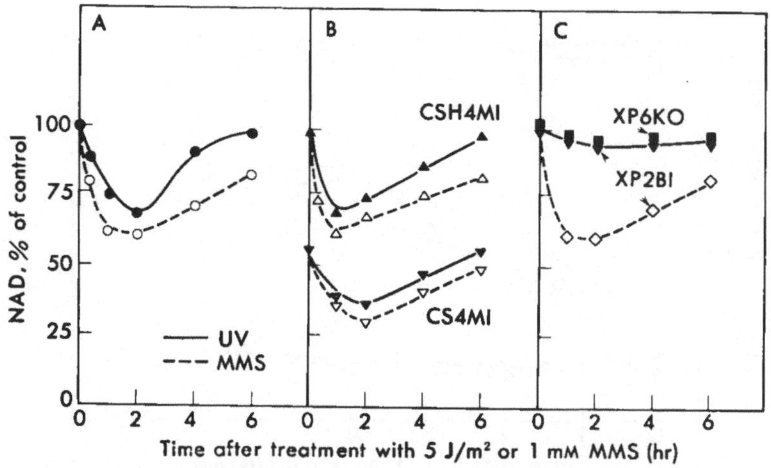


FIG. 1. The reduction and recovery kinetics of cellular NAD levels in response to 254 nm UV (5 J/m²) and MMS (1 mM). NAD content in cells in log-growth phase was determined by the enzyme cycling assay method (*3*). CSH4MI is heterozygous (CS4MI's mother). A, NHSF6; B, CS; C, XP.

response to the UV-induced decline / recovery curve (Fig. 1 B). Such a CS4MI response supports our previous notion that CS cells are not detectably deficient in intrinsic poly(ADP-ribose) polymerase activity (*20*). However, excision-defective XP6KO group-A and XP2BI group-G cells showed a very little NAD reduction in response to 5 J/m²-induced DNA damage (Fig. 1 C), in keeping with previous results (*17–19*). On the other hand, treatment with a low 1-mM concentration of methyl methanesulfonate (MMS), which induces single-strand breaks, effectively reduced the NAD level to a similar extent which then caused a slower recovery in all the strains used.

A greater stimulation of poly(ADP-ribose) synthesis by alkylating agents is a well-known phenomenon. Figure 2 shows the time-course changes in the enzyme activity after irradiation with 5 J/m² when assayed in the permeabilized cells (*8*, *20*). In normal (NHSF6, TIG-1) and CS (GM739, CS2MI, and CS3MI) cells, the relative increase in enzyme activity was rather low, within 2-fold overall (*20*): a little higher during the initial 2 hr and significantly higher in the later period of 2 to 6 hr. This apparent time-course change does not well reflect the NAD change in Fig. 1. Therefore, we assume that *in vivo* poly(ADP-ribosyl)ation due to rapid synthesis before assay in the permeabilieed cells may cocur to a significant extent during the early chromatin relaxation / DNA repair period (thus the low [$^3$H]NAD incorporation in Fig. 2), and poly(ADP-ribose) synthesized turns over rapidly to a normal level during the later chromatin reconstruction period (thus the higher [$^3$H]NAD incorporation in Fig. 2). The increase in the enzyme assay under our conditions is therefore apparent, while the NAD change *in vivo* is a faithful indicator of poly(ADP-ribose) synthesis *in vivo*, as pointed out by Jacobson *et al.* (*3*). It is also characteristic that repair-defective XP6KO and XP2BI cells display no or a little lower poly(ADP-ribose) polymerase activity (Fig. 2), as this correlated well with no apparent change in the cellular NAD level after irradiation (Fig. 1 C).

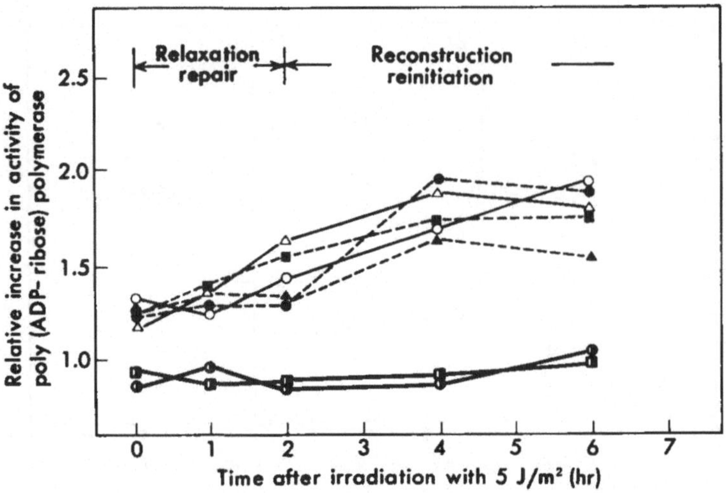


FIG. 2. Time-course changes in poly(ADP-ribose) polymerase activity after UV irradiation with 5 J/m², as assayed in the permeabilized cell system of Berger *et al.* (*6, 17*) as described previously (*20*). ○ NHSF6; ● CS2MI; △ TIG-1; ■ GM739; ▲ CS3MI; ◑ XP6KO; ◨ XP2BI.

TABLE 1. T4-endonuclease V-induced Poly(ADP-ribose) Synthesis

| Cell strain | T4-endo-nuclease V, (0.1 unit/assay) | Poly(ADP-ribose) polymerase activity, (cpm/mg protein/30 min[a]) | | Relative increase |
|---|---|---|---|---|
| | | 0 J/m² | 10 J/m² | |
| NHSF6 (normal) | − | 2,896 | 3,128 | 1.1 |
| | + | 2,938 | 8,210 | 2.8 |
| XP6KO group-A | − | 2,923 | 3,040 | 1.0 |
| | + | 2,822 | 7,361 | 2.6 |

[a] Mean of triplicate assays. Standard error fell within 8% of each mean value.

XP6KO group-A cells were used as a model strain to see whether enzymatic incision breaks activate poly(ADP-ribose) synthesis. Pyrimidine-dimer specific T4-endonuclease V (0.1 unit/assay, courtesy of Dr. M. Sekiguchi) was applied to the permeabilized cells during a 30-min assay incubation. Table 1 shows that immediately after 10 J/m², the absence of T4-endonuclease caused only a small or no increase in poly(ADP-ribose) synthesis in NHSF6 (normal) and XP6KO cells, respectively, while its presence stimulated such synthesis 2.6–2.8-fold in both strains.

The effect of inhibitors of poly(ADP-ribose) polymerase on UDS was examined by incubating repair-proficient NHSF6 and CS20S cells with [$^3$H]thymidine (5 $\mu$Ci/ml, 5 Ci/mmole)-plus-2 mM hydroxyurea in the presence of sufficient 3-ABA, 5-methylnicotinamide or theophylline for 4 hr after 10 J/m². Table 2 shows the results demonstrating that neither of the inhibitors suppressed or enhanced repair incorporation, being accompanied with no apparently altered NAD pool as a consequence of the enzyme inhibition. In control "none", the NAD level was once minimally decreased at 2 hr (see Fig. 4 A) and remained constant at about 40% at 4 hr during repair incubation (Table 2, Fig. 4 A).

TABLE 2. Effects of Inhibitors of Poly(ADP-ribose) Polymerase on Unscheduled DNA Synthesis after 10 J/m² 254 nm UV

| Cells and treatment | Unscheduled DNA synthesis per culture | | NAD+ |
|---|---|---|---|
| | cpm[a] | % | % of unirrad. |
| NHSF6 (normal fibroblasts) | | | |
| None | 2,810±250 | 100 | 42 |
| 3-ABA, 5 mM | 2,630±317 | 94 | 98 |
| 5-Methylnicotinamide, 5 mM | 2,892±193 | 103 | 97 |
| Theophilline, 5 mM | 2,610± 81 | 93 | 98 |
| CS20S (Cockayne fibroblasts) | | | |
| None | 2,710±203 | 100 | nd[b] |
| 3-ABA, 5 mM | 2,760±310 | 102 | nd |
| 5-Methylnicotinamide, 5 mM | 2,671±276 | 99 | nd |
| Theophylline, 5 mM | 2,873±117 | 106 | nd |

[a] Incorporation into acid-insoluble DNA of 5 μCi/ml of [³H]thymidine (5 Ci/mmol) in the presence of 2 mM hydroxyurea for 4 hr after irradiation with 10 J/m², corrected for the respective unirradiated control. [b] nd, not detected.

TABLE 3. Effects of Post-treatment with 3-ABA on UV and MMS survivals

| Treatment | Survival parameters[a] | Normal | | Cockayne | |
|---|---|---|---|---|---|
| | | NHSF6 | TIG-1 | CS2OS(A) | GM1428(B) |
| UV | n | 1.5 | 2.0 | 1.0 | 1.0 |
| | Do, J/m² | 5.0 | 4.5 | 1.0 | 1.0 |
| UV+5 mM 3-ABA | n | 1.5 | 2.0 | 1.0 | 1.0 |
| | Do | 5.0 | 4.5 | 0.9 | 1.0 |
| MMS | n | 1.2 | 1.2 | 1.2 | 1.2 |
| | Do, μM·hr[b] | 65 | 65 | 65 | 65 |
| MMS+5 mM 3-ABA | n | 1.2 | 1.2 | 1.2 | 1.2 |
| | Do μM·hr[b] | 65 | 65 | 65 | 65 |

[a] Extrapolation number (n) is obtained at the intercept when the exponential part of the survival curve is extrapolated to the ordinate, and mean lethal dose (Do) is the dose which reduces survival in the exponential region of the survival curve to 1/*e* (=37%). [b] Treatment with MMS for 1 hr at 37°C in phosphate-buffered saline.

The above findings clearly indicate that incision breaks during the early repair phase reduce the cellular NAD level to convert NAD to poly(ADP-ribose) in repair-proficient cells. However, the subsequent repair synthesis may not require extensive ADP-ribosylation as evidenced by the inhibitor experiment, since excision repair is rapid and coordinated. It should be kept in mind that we have little knowledge of the *in vivo* fine ADP-ribosylation for rearrangement of nucleosomes during excision repair of UV damage.

Table 3 shows that the continuous post-treatment with 5 mM 3-ABA produced no synergistic killing effect on UV and MMS survivals of normal human (NHSF6, TIG-1) and Cockayne (CS2OS group-A), GM1428 group-B) fibroblasts. Further, CS cells with a lower level of NAD can perform normal rejoining of X-ray breaks and survive MMS and X-rays in the normal fashion (*20–22*). Therefore, we suggest that in human fibroblasts, the inhibitors of poly(ADP-ribose) polymerase do not seriously

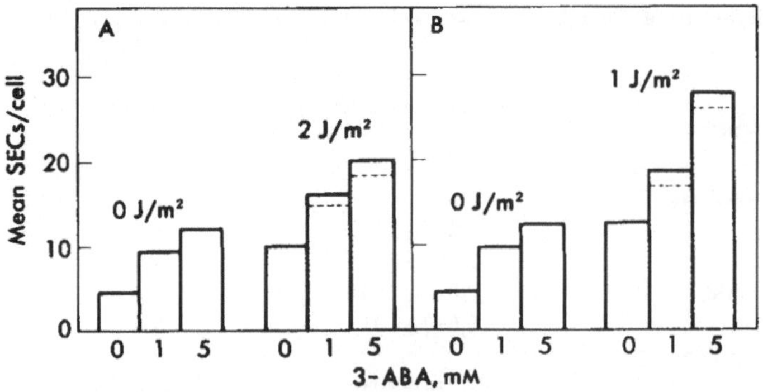


FIG. 3. Inductions of SCEs by 3-ABA with or without UV irradiation. NHSF6 (A) and CS2OS (B) cells unirradiated or irradiated 2 and 1 J/m², respectively, and incubated for 72 hr in the bromodeoxyuridine-deoxycytidine medium (*20*) in the absence or presence of 3-ABA. Each frequency is the mean of 50 cells scored. --- the expected synergistic effect.

affect the repair of UV and MMS lesions within survival dose range. This is a clear distinction from the inhibition observed in L1210 lymphoma cells treated with dimethyl sulfate (*4*). However, a high supralethal dose of MMS (>1 mM) causes extensive DNA breaks, whose repair in human fibroblasts is delayed or inhibited by the poly(ADP-ribose) inhibitors (*9*, *10*). In this regard, 3-ABA increases synergistically the sister chromatid exchange (SCE) frequency induced by MMS (*24*). Figure 3 displays that 1–5 mM 3-ABA exerts no synergistic effect on UV (1–2 J/m²)-induced SCE formation in either NHSF6 or CS2OS cells, although 3–ABA itself probably raised the SCE frequency by perturbing replicative DNA synthesis (see below), in accord with previous results with 3-ABA (*24*) and nicotinamide (*25*). This phenomenon further supports a view of inactive involvement of poly(ADP-ribose) in the repair of UV-induced pyrimidine dimers.

### *DNA Synthesis, NAD Level, and Poly(ADP-ribose) Synthesis after UV Irradiation of Human Fibroblasts*

Figure 4 shows the relationship between the decline / recovery curves of cellular NAD level and DNA-synthetic rate in NHSF6 and GM739 CS cells. In Fig. 4 A, the *in vivo* NAD level in NHSF6 cells decreased abruptly to dose-dependent minimums around 2 hr, and subsequently increased to reach control level at 6 hr after 5 J/m², but not substantially after 7 and 10 J/m². Such dose and time-dependent NAD patterns correlate surprisingly well with those of the DNA synthetic rate in irradiated NHSF6 cells (Fig. 4 B). The continued NAD depletion and lack of substantial recovery of DNA synthesis in normal cells given higher UV doses of more than Do (see Table 3) implies that alternately occurring incision breaks in replication domains and concurrent extensive poly(ADP-ribose) synthesis interrupt with repair-coupled reinitiation of replicon synthesis over a long period of time.

On the other hand, GM739 CS cells also exhibited a 50% lower NAD level (Fig. 4 C), which a UV dose of 5 J/m² reduced further at 2 hr, although subsequent recovery occurred to the initial low level in that strain (Fig. 4 C). However, the rate of DNA synthesis in the CS cells after 5 J/m² absolutely failed to recover (Fig. 4 D),

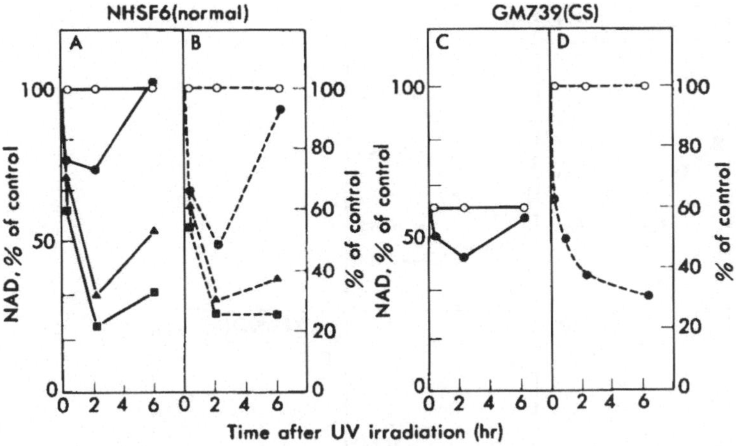


FIG. 4. Relation between the decline/recovery curves of cellular NAD (A, C) and rate of DNA synthesis (B, D) after UV irradiation. Log-growth phase cells were irradiated with 0 (○—○, ○---○), 5 (● ●, ●---●), 7 (▲—▲, ▲---▲), and 10 J/m² (■ ■, ■---■) and incubated. Rate of DNA synthesis was determined by pulse-labelling with [³H]thymidine (5 μCi/ml, 5 Ci/mmol) for 30 min at indicated times, and expressed as % of control. NAD was determined as described in Fig. 1.

as described previously (*20, 21, 23*). Thus, CS cells cannot resume the repair-coupled reinitiation of replicon synthesis due specifically to a low NAD level. In addition, Fig. 4 infers that an approximately 40% reduced NAD level from normal may be a critical level with respect to reinitiation of replicon synthesis once suppressed by UV damage in repair-proficent cells.

Nicotinamide (Nam) is a precursor of NAD and a potent inhibitor of poly(ADP-ribose) polymerase (*4–7*). Figure 5 illustrates the effects of 1 mM NAD and/or 10 mM Nam on the rate of replicative synthesis in unirradiated and 5 J/m²-irradiated NHSF6, GM739 CS, and XP6KO group-A cells. Nam inhibited progressively, but partially, the DNA synthesis until 6 hr when tested in unirradiated NHSF6 and GM739 cells (Fig. 5 A and C). Such an inhibition saturated at 40–50%. Thus, gross replicative DNA synthesis seems to consist of poly(ADP-ribose)-sensitive and resistant fractions. In the NHSF6 cells receiving 5 J/m² at 0 hr, an exogenous supply of 1 mM NAD caused slight enhancement of recovery of DNA synthesis which was inhibited by the copresence of 10 mM Nam (Fig. 5 B). In GM739 cells, the NAD addition rescued an otherwise defective recovery of DNA synthesis (Fig. 5 D) due to elevation by incorporated NAD (*20*). Such a normalizing effect of NAD was completely abolished by the copresence of Nam (Fig. 5 D). Therefore, the reinitiation of replicon synthesis once suppressed by UV damage is a sensitive step to the inhibition of poly(ADP-ribose) synthesis. According to Thraves and Smulson (*26*), poly(ADP-ribose) polymerase and ADP-ribosylated chromatin proteins are enriched in the regions of DNA synthesis during S phase, although the total enzyme activity is low in this phase (*6*). This polymer seems to function as a regulator. Further, we observed the similar results with 10 mM 3-ABA, but 1–5 mM concentrations were less effective in inhibiting replicative DNA synthesis (unpublished data). In this regard, 2 mM 3-ABA has no effect on nucleoid sedimentation of relaxation and supercoiling due to inci-

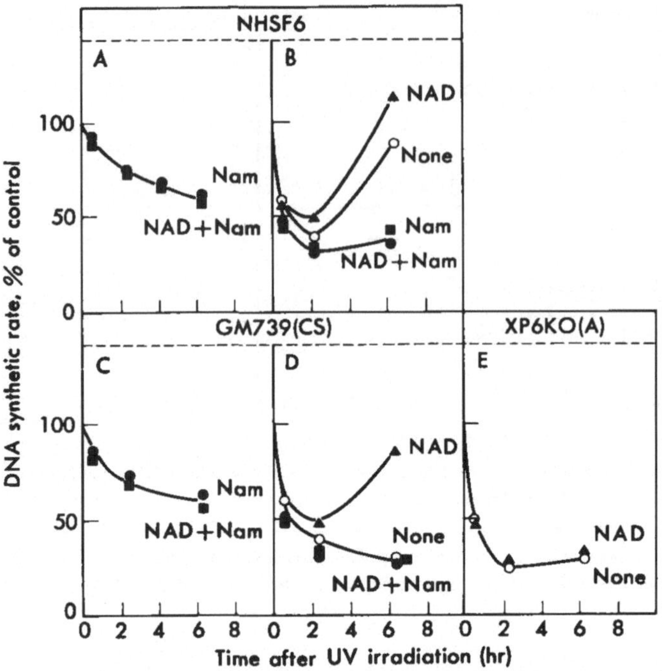


FIG. 5. Effects of NAD and/or Nam on the rate of DNA synthesis in unirradiated and UV-irradiated cells. NHSF6, GM739 CS, and XP6KO group-A cells were unirradiated or irradiated with 5 J/m², and incubated in the absence or presence of 1 mM NAD and/or 10 mM Nam, followed by pulse-labelling for 30 min at indicated times. ○ none; ▲ NAD; ● Nam; ■ NAD+Nam. A, 0J/m²; B, 5 J/m²; C, 0 J/m²; D, 5 J/m²; E, 5 J/m².

sion breaks and repair, respectively, in UV-irradiated normal human fibroblasts (*27*), and CS cells completely reconstruct the chromatin after UV irradiation (*28*). Therefore, further studies remain to clarify the specific mechanism by which poly(ADP-ribose) may trigger DNA synthesis in irradiated CS cells.

XP6KO group-A cells also failed to recover post-UV DNA synthesis after a lethal dose as high as 5 J/m² (Fig. 5 E). This is quite different from the complete block to replication by lesions remaining absolutely unexcised. In the XP cells, NAD exerted no effect on post-UV DNA synthesis (Fig. 5 E).

ACKNOWLEDGMENTS

This work was supported in part by a Grant-in-Aid for Cancer Research from the Ministry of Education, Science and Culture, Japan. We thank Dr. M. Ikenaga for supply of CS cells and A. Katanoda for her skillful assistance.

REFERENCES

1. Hayaishi, O. and Ueda, K. Poly(ADP-ribose) and ADP-ribosylation of proteins. Annu. Rev. Biochem., *46*: 95–116, 1977.
2. Sugimura, T., Miwa, M., Saito, H., Kanai, Y., Ikejima, M., Terada, M., Yasuda, M., and Utakoji, T. Studies of nuclear ADP-ribosylation. Adv. Enzyme Regul., *18*: 195–220, 1980.

3. Jacobson, M. K., Sims J. L., Juarez-Salinas, H., Levi, U., Barton, R. A., and Jacobson E. L. Mechanism of carcinogen-induced alteration of NAD metabolism. *In;* M. E. Smulson and T. Sugimura (eds.), Novel ADP-ribosylation of Regulatory Enzyme and Proteins, pp. 239–248, Elsevier/North-Holland Publ. Co., Amsterdam, 1980.
4. Durkacz, B., Omidiji, O., Gray, D. A., and Shall, S. (ADP-ribose)$_n$ participates in DNA excision repair. Nature, *283*: 593–596, 1980.
5. Skidmore, C. J., Davies, M. I., Goodwin, P. M., Omidiji, O., Zia'ee, A., and Shall, S. Poly(ADP-ribose) polymerase—an enzyme sensitive to DNA damage. *In*; M. E. Smulson and T. Sugimura (eds.), Novel ADP-ribosylation of Regulatory Enzyme and Proteins, pp. 197–204, Elsevier / North-Holland Publ. Co., Amsterdam, 1980.
6. Berger, N. A., Sikorski, G. W., and Petzold, S. J. Association of poly(ADP-ribose) synthesis with DNA strand breaks in replication and repair *In;* M. E. Smulson and T. Sugimura (eds.), Novel ADP-ribosylation of Regulatory Enzymes and Proteins, pp. 185–195, Elsevier / North-Holland Publ. Co., Amsterdam, 1980.
7. Nduka, N., Skidmore, C. J., and Shall, S. The enhancement of cytotoxcity of N-methyl-N-nitorosourea and of $\gamma$-radiation by inhibitors of poly(ADP-ribose) polymerase. Eur. J. Biochem., *105*: 525–530, 1980.
8. Benjamin, R. C. and Gill, D. M. Poly(ADP-ribose) synthesis *in vitro* programmed by damaged DNA. J. Biol. Chem., *255*: 10502–10508, 1980.
9. Cleaver, J. E., Bodell, W. J., Morgan, W. F., and Zelle, B. Difference in the regulation by poly(ADP-ribose) of repair of DNA damage from alkylating agents and ultraviolet light according to cell type. J. Biol. Chem., in press.
10. James, M. R. and Lehmann, A. R., The role of (ADP-ribose)$_n$ in DNA repair in human fibroblasts. Prog. Mutat. Res., *4*: 177–182, 1982.
11. Jacobson, E. L., Lange, R. A., and Jacobson, M. K. Pyridine nucleotide synthesis in 3T3 cells. J. Cell. Physiol., *99*: 417–426, 1979.
12. Bussee, D. L., Rankin, P., Mitschell, V. R., Cantrell, E. T., and Jacobson, M. K. Correlation of carcinogen-induced unscheduled DNA synthesis and NAD reduction in fresh human lymphocytes. Cancer Lett., *11*: 153–160, 1980.
13. Berger, N. A. and Sikorski, G. W. Nicotinamide stimulates repair of DNA damage in human lymphocytes. Biochem. Biophys. Res. Commun., *95*: 67–72, 1980.
14. Miwa, M., Kanai, M., Kondo, T., Hoshino, H., Ishihara, K., and Sugimura, T. Inhibitors of poly(ADP-ribose) polymerase enhance unscheduled DNA synthesis in human peripheral lymphocytes. Biochem. Biophys. Res. Commun., *100*: 463–470, 1981.
15. Bohr, V. and Klenow, H. 3-Aminobenzamide stimulates unscheduled DNA synthesis and rejoining of strand breaks in human lymphocytes. Biochem. Biophys. Res. Commun., *102*: 1254–1261, 1981.
16. Creissen, D. and Shall, S. Regulation of DNA ligase activity by poly(ADP-ribose). Nature, *296*: 271–272, 1982.
17. Berger, N. A., Sikorski, G. W., Petzold, S. J., and Kurohara, K. K. Defective poly-(adenosine diphosphoribose) synthesis in xeroderma pigmentosum. Biochemistry, *19*: 289–293, 1980.
18. Berger, N. A. and Sikorski, G. W. Poly(adenosine diphosphoribose) synthes's in ultraviolet-irradiated xeroderma pigmentosum cells reconstituted with *Micrococcus lueteus* UV-endonuclease. Biochemistry, *20*: 3610–3614, 1981.
19. McMurry, L. S. and Jacobson, M. K. Poly(ADP-ribose) synthesis following DNA damage in cells heterozygous and homozygous for the xeroderma pigmentosum genotype. J. Biol. Chem., *256*: 551–553, 1981.

20. Fujiwara, Y., Goto. K., and Kano, Y. Ultraviolet hypersensitivity of Cockayne's syndrome fibroblasts, effects of nicotinamide adenine dinucleotide and poly(ADP-ribose) synthesis. Exp. Cell Res., *139*: 207–215, 1982.
21. Fujiwara, Y., Ichihashi, M., Kano, Y., Goto, K., and Shimizu, K. A new human photosensitive subject with a defect in the recovery of DNA synthesis after ultraviolet irradiation. J. Invest. Dermatol., *77*: 256–263, 1981.
22. Fujiwara, Y., Kano, Y., Paul, P., Goto, K., Yamamoto, K., and Miyazaki, N. Excision and crosslink repair of DNA and sister chromatid exchanges in cultured human fibroblasts with different repair capacities. Gann Monogr. Cancer Res., *27*: 33–44, 1981.
23. Lehmann, A. R., Kirk-Bell, S., and Mayne, L. Abnormal kinetics of DNA synthesis in ultraviolet-irradiated cells from patients with Cockayne syndrome. Cancer Res., *39*: 4237–4241, 1979.
24. Morgan, W. F. and Cleaver, J. E. 3-Aminobenzamide synergistically increases sister chromatid exchanges in cells exposed to methyl methanesulfonate but not to ultraviolet light. Mutat. Res., *104*: 361–366, 1982.
25. Utakoji, T., Hosoda, K., Umezawa, K., Sawamura, M., Matsushima, T., Miwa, M., and Sugimura, T. Induction of sister chromatid exchanges by nicotinamide in Chinese hamster lung fibroblasts and human lymphoblastoid cells. Biochem. Biophys. Res. Commun., *90*: 1147–1152, 1979.
26. Thraves, P. and Smulson, M. Poly(ADP-ribose) and DNA repair studies at the polynucleosomal level of chromatin. Prog. Mutat. Res., *4*: 147–166, 1982.
27. Charles, W. C. and Cleaver, J. E. Comparison of nucleoid and alkaline sucrose gradients in the analysis of inhibitors of DNA repair in human fibroblasts. Biochem. Biophys. Res. Commun., *107*: 250–257, 1982.
28. Cleaver, J. E. Normal reconstruction of DNA supercoiling and chromatin structure in Cockayne syndrome cells during repair of damage from ultraviolet light. Am. J. Hum. Genet., *34*: 566-575, 1982.

ADP-RIBOSYLATION, DNA REPAIR AND CANCER, M. MIWA ET AL. (EDS.),
JAPAN SCI. SOC. PRESS. TOKYO/VNU SCIENCE PRESS, UTRECHT, PP. 219-226, 1983

# Poly(ADP-ribose) Polymerase Mediates the Suicide Response to Massive DNA Damage: Studies in Normal and DNA-repair Defective Cells

N. A. BERGER, J. L. SIMS, D. M. CATINO, and S. J. BERGER

*Hematology/Oncology Division of the Department of Medicine and the Department of Pharmacology, Washington University School of Medicine and The Jewish Hospital of St. Louis, St. Louis, Missouri 63110, U.S.A.*

*Abstract:* Treatment of cells with DNA damaging agents results in a dose dependent decrease in $NAD^+$ and ATP pool sizes. The decrease in $NAD^+$ is associated with the activation of poly(ADP-ribose) polymerase and the decrease in ATP is consequent to the fall in $NAD^+$. Depletion of both $NAD^+$ and ATP can be blocked or retarded by inhibitors of poly(ADP-ribose) polymerase. Both the stimulation of poly(ADP-ribose) synthesis and the effect of enzyme inhibitors have been confirmed in intact cells by using enzymatic cycling techniques to measure the disappearance of $NAD^+$ and high pressure liquid chromatography (HPLC) to measure fluctuations in polymer levels. As a consequence of the depletion of $NAD^+$ and ATP pools, cells exhibit a marked impairment in their ability to conduct all energy dependent functions. Thus cells treated with high levels of DNA damaging agents exhibit severe suppression of DNA replication and repair, RNA synthesis and protein synthesis. The use of inhibitors of poly-(ADP-ribose) polymerase to prevent the depletion of $NAD^+$ and ATP partially restores the cells' ability to conduct DNA, RNA, and protein synthesis. This preservation of the $NAD^+$ and ATP pools accounts for the recent observations that inhibitors of poly(ADP-ribose) stimulate the level of DNA repair synthesis in cells treated with high levels of DNA damaging agents. We have also examined cells from patients with several of the disorders of DNA repair and have found that cells from patients with Fanconi's anemia have lower than normal $NAD^+$ levels. When cells from these patients are treated with DNA damaging agents their $NAD^+$ pools are depleted to levels that are lower than those which occur in cells from normal donors. An impaired ability to maintain nucleotide pools and energy dependent functions may contribute to the decreased survival that Fanconi's anemia cells exhibit following DNA damage. Thus in eukaryotes, high levels of DNA damage activate poly(ADP-ribose) polymerase to a degree that depletes cellular $NAD^+$ levels and subsequently depletes ATP levels causing a decrease in energy dependent functions which can consequently lead to cell death before DNA repair can be accomplished. Such a suicide mechanism may be of benefit to an organism where it would be preferrable for cells with severely damaged DNA to die rather than risk repair with a high level of infidelity.

When cells are treated with sublethal doses of DNA-damaging agents, they alter their metabolic and synthetic processes to repair the damage and restore cellular functions. When cells are treated with massive doses of DNA damaging agents, they undergo rapid reductions in glycolysis (*1*), and macromolecular synthesis followed by cell death (*2*). Rapid cell death occurs before the genetic effects of DNA damage become manifest and before DNA repair can be completed. This process may be beneficial to multicellular organisms, since it might be preferable for cells with severely damaged DNA to die quickly rather than risk repair with a high level of infidelity. Our studies indicate that the activation of poly(ADP-ribose) polymerase by DNA damage and the consequent depletion of cellular $NAD^+$ levels may be the primary means of effecting this suicide response to massive DNA damage (*2*).

*Relation of the Suicide Response to Cellular $NAD^+$ Levels*

Figure 1 illustrates the failure of the repair pathway in L1210 cells treated with high doses of DNA damaging agents. The upper curve shows that the level of replicative DNA synthesis decreased as the cells were treated with increasing doses of N-methyl-N'-nitro-N-nitrosoguanidine (MNNG). The lower curve shows that, in the presence of 10 mM hydroxyurea, doses of MNNG as high as 2 μg/ml stimulated the level of DNA repair synthesis, whereas higher doses depressed it. If repair synthesis stopped increasing after 2 μg/ml MNNG because the system was saturable, then the level of repair synthesis should have remained constant in response to the higher

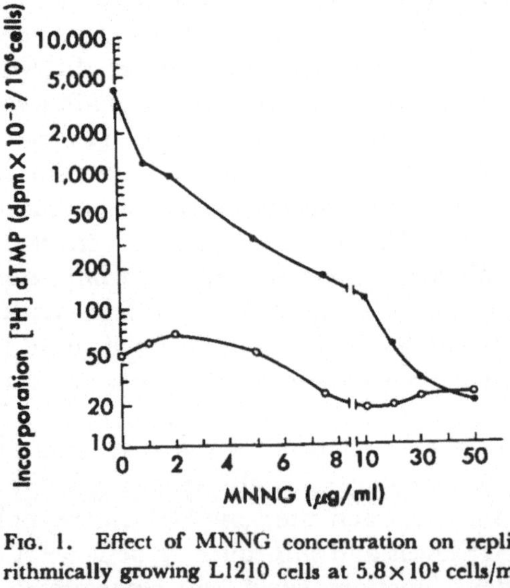


FIG. 1. Effect of MNNG concentration on replicative and repair synthesis of DNA. Logarithmically growing L1210 cells at $5.8 \times 10^5$ cells/ml were treated with the indicated concentrations of MNNG, then portions of each treatment group were adjusted to contain 10 mM hydroxyurea. After 15 min, [$^3$H]thymidine was added to each group of cells at a final concentration of 0.5 μCi/ml and incubations were continued for 6 hr at 37°C. [$^3$H]thymidine incorporation into trichloroacetic acid (TCA) precipitable material was determined for the 6 hr incubation period. ● replicative DNA synthesis measured in the absence of hydroxyurea; ○ repair DNA synthesis measured in the presence of 10 mM hydroxyurea.

TABLE 1. Effect of Nicotinamide on DNA Synthesis in L1210 Cells

| Treatment | Incorporation [$^3$H]dTMP (dpm/$10^6$ cells/6 hr) | |
|---|---|---|
| | −HU | +HU |
| Control (undamaged) | 1,450,000 | 36,000 |
| Control+nicotinamide | 1,380,000 | 38,000 |
| 20 μg/ml MNNG | 65,000 | 19,000 |
| 20 μg/ml MNNG+nicotinamide | 175,000 | 93,000 |

Logarithmically growing L1210 cells were pretreated for 15 min with 10 mM hydroxyurea (+HU) or with an equivalent amount of saline (−HU). At time zero, the indicated cells were treated with 20 μg/ml MNNG. The indicated cultures were then adjusted to contain 2 mM nicotinamide. [$^3$H]thymidine was added to label DNA. After a 6 hr incubation at 37°C, the incorporation of [$^3$H]dTMP into TCA precipitates was determined as described (*2*).

doses of MNNG. The results thus suggest that something occurs during the response to massive DNA damage that depresses DNA repair synthesis.

It has previously been shown that poly(ADP-ribose) polymerase is activated by the DNA strand breaks that occur in response to DNA damage (*3–5*). At sublethal levels of DNA damage, activation of this enzyme leads to synthesis of poly(ADP-ribose) which presumably modifies chromatin structure to facilitate the DNA repair process (*6, 7*). Under these conditions, $NAD^+$ deficiency or poly(ADP-ribose) polymerase inhibitors interfere with the DNA repair process and potentiate the cytocidal effects of the DNA damage (*8*).

Poly(ADP-ribose) polymerase is also activated in response to massive DNA damage and enzyme inhibitors retard the rates of polymer synthesis and $NAD^+$ lowering (*9*). Table 1 shows that poly(ADP-ribose) polymerase inhibitors also retard the damage induced decrease in DNA synthesis. Thus, treatment of L1210 cells with MNNG reduced the level of [$^3$H]dTMP incorporation from 1.45 million dpm/$10^6$ cells/6 hr to 65,000 dpm/$10^6$ cells/6 hr. When control cells were treated with 2 mM nicotinamide there was essentially no effect on the level of DNA synthesis. In contrast, when cells treated with 20 μg/ml MNNG were incubated with 2 mM nicotinamide, the level of DNA synthesis was stimulated or, alternatively, the drastic fall in DNA synthesis was partially prevented. When the level of DNA repair synthesis was assessed in the same cells by incubation in the presence of 10 mM hydroxyurea, it was apparent that MNNG suppressed the level of DNA repair synthesis but nicotinamide stimulated it.

The effect of nicotinamide in cells subjected to massive DNA damage shows that inhibitors of poly(ADP-ribose) polymerase can partially prevent the suppression of DNA repair synthesis that occurs as part of the suicide response. To define further the suicide pathway, we evaluated the effect of increasing doses of MNNG on $NAD^+$ and ATP levels and then determined the effect of poly(ADP-ribose) polymerase inhibitors on these levels. Fingure 2 shows that increasing doses of MNNG resulted in a dose dependent depletion of cellular $NAD^+$ levels and this was accompanied by a dose dependent depletion of ATP. Thus, within 6 hr after treatment with 2 μg/ml MNNG, $NAD^+$ was decreased to 36% of starting levels and ATP was decreased to 65% of its initial level. Within 6 hr after treatment with 20 μg/ml MNNG, $NAD^+$ was drastically depleted to less than 0.01% of its starting level and

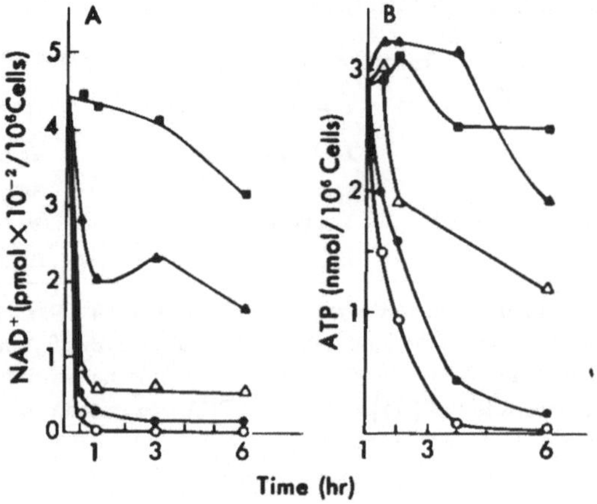


FIG. 2. Effect of MNNG concentration on rates of NAD+ (A) and ATP (B) lowering. Logarithmically growing L1210 cells were treated with the indicated concentration of MNNG. At the indicated times, duplicate samples were removed from each flask and NAD+ and ATP were extracted and measured by enzymatic cycling methods. Final concentrations of MNNG were 0 (■), 2 μg/ml (▲), 5 μg/ml (△), 10 μg/ml (●), and 20 μg/ml (○).

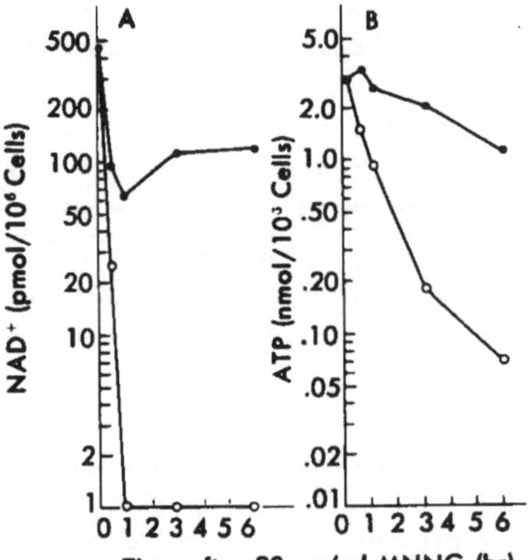


FIG. 3. Effect of Nicotinamide on MNNG-induced NAD+ (A) and ATP (B) depletion. Logarithmically growing L1210 cells were treated with 20 μg/ml MNNG and then a portion of the culture was adjusted to contain 2 mM nicotinamide (+Nam). The cells were then incubated at 37° C for 6 hr. Samples were removed at the indicated times for determinations of NAD+ and ATP levels. ○ cells treated with MNNG alone; ● cells treated with MNNG and then adjusted to contain 2 mM nicotinamide.

ATP was depleted to 2.4% of its initial level. Figure 3 shows that depletion of both NAD+ and ATP can be lessened by blocking poly(ADP-ribose) polymerase with 2 mM nicotinamide: when cells were treated with 20 μg/ml MNNG and then incubated in the presence of 2 mM nicotinamide, NAD+ levels remained at 26% and ATP remained at 38% of their pretreatment levels. Similar results have been obtained with benzamide.

TABLE 2. Relation of $NAD^+$ and ATP Pool Sizes to DNA Repair Synthesis

| MNNG ($\mu$g/ml) | Nicotinamide (2 mM) | $NAD^+$ (pmol/$10^6$ cells) | ATP (pmol/$10^6$ cells) | Unscheduled DNA Synthesis (dpm[$^3$H]dTMP/ $10^6$ cells) |
|---|---|---|---|---|
| 0 | − | 312 | 2,500 | 62,000 |
| | + | 524 | 2,000 | 64,000 |
| 2 | − | 162 | 1,900 | 64,000 |
| | + | 492 | 2,300 | 70,000 |
| 5 | − | 57 | 1,200 | 70,000 |
| | + | 335 | 2,000 | 70,000 |
| 10 | − | 6.5 | 134 | 40,000 |
| | + | 134 | 1,600 | 104,000 |
| 20 | − | 0.04 | 70 | 31,000 |
| | + | 118 | 1,100 | 91,000 |

Logarithmically growing L1210 cells were pretreated for 15 min with 10 mM hydroxyurea. At time zero, the indicated concentrations of MNNG were added and where indicated (+), the cultures were adjusted to contain 2 mM nicotinamide and then [$^3$H]dThd was added to label DNA. After 6 hr incubation at 37°C, samples were removed for measurements of $NAD^+$ and ATP by enzymatic cycling methods (*2*). The pretreatment levels of these nucleotides were 449 pmol $NAD^+/10^6$ cells and 2,900 pmol ATP/$10^6$ cells. Unscheduled DNA synthesis was determined as the amount of [$^3$H]dTMP incorporated into TCA precipitable material during the 6 hr incubation in the presence of 10 mM hydroxyurea (*2*).

TABLE 3. Effect of Poly(ADP-ribose) Polymerase Inhibitors on Macromolecular Synthesis

| Treatment | [$^3$H]dTMP | [$^3$H]UMP (dpm/$10^6$ cells/6 hr) | [$^3$H]leu |
|---|---|---|---|
| 20 $\mu$g/ml MNNG | 18,300 | 3,500 | 5,100 |
| 20 $\mu$g/ml MNNG+2 mM Nam | 77,000 | 13,300 | 20,900 |
| 20 $\mu$g/ml MNNG+2 mM Bam | 109,000 | 18,700 | 30,000 |

Logarithmically growing L1210 cells were treated with hydroxyurea and MNNG as described in the legend to Table 2. Where indicated, portions were adjusted to contain 2 mM nicotinamide (Nam) or 2 mM benzamide (Bam). Different samples were labeled to assess DNA, RNA, and protein synthesis by the addition of [$^3$H]dThd, [$^3$H]Urd or [$^3$H]leu. After a 6 hr incubation at 37 C, incorporation of label into TCA precipitates was determined as described (*2*).

Table 2 shows the relation of $NAD^+$ and ATP pool sizes to the level of DNA repair synthesis. In this experiment, MNNG concentrations up to 5 $\mu$g/ml resulted in increasing amounts of DNA repair synthesis whereas higher doses suppressed DNA repair. The doses of MNNG that suppressed DNA repair are exactly those that caused severe depletion of $NAD^+$ and ATP. When the MNNG treated cells were incubated with 2 mM nicotinamide, its ability to block the drastic depletion of $NAD^+$ and ATP resulted in stimulation of DNA repair synthesis. Thus, depletion of cellular $NAD^+$ and ATP pools, consequent to massive DNA damage, contributes to a rapid reduction in DNA synthesis and treatment of these cells with nicotinamide appears to stimulate the level of DNA repair synthesis by preserving the $NAD^+$ and ATP pools.

Since RNA and protein synthesis are also ATP-requiring processes, we reasoned that the use of poly(ADP-ribose) polymerase inhibitors to preserve ATP pools should

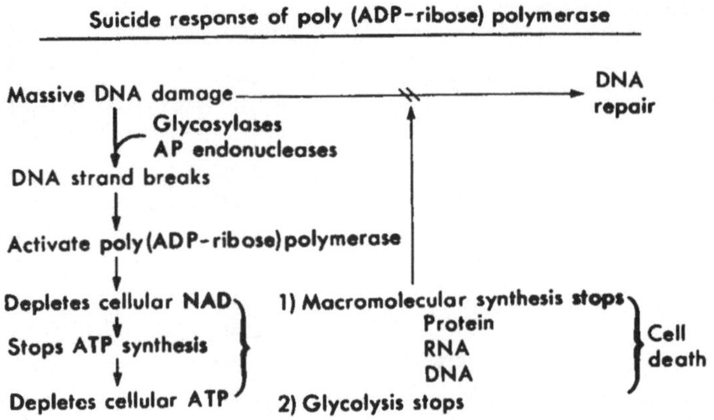


FIG. 4. Schematic proposal for the involvement of poly(ADP-ribose) polymerase in the suicide response to massive DNA damage.

stimulate synthesis of these macromolecules in massively damaged cells. Table 3 shows that, as predicted, both nicotinamide and benzamide stimulated the level of DNA, RNA, and protein synthesis in cells treated with 20 $\mu$g/ml MNNG. It is noteworthy that benzamide, a more potent inhibitor of poly(ADP-ribose) polymerase than nicotinamide (*9*), caused higher levels of stimulation of DNA, RNA, and protein synthesis.

Figure 4 summarizes the suicide response to massive DNA damage mediated by poly(ADP-ribose) polymerase. In this scheme, the DNA damage leads to spontaneous DNA strand breaks in addition to those brought about by glycosylases and AP endonucleases specific for the types of DNA damage. The DNA strand breaks activate poly(ADP-ribose) polymerase which in turn depletes cellular $NAD^+$ pools. Since $NAD^+$ is required at several steps in the generation of ATP, depletion of $NAD^+$ results in depletion of cellular ATP. As a consequence of $NAD^+$ and ATP depletion, there is a rapid decrease in energy dependent functions such as glycolysis and synthesis of DNA, RNA, and protein. Shutdown of these vital processes leads to rapid cell death before the genetic effects of DNA damage become manifest and before DNA repair can be completed.

### *$NAD^+$ Deficient Cells*

Considering that $NAD^+$ and poly(ADP-ribose) polymerase are involved in the normal process of DNA repair as well as in the suicide response to massive DNA damage, it is likely that individuals with abnormally low $NAD^+$ pools might be deficient in some aspects of DNA repair and might also show increased susceptibility to the lethal effects of cytotoxic agents. Cells from such individuals could be defective in DNA repair without showing abnormalities in any of the specific enzymes involved in the repair process. Further, since $NAD^+$ pool sizes appear to be involved in differentiation processes (*10*), individuals with abnormalities in $NAD^+$ might also show developmental abnormalities.

We surveyed poly(ADP-ribose) polymerase activity in cells from patients with genetic disorders of DNA repair by assaying the $NAD^+$-lowering response to DNA damage. Cells from patients with xeroderma pigmentosum, Cockayne's syndrome, Fanconi's anemia, ataxia telangiectasia, and Bloom's syndrome all showed $NAD^+$ lowering in response to treatment with MNNG. In the course of these studies, we noted that fibroblasts from many patients with Fanconi's anemia had lower $NAD^+$ levels than were observed in cells from normal donors (*11*). Freshly prepared lymphocytes from normal individuals and from patients with Fanconi's anemia all had low cellular $NAD^+$ levels. When lymphocytes are stimulated to proliferate by phytohemagglutinin, they undergo marked amplification of their pyridine nucleotide pools. Cells from most of the patients with Fanconi's anemia also amplified their $NAD^+$ pools, however they reached lower levels than those achieved by normal cells. When lymphocytes from patients with Fanconi's anemia were treated with DNA damaging agents, the $NAD^+$ pools were lowered even further than they were in normal cells (*11*), suggesting that Fanconi's cells may be more likely to undergo the suicide response following activation of poly(ADP-ribose) polymerase.

Earlier studies showed that UV irradiation of Faconi's anemia cells resulted in development of DNA strand breaks just as it did in normal cells (*12*). However, normal cells repaired the strand breaks whereas the breaks were persistent in the Fanconi's cells (*12*). This abnormality is similar to that seen in cells with low $NAD^+$ levels induced by nicotinamide starvation (*8*). Patients with Fanconi's anemia have also been found to show increased susceptibility to the lethal effects of chemotherapy and this may be an exaggeration of the suicide response due to pre-existing deficiencies of cellular $NAD^+$ pools. It is interesting too that patients with Fanconi's anemia have developmental anomalies of the skeletal and renal systems in addition to their pancytopenia and stunted growth. Since $NAD^+$ pool sizes and poly(ADP-ribose) polymerase are important in differentiation (*10*) as well as in DNA repair (*6, 8*), it may be that abnormalities in cellular $NAD^+$ levels could be etiologically related to many of the abnormalities seen in Fanconi's anemia. We have recently demonstrated that growth of Fanconi's cells in the presence of high concentrations of nicotinamide raises their $NAD^+$ levels and it will be interesting to determine whether such changes can eliminate the repair defect or improve the clinical status of patients.

## ACKNOWLEDGMENTS

These studies were supported by NIH grants CA24986 and GM26463. J. L. Sims is a recipient of a training fellowship, part of an institutional NSRA training grant T32CA09118. N. A. Berger is a Leukemia Society of America scholar.

## REFERENCES

1. Dixon, M. and Needham, D. M. Biochemical research on chemical warfare agents. Nature, *159*: 432–438, 1946.
2. Sims, J. L., Berger, S. J., and Berger, N. A. Poly(ADP-ribose) polymerase inhibitors preserve $NAD^+$ and ATP pools in DNA-damaged cells: Mechanism of stimulation of unscheduled DNA synthesis. Biochemistry, 1983, in press.
3. Benjamin, R. C. and Gill, D. M. Poly(ADP-ribose) Synthesis *in vitro* programmed by damaged DNA: A comparison of DNA molecules containing different types of strand breaks. J. Biol. Chem., *255*: 10502–10508, 1980.

4. Ohgushi, H., Yoshihara, K., and Kamiya, T. Bovine thymus poly(adenosine diphosphate ribose) polymerase, physical properties and binding to DNA. J. Biol. Chem., *255*: 6205–6211, 1980.
5. Cohen, J. J., Catino, D. M., Petzold, S. J., and Berger, N. A. Activation of poly-(adenosine diphosphate ribose) polymerase by SV40 minichromosomes: Effects of DNA damage and histone H-1. Biochemistry, *21*: 4931–4940, 1982.
6. Berger, N. A. and Sikorski, G. W. Poly(adenosine diphosphoribose) synthesis in ultraviolet-irradiated xeroderma pigmentosum cells reconstituted with *Micrococcus luteus* UV endonuclease. Biochemistry, *20*: 3610–3614, 1981.
7. Poirier, G. G., de Murcia, G., Jongstra-Bilen, J., Niedergang, C., and Mandel, P. Poly(ADP-ribosyl)ation of polynucleosomes causes relaxation of chromatin structure. Proc. Natl. Acad. Sci. U.S., *79*: 3423–3427, 1982.
8. Durkacz, B. W., Omidiji, O., Gray, D. A., and Shall, S. $(ADP\text{-ribose})_n$ participates in DNA excision repair. Nature, *283*: 593–596, 1980.
9. Sims, J. L., Sikorski, G. W., Catino, D. M., Berger, S. J., and Berger, N. A. Poly-(ADP-ribose) polymerase inhibitors stimulate unscheduled DNA synthesis in normal human lymphocytes. Biochemistry, *21*: 1813–1821, 1982.
10. Caplan, A. I. and Ordahl, C. P. Irreversible gene repression model for control of development. Science, *201*: 120–130, 1978.
11. Berger, N. A., Berger, S. J., and Catino, D. M. Abnormal $NAD^+$ levels in cells from patients with Fanconi's anemia. Nature, *299*: 271–273, 1982.
12. Poon, P. K., O'Brien, R. L., and Parker, J. W. Defective DNA repair in Fanconi's anemia. Nature, *250*: 223–225, 1974.

ADP-RIBOSYLATION, DNA REPAIR AND CANCER, M. MIWA ET AL. (EDS.),
JAPAN SCI. SOC. PRESS, TOKYO/VNU SCIENCE PRESS, UTRECHT, PP. 227-242, 1983

# Influence of Inhibitors of Poly(ADP-ribose) Polymerase on DNA Repair, Chromosomal Alterations, and Mutations

A. T. NATARAJAN, A. A. van ZEELAND, and T.S.B. ZWANENBURG

*Department of Radiation Genetics and Chemical Mutagenesis, State University of Leiden, Sylvius Laboratory, Wassenaarseweg 72, 2333 AL Leiden and J. A. Cohen Institute, Interuniversity Institute for Radiopathology and Radiation Protection, Leiden, The Netherlands*

*Abstract:* The influence of inhibitors of poly(ADP-ribose) polymerase such as 3-aminobenzamide (3AB) and benzamide (B) on the spontaneously occurring as well as mutagen induced chromosomal aberrations, sister chromatid exchanges (SCEs) and point mutations has been studied. In addition, we have measured the influence of 3AB on DNA repair following treatment with physical and chemical mutagens.

Post treatment of X-irradiated mammalian cells with 3AB increases the frequencies of induced chromosomal aberrations by a factor of 2 to 3. Both acentric fragments and exchanges increase indicating that the presence of 3AB slows down the repair of DNA strand breaks (probably DNA double strand breaks), thus making breaks available for interaction with each other to give rise to exchanges.

3AB, when present in the medium containing bromodeoxyuridine(BrdUrd) during two cell cycles, increases the frequencies of SCEs in Chinese hamster ovary cells (CHO) in a concentration dependent manner leading to about a 10-fold increase at 10 mM concentration. Most 3AB induced SCEs occur during the second cell cycle, in which DNA containing bromouridine (BU) is used as template for replication. BU containing DNA appears to be prone to errors during replication. The extent of increase in the frequencies of SCEs by 3AB is correlated with the amount of BU incorporated in the DNA of the cells. The frequencies of spontaneously occurring DNA single strand breaks in cells grown in BrdUrd containing medium are higher than in the cells grown in normal medium and this increase depends on the amount of BU incorporated in the DNA of these cells.

We have studied the extent of increase in the frequencies of SCEs due to 1 mM 3AB in several human cell lines, including those derived from patients suffering from genetic diseases such as ataxia telangiectasia (A-T), Fanconi's anemia (FA), and Huntington's chorea. None of these syndromes showed any increased response when compared to normal cells. 3AB, however, increased the frequencies of spontaneously occurring chromosomal aberrations in A-T and FA cells.

3AB does not influence the frequencies of SCEs induced by UV or mitomycin C (MMC) in CHO cells. However, it increases the frequencies of SCEs induced by ethyl methanesulfonate (EMS) and methyl methanesulfonate (MMS).

Under the conditions in which 3AB increases the frequencies of spontaneously occurring as well as induced SCEs, it does not increase the frequencies of point mutations in hypoxanthine-guanine phosphoribosyltransferase (HGPRT) locus. 3AB does not influence the amount of repair replication following dimethylsulphate (DMS) treatment of human fibroblasts, or UV irradiated human lymphocytes. However, the amount of unscheduled DNA synthesis (UDS) in UV-irradiated lymphocytes is strongly enhanced by 3AB. The results can be explained, if one assumes that 3AB is able to lower the endogenous synthesis of dTTP in the cell.

When lesions are induced in the DNA by mutagenic agents, they are subjected to cellular repair. Unrepaired and misrepaired lesions lead to biological effects such as cell killing, point mutations and chromosomal alterations (aberrations and sister chromatid exchanges (SCEs)). It is difficult to correlate any particular type of lesion to a specific biological effect. However, in specific cases this can be done. For example, shortwave UV induced cyclobutane dimers appear to be responsible for most of the biological effects induced by UV, which can be demonstrated by photoreactivation (PR) experiments, *i.e.* with removal of dimers by PR, there is a concomitant reduction in the frequencies of chromosomal aberrations, SCEs, and point mutations (*1–3*). Similarly, DNA double strand breaks appear to be involved in the production of chromosomal aberrations induced by ionizing radiations (*4*, *5*). However, such direct correlations between biological effects and lesions induced by chemical mutagens or carcinogens are difficult in view of the number and types of lesions involved (*6*).

The frequenices of induced chromosomal alterations and point mutations increase with the dose and an inhibition of DNA repair following mutagen treatment usually increases the biological effects. Similarly, by preventing the cells from dividing following mutagen treatment, one can allow them to repair premutational damage, thus reducing the biological effects induced. Comprehensive studies involving quantification of primary DNA lesions, their repair and biological effects will enable one to understand to some extent the complex processes involved in the manifestation of specific events that follow the treatment of cells with mutagenic carcinogens. Evidence accumulated in the last few years has demonstrated that poly(ADP-ribose) polymerase is involved in the repair and replication of DNA. In the present paper we review the results obtained in our laboratory on the influence of inhibitors of poly-(ADP-ribose) polymerase on the induction of chromosomal aberrations, SCEs, and point mutations by physical and chemical mutagens as well as associated DNA repair.

*Influence of 3-Aminobenzamide (3AB) on the Frequencies of X-ray Induced Chromosomal Aberrations*

Since it is well documented that: a) most of the chromosomal aberrations induced by ionizing radiations arise from DNA double strand breaks (*5*), and b) induction of DNA double strand breaks stimulates poly(ADP-ribose) polymerase synthesis in cells, implicating its role in the repair of DNA damage (*7*), we studied the effects of 3AB, a potent inhibitor of poly(ADP-ribose) polymerase on the frequencies of

TABLE 1. Chromosomal Aberrations Induced by 200 rad X-rays in Peripheral Blood Lymphocytes of Different Donors

| Treatment | Chromosome aberrations/100 cells | | | | | | | | | | | |
|---|---|---|---|---|---|---|---|---|---|---|---|---|
| | Donor 1 | | | Donor 2 | | | Donor 3 | | | Donor 4 | | |
| | F″ | EX | T | F″ | EX | T | F″ | EX | T | F″ | EX | T |
| 200 rad | 22 | 18 | 40 | 21 | 19 | 40 | 27 | 19 | 46 | 19 | 19 | 38 |
| 200 rad+3 AB | 61 | 37 | 98 | 62 | 49 | 111 | 56 | 45 | 101 | 75 | 44 | 139 |

Lymphocytes were irradiated at $G_0$ stage and incubated in a medium with or without 3 mM 3-aminobenzamide for 6 hr before mitogen stimulation. (Cells harvested at 48 hr culture time).
F″, chromosome breaks; EX, dicentries and rings; T, total.

X-ray induced chromosomal aberrations, In peripheral human blood lymphocytes when X-irradiated with 200 rad at $G_0$ stage and incubated in a medium containing 3 mM 3AB for 6 hr prior to mitogenic stimulation, or during the entire period of 48 hr of culture, the frequencies of chromosomal aberrations increase 2 to 3-fold (*8*). The increase due to 3AB posttreatment was similar, irrespective of the duration indicating that most of the aberrations induced by X-rays are fixed immediately after irradiation (*8*). In Table 1, data on the potentiation of 3AB on the frequencies of X-ray induced chromosome aberrations in 4 different donors are presented. There is no significant inter-donor difference for the potentiating effect of 3AB which is in the range of 2 to 3-fold over the X-irradiated controls. This effect of 3AB can be interpreted as due to slowing down or partial prevention of rejoining DNA double strand breaks by this inhibitor, thus allowing possibilities for interaction between breaks to generate exchanges.

Treatment of Chinese hamster ovary cells (CHO) in $G_2$ with 3AB following X-irradiation also increases the frequencies of induced aberrations by a factor of 2 to 3. Both types of aberrations, breaks and exchanges, increase to a similar extent (*8*). The other inhibitors of DNA repair such as ara C, aphidicolin and hydroxyurea which act by different mechanisms, when given immediately after X-irrdiation of $G_2$ cells increase only chromatid breaks and not exchanges (*8–11*). 3AB is different in this respect, as this inhibitor does not completely prevent the rejoining of breaks, but delays their repair, thus leading to an increase in the yield of chromosomal exchanges as well.

### *Influence of 3AB on the Frequencies of Shortwave UV (254 nm) Induced Chromosomal Aberrations*

Though a preliminary study in our laboratory (*8*) indicated a potentiating effect of 3AB on UV-induced chromosomal aberrations, extensive studies conducted later have demonstrated that this inhibitor does not influence the frequencies of UV-induced chromosomal aberrations in CHO cells. The results of an experiment involving a multiple fixation schedule in which UV-irradiated cells were posttreated with 3 mM 3AB for 3 to 24 hr and fixed are presented in Table 2. Unlike X-rays which produce DNA strand breaks directly thus leading to chromosomal aberrations at all stages of the cell cycle, UV-induced chromosomal aberrations are formed by a dif-

TABLE 2. Influence of 3AB (3 mM) Posttreatment on the Frequencies of Shortwave UV Induced Chromosomal Aberrations at Different Fixation Times (UV dose, 5 J/m²) in CHO Cells

| Treatment | Recovery time (hr) | Aberrations/100 cells | | |
|---|---|---|---|---|
| | | Breaks | Exchanges | Total |
| UV | 3 | 12 | 2 | 14 |
| UV+3AB | 3 | 14 | 1 | 15 |
| UV | 6 | 22 | 2 | 24 |
| UV+3AB | 6 | 19 | 3 | 22 |
| UV | 15 | 37 | 114 | 151 |
| UV+3AB | 15 | 57 | 107 | 164 |
| UV | 18 | 105 | 128 | 233 |
| UV+3AB | 18 | 99 | 131 | 230 |
| UV | 21 | 69 | 137 | 206 |
| UV+3AB | 21 | 59 | 111 | 170 |
| UV | 24 | 97 | 217 | 314 |
| UV+3AB | 24 | 84 | 195 | 279 |

ferent mechanism, closely associated with DNA synthesis. Other inhibitors of DNA repair such as ara C, aphidicolin, hydroxyurea, have profound influence on the frequencies of UV-induced chromosomal aberrations (*8*). The present results suggest that inhibitors of poly(ADP-ribose) polymerase do not involve the repair of DNA lesions induced by shortwave UV in CHO cells.

## *SCEs*

SCEs are the outcome of exchanges of replicated DNA at homologous loci, a process involving DNA double strand breaks and rejoining of such breaks between sister chromatids. They occur spontaneously at very low frequencies. To detect SCEs, sister chromatids have to be differentiated. This can be achieved by growing cells for two cycles in a medium containing 5-bromodeoxyuridine (BrdUrd) or by growing cells for one cycle in BrdUrd and the second cycle in a medium containing thymidine (TdR). Both these procedures will lead to sister chromatids with different constitution (BB-BT or TT-TB) thus enabling sister chromatid differentiation following appropriate staining procedures (*12*). SCEs are efficiently induced by 'S' dependent agents, such as UV and alkylating agents. X-rays and Bleomycin which induce DNA strand breaks directly are poor inducers of SCEs. In view of the fact that most of the potent alkylating agents efficiently induce SCEs, attempts have been made to directly link their induction to the induction of point mutations (*13*).

Oikawa and his coworkers (*14*) were the first to report that several inhibitors of poly(ADP-ribose) polymerase induce SCEs. This ability was correlated with the extent of inhibition of poly(ADP-ribose) polymerase. A dose-dependent increase in the frequencies of SCEs in CHO cells by 3AB was demonstrated by us (*9*). Interestingly, even under conditions in which very high frequencies of SCEs were obtained due to 3AB there was no increase in the spontaneous frequencies of chromosomal aberrations in these cells.

Further to elucidate the mechanism of induction of SCEs by 3AB or benzamide (B) experiments were designed to answer the following questions: a) do the inhibitors

TABLE 3. Influence 3 mM 3AB on the Frequencies of SCEs in CHO Cells

| Cell cycle 1 | Cell cycle 2 | SCEs/cell |
|---|---|---|
| BrdUrd | BrdUrd (30%) | 4.28 |
| BrdUrd | BrdUrd (70%) | 5.16 |
| BrdUrd | BrdUrd (100%) | 6.48 |
| BrdUrd+3AB | BrdUrd (30%) | 7.04 |
| BrdUrd | BrdUrd (30%)+3AB | 11.00 |
| BrdUrd+3AB | BrdUrd (70%) | 8.80 |
| BrdUrd | BrdUrd (70%)+3AB | 20.92 |
| BrdUrd+3AB | BrdUrd (100%) | 9.16 |
| BrdUrd | BrdUrd (100%)+3AB | 30.64 |

The cells were grown in medium containing different concentrations of BrdUrd to yield 30 or 70 or 100% substitution in their DNA.

affect the frequencies of SCEs in the first and second cell cycles in a similar way? and b) will growing cells in BrdUrd or thymidine containing medium during the second cycle have an influence on the extent of SCEs induced by 3AB or B? The results clearly demonstrated that: a) 3AB or B when present only during the first cell cycle increased the frequencies of SCEs about 2-fold, b) when present during the second cell cycle they increased the frequencies of SCEs 7 to 10-fold, c) when the inhibitor was present during both cell cycles the effect was additive, and d) the increase in the second cycle was independent of the precursors present in the medium, *i.e.*, whether it contained BrdUrd or not (*9*). The dramatic increase in the frequencies of SCEs during the second cycle in the presence of 3AB or B was interpreted as being due to errors in replication of DNA, when bromouridine (BU) containing DNA strand was used as template, which happens to be the only difference between the first and second cell cycles in these experiments. In order to confirm this finding further, CHO cells were cultured in medium containing BrdUrd for 4 cell cycles, so as to substitute both strands of DNA optimally and grown further for two cycles in medium containing thymidine, or first cycle in thymidine and the second cycle in BrdUrd, in order to obtain sister chromatid differentiation. In this case the presence of the 3AB during the first cell cycle maximally potentiated the frequencies of SCEs, supporting the conclusion that replication using BU containing strand as template is responsible for the increased frequecies of SCEs in cells grown in a medium containing 3AB. The direct involvement of incorporated BU in the formation of SCEs was further elucidated by growing cells in medium containing different concentrations of BrdUrd to get 30%, 70%, and 100% substitution. When such cells were challenged with 3AB during two cell cycles, the potentiation due to 3AB was dependent upon the BU content of the DNA (Table 3) (*15*). We have interpreted these results as indicating that BU containing DNA is fragile and during replication is subjected to spontaneous breaks. The repair of such breaks is slowed down by 3AB, thus allowing the possibility of formation of SCEs. We have measured the frequency of single strand breaks in DNA of cells from three different strains grown in medium containing different concentrations of BrdUrd. The measurements made with alkaline sucrose gradients, which include SSBs as well as alkali labile sites, demonstrated

TABLE 4. Single Strand Breaks and Alkali-labile Sites in BrdUrd Substituted Cells

| Substitution (%) | SSBs/$10^9$ daltons | | |
|---|---|---|---|
| | CHO | L5178Y | Human foreskin |
| 0 | — | — | — |
| 30 | 1.5 | 0.5 | 1.9 |
| 100 | 4.8 | 4.0 | 4.5 |

Exponentially growing cells were incubated for 24 hr in medium containing 10 $\mu$M thymidine (0), or 7 $\mu$M thymidine+3 $\mu$M BrdUrd (30%), or 10 $\mu$M BrdUrd (100%) in the presence of 1 $\mu$M FdUrd and 100 $\mu$M deoxycytidine. Determination of SSBs was made on alkaline sucrose gradients.

an increase in SSBs with increasing substitution of BU in the DNA (Table 4). Preliminary studies have indicated also that the activity of poly(ADP-ribose) polymerase (measured by the extent of incorporation of [$^3$H]ADP-ribose in the chromatin) increases in cells with increased substitution of BU in the cellular DNA. These observations link incorporated BU in the DNA to the origin of spontaneous occurrence of SCEs, and modification of poly(ADP-ribose) polymerase levels in cells will alter the frequencies of the SCEs. Results of the SCE test should be carefully evaluated as one has to employ cells with BU substituted chromosomes for these studies.

*Potentiation of the Frequencies of SCEs by 3AB in Different Cell Lines*

We have compared the frequencies of SCEs induced by 1 mM 3AB given for two cycles in a medium containing 5 $\mu$M BrdUrd in normal as well as transformed cells. In CHO cells, 3 mouse cell lines, including one ascites tumor line, and one teratocarcinoma cell line (PCC3). the base line frequencies of SCEs increased by a factor of 3 due to 3AB (Table 5). However, two normal human fibroblast cell lines

TABLE 5. Comparison of Potentiating Effect of 3AB (1 mM) Supplemented in the Growth Medium for Two Cell Cycles on the Base Line Frequencies of SCEs in Different Cell Lines

| Cell line | | SCEs/cell |
|---|---|---|
| Chinese hamster | | |
| CHO | Control | 6.42 |
| | 3 AB | 18.15 |
| Mouse | | |
| PG 19 | Control | 6.45 |
| | 3 AB | 20.84 |
| PCC3 | Control | 6.43 |
| | 3 AB | 18.20 |
| Ascites tumor cells | | |
| | Control | 13.32 |
| | 3 AB | 35.58 |
| Human foreskin fibroblasts | | |
| VH 8 | Control | 9.00 |
| | 3 AB | 11.32 |
| VH 16 | Control | 8.10 |
| | 3 AB | 11.00 |

TABLE 6. Induction of SCEs by 3AB in Fibroblasts from Normal Individuals and Patients with Mutagen Sensitive Genetic Disorders

| | Normal | | FA | | A-T | | Huntington's disease | |
|---|---|---|---|---|---|---|---|---|
| | I | II | I | II | I | II | I | II |
| Control | 9.0±1.8 | 8.1±1.9 | 8.9±1.6 | 8.5±1.7 | 8.8±1.7 | 8.2±1.6 | 7.7±1.6 | 10.0±1.8 |
| 1 mM | 11.3±1.8 | 11.0±1.9 | 11.4±1.7 | 12.6±1.6 | 11.0±1.8 | 12.8±1.7 | 11.2±2.0 | 13.2±1.9 |
| 3 mM | 18.6±2.1 | 15.9±1.8 | 17.4±2.0 | 16.2±2.0 | 17.1±2.1 | 13.9±1.7 | 12.9±1.7 | 18.1±1.9 |

Cells were grown for two cycles in a medium containing 5 μM BrdUrd and appropriate concentration of 3AB.

TABLE 7. Frequencies of Chromosomal Aberrations in Fibroblasts from Normal Individual and Patients with A-T and FA Grown for 72 hr in Medium Containing BrdUrd (5 μM) and 3AB

| Cell type | Treatment | Aberrations/100 cells | | | | |
|---|---|---|---|---|---|---|
| | | Gaps | Breaks | Iso-breaks | Exchanges | Total |
| Normal | Control | 2 | 1 | 0 | 0 | 3 |
| | 3 mM 3AB | 3 | 1 | 0 | 0 | 4 |
| FA | Control | 11 | 6 | 2 | 0 | 19 |
| | 1 mM 3AB | 13 | 23 | 3 | 0 | 39 |
| | 3 mM 3AB | 11 | 30 | 4 | 0 | 45 |
| A-T | Control | 2 | 11 | 1 | 0 | 14 |
| | 1 mM 3AB | 4 | 13 | 3 | 1 | 21 |
| | 3 mM 3AB | 7 | 29 | 11 | 0 | 47 |

responded with much less potentiation (<2) by 3AB. This may indicate that transformed cells may be more susceptible to the potentiation of SCEs by 3AB, either due to increased incorporation of BrdUrd in the DNA or different intra-cellular levels of poly(ADP-ribose) polymerase. This problem is currently being investigated in our laboratory.

The influence of 3AB on the frequencies of SCEs in two different cell lines of normal individuals as well as patients of inherited diseases such as ataxia telangiectasia (A-T), Fanconi's anemia (FA), and Huntington's disease has been investigated. The results presented in Table 6 show that all cell lines irrespective of their genetic constitution respond in a remarkably similar manner, with low increase in the base line frequencies of SCEs. However, under these growth conditions, *i.e.*, presence of BrdUrd and 3AB in the medium, cells such as A-T and FA which have low but significant levels of spontaneously occurring chromosomal aberrations, respond to 3AB with increased aberration frequency in a dose dependent manner (Table 7). This may be interpreted as an interference of 3AB on the repair of spontaneously occurring DNA breaks in these syndromes.

### *Influence of Inhibition of Poly(ADP-ribose) Polymerase on Mutagen Induced SCEs*

We have estimated the extent of modification by inhibitors of poly(ADP-ribose) polymerase—3AB or B—on the frequencies of SCEs induced by several mutagens

TABLE 8. Influence of Inhibitors of Poly (ADP-ribose) Polymerase on the Yield of Induced SCEs in CHO Cells

| Treatment | SCEs/cell | Increase over control |
|---|---|---|
| Control | 6.0 | — |
| 3AB (3 mM) | 27.2 | 21.2 |
| UV (254 nm) (5 J/m²) | 28.1 | 22.1 |
| UV (254 nm)+3AB (3 mM) | 54.6 | 5.3 (12.2%) |
| Control (BU substituted) | 6.3 | — |
| 3AB (3 mM) | 28.0 | 21.7 |
| UV (254 nm) (1.5 J/m²) | 64.7 | 58.4 |
| UV (254 nm)+3AB (3 mM) | 63.7 | −22.4 (−28.0%) |
| Control | 7.8 | — |
| B | 39.8 | 32.0 |
| MMS (1 mM/1 hr) | 32.1 | 24.3 |
| MMS+B | 91.1 | 27.0 (48.0%) |
| Control | 9.6 | — |
| 3AB (3 mM) | 23.4 | 13.8 |
| EMS | 26.5 | 16.9 |
| EMS+3AB (3 mM) | 74.9 | 34.6 (112.7%) |
| Control | 8.2 | — |
| B | 35.1 | 26.9 |
| ENU (5 mM/1 hr) | 29.7 | 21.5 |
| ENU+B | 65.2 | 8.6 (17.8%) |
| Control | 7.1 | — |
| B | 42.2 | 35.1 |
| MMC (0.018 μg/ml/1 hr) | 30.7 | 23.6 |
| MMC+B | 59.7 | 1.0 (1.7%) |

Following mutagen treatment, the cells were incubated for 2 cycles in medium containing 3AB and 5 μM BrdUrd.

(Table 8). In these experiments mutagen-treated CHO cells were grown for two cycles in medium containing 5 μM BrdUrd and 3 mM 3AB or B. There is no or very weak potentiation on SCEs induced by shortwave UV or mitomycin C (MMC). However, with monofunctional alkylating agents such as ethyl methanesulfonate (EMS) and methyl methanesulfonate (MMS), there is a strong potentiation, whereas with ethyl nitrosourea (ENU) the potentiation is weak. Thus, these inhibitors do not always increase the frequencies of induced SCEs, and the modification depends upon the type of mutagen used, the types of lesions induced and the repair processes which follow. This would mean that the response to 3AB potentiation will vary between cell lines with different DNA repair capacity. In contrast to CHO cells which are poor in excision repair, normal human cells which are proficient in such repair respond to 3AB with strong potentiation (more than double) of UV-induced SCEs (*16*).

When cells are treated with a mutagen in the presence of 3AB or B, the response to induction of SCEs varies depending upon the mutagen and the inhibitor used. In the case of monofunctional alkylating agents there is no influence of 3AB on the induction of SCEs. However, when shortwave UV is used, 3AB and not B protects the cells from the lethal effects. With long-wave UV irradiation of BU substituted cells, there is a potentiation by 3AB and not B for induction of biological effects (*17*). This modification is not related to inhibition of poly(ADP-ribose) polymerase. 3AB protects against shortwave UV by direct absorption, whereas with the long wave UV we do not know at present the exact mechanism involved.

### *Inhibitors of Poly(ADP-ribose) Polymerase and Point Mutations*

Since SCEs have been considered as an indicator of mutagenesis, it was of interest to test inhibitors of poly(ADP-ribose) polymerase on their capacity to induce point mutations. B and 3AB do not induce mutations in hypoxanthine-guanine phosphoribosyltransferase (HGPRT) locus in CHO cells either when treated alone or in the presence of BrdUrd for two cell cycles, a condition which induces about 10-fold increase in the frequencies of SCEs (*9*). The influence of 3AB posttreatment on the induction of mutations at the HGPRT locus by EMS, ENU, and shortwave UV has been studied. In all cases posttreatment has no influence on the frequencies of mutations (*17*). On the other hand, posttreatment with 3AB of EMS treated cells clearly increases the frequencies of induced SCEs (Table 9). 3AB does not effectively influence UV induced SCEs. When 3AB is given during treatment, there is no effect on EMS induced mutations or SCEs. However, when cells are irradiated with shortwave UV in the presence of 3AB there is protection for induction of SCEs and mutations due to the absorption of UV by 3AB molecules.

From the results presented above it appears that SCEs and point mutations are not always a related phenomenon and one has to be cautious in using the exclusively SCE inducing effect of a chemical as an indicator of mutagenicity (*9*).

TABLE 9. Influence of 3AB on the Frequencies of SCEs and Point Mutations ($HGPRT^-$) Induced in CHO Cells by EMS

| Treatment | Frequencies of: | | Mutation frequency ($10^{-5}$) |
|---|---|---|---|
| | SCEs | Induced SCEs | |
| Control | 9.6 | — | 4.3 |
| 3AB (3 mM) | 23.4 | 13.8 | — |
| EMS (4 mM) | 26.5 | 16.9 | 29.4 |
| EMS (4 mM)+3AB | 74.9 | 48.4 | 30.5 |
| EMS (6 mM) | 35.7 | 26.1 | 40.9 |
| EMS (6 mM)+3AB | 84.7 | 61.3 | 41.2 |

EMS treatment for 2 hr at 37°C followed by 28 hr incubation in medium containing 3 mM 3AB.

### *Effects of 3AB on DNA Repair in Human Cells*

It has been shown by several investigators (*18*, *19*) that 3AB is able to increase the amount of excision repair measured as unscheduled DNA synthesis in human lymphocytes treated with UV-light or alkylating agents as well as in diploid human fibroblasts treated with dimethylsulphate (DMS). In addition, it has been observed that 3AB inhibits the repair of single strand breaks or alkali labile sites induced by DMS in mouse L1210 cells and in human fibroblasts (*19*, *20*). A possible explanation which has been put forward (*18*, *19*) is that the inhibition of poly(ADP-ribose) polymerase prevents the rejoining step in excision repair. Under these conditions nick translation at repair sites might take place resulting in a larger size of repair patches. If the number of repair sites is not changed the amount of incorporation of repair label will be increased. We have carried out experiments in which we investigated the effect of 3AB on the size of repair patches induced by UV-irradiation or DMS in human diploid fibroblasts as well as in human lymphocytes. Repair replication was determined using the CsCl density gradient technique (*21*). Human fibroblasts (VH-16) were cultured for 2–3 days in medium containing 0.3 μCi/ml $^{32}$P (as orthophosphate) and subsequently grown to confluency in medium without radioactivity (MEM plus 15% fetal calf serum). One hour prior to the mutagenic treatment, BrdUrd ($10^{-5}$ M) and FUdR ($10^{-6}$ M) were added to the medium. At the time of treatment the medium was removed and the monolayers were rinsed with phosphate buffered saline. The cells were then either irradiated with shortwave UV-light or incubated for 20 min in medium without serum containing 0–600 μM DMS, BrdUrd ($10^{-5}$ M), FUdR ($10^{-6}$ M). 3AB and/or hydroxyurea were included where indicated. Following treatment the cells were incubated in medium containing 5 μCi/ml [$^3$H]BrdUrd (25 Ci/mmole), $10^{-5}$ M unlabeled BrdUrd, and $10^{-6}$ M FUdR. 3AB and/or hydroxyurea were present where indicated. After a 4 hr repair period the cells were lysed and neutral plus alkaline CsCl density gradients were used to dertermine the amount of repair replication (*21*). In those cases where the size of the repair patches was analyzed, two neutral CsCl gradients were used prior to sonification of the DNA and analysis on an alkaline CsCl gradient. In each experiment the specific activity of the DNA expressed in cpm $^{32}$P/μg DNA was determined and the amount of repair replication was expressed as cpm $^3$H/μg DNA.

Figure 1 shows the results obtained after UV-irradiation. The presence of 3 mM 3AB in the medium during the repair period had no effect on the amount of repair replication observed in these cells. The alkaline CsCl gradients on which the size of the UV-induced repair patches was determined is shown in Fig. 2. Sonification of the DNA causes a shift in the [$^3$H]BrdUrd labeled DNA to a position in the gradient corresponding with a higher density than the [$^{32}$P]labeled parental DNA. This is

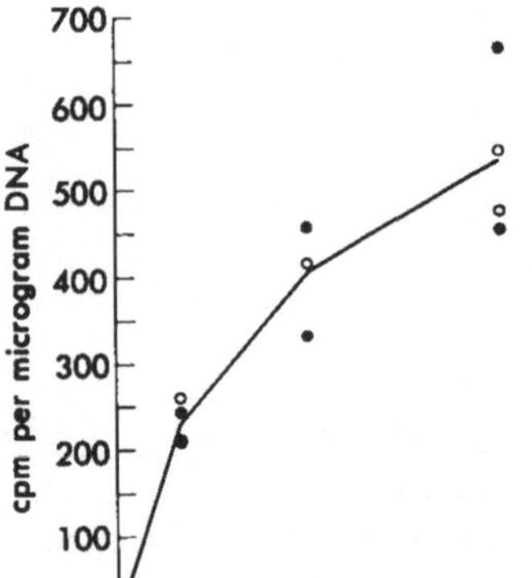


FIG. 1. The influence of 3 mM 3AB on UV-induced DNA repair replication in diploid human fibroblasts (VH-12). Cells were confluent and repair was measured over a 4 hr period. ○ no 3AB; ● 3 mM 3AB.

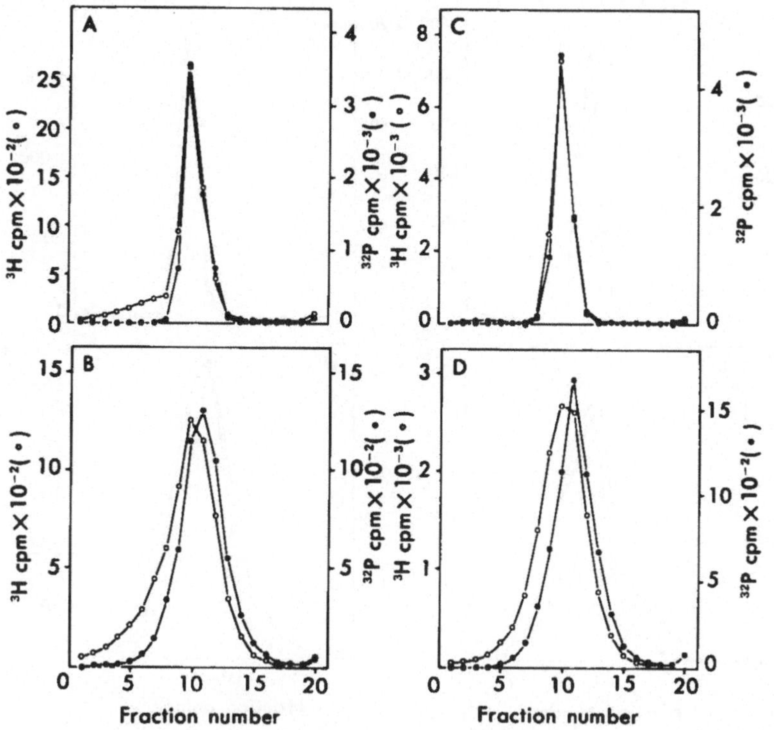


FIG. 2. Alkaline CsCl gradients of DNA from UV-irradiated VH-16 cells treated as described in Fig. 3. The DNA was purified on two neutral CsCl gradients and subsequently analyzed with and without sonification on alkaline CsCl gradients as described. A, 0 mM 3AB−sonification. B, 0 mM 3AB+sonification; C, 3 mM 3AB−sonification; D, 3 mM 3AB+sonification.

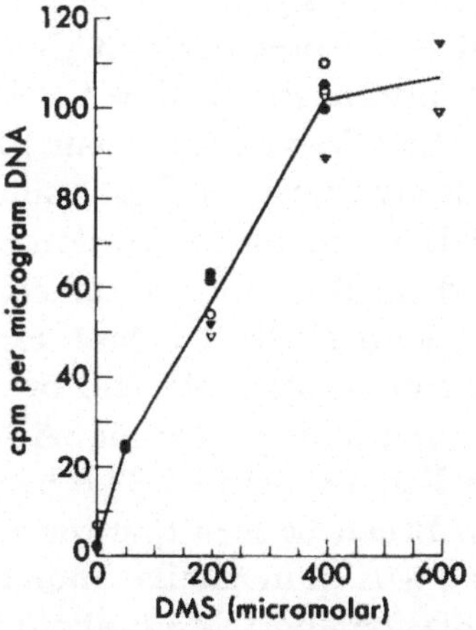


FIG. 3. Influence of 3 mM 3AB in DMS-induced DNA repair replication in human diploid fibroblasts (VH-12). Cells were confluent and repair was measured over a period of 4 hr. DMS treatment was for 20 min in medium without serum. No 3AB (○). 3 mM 3AB (●) present during DMS treatment and repair period. 10 mM hydroxyurea (▽) added 30 min before DMS treatment and present until the end of the repair period. 10 mM hydroxyurea and 3 mM 3AB (▼) added 30 min before the DMS treatment and present until the end of the repair period.

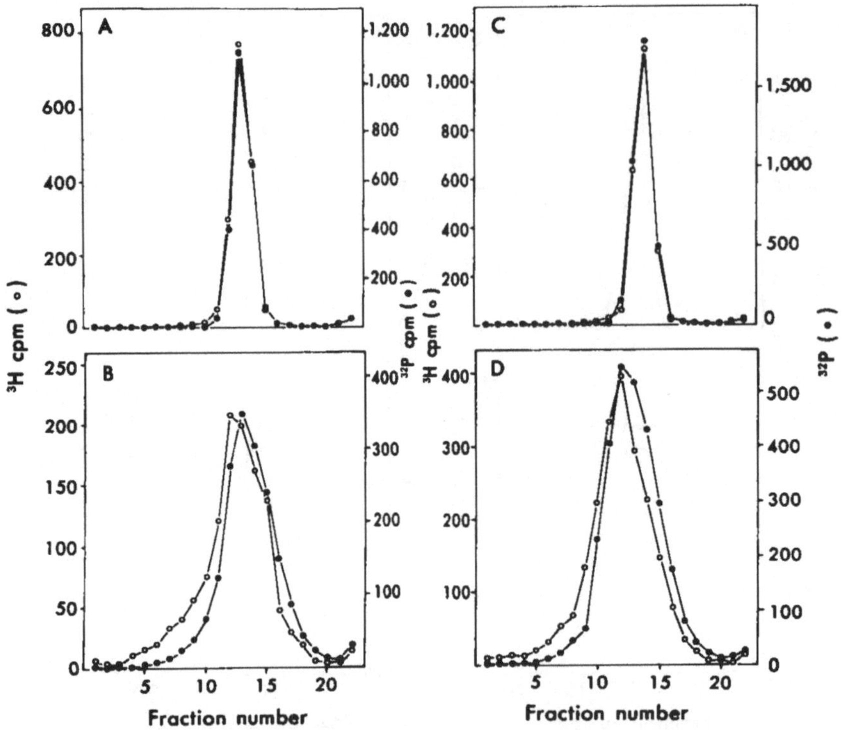


FIG. 4. Alkaline CsCl gradients of DNA from VH-16 cells treated as described in the experiments without hydroxyurea, described in Fig. 3. The DNA was purified and analyzed as described in Fig. 2. A, 0 mM 3AB−sonification; B, 0 mM 3AB+sonification; C, 3 mM 3AB−sonification; D, 3 mM 3AB+sonification.

because the [$^3$H]BrdUrd-labeled repair patch becomes a significant part of the total length of the DNA fragment. The results show that the presence of 3AB during the repair period does not effect the shift of the repair label in the alkaline CsCl gradients and therefore 3AB does not influence the size of the UV-induced repair patch. The results obtained with DMS in the same cell strain are shown in Figs. 3 and 4. Here also there is no effect of 3AB on the amount of DNA repair replication. In Fig. 3 the data of several experiments are pooled. As indicated, in some instances 3AB has been added to the cells 30 min before treatment with DMS. In those experiments also hydroxyurea (10 mM) was present from that time on until the end of the repair period. In none of the experiments there was a potentiating effect of 3AB on DNA repair replication. The alkaline CsCl gradients in Fig. 4 show that 3AB has no effect on the size of DMS induced DNA repair patches. It can be seen that the shift of the repair label in the gradients shown in Figs. 2 and 4 is quite similar indicating that the size of the repair patches induced by UV-irradiation and DMS is about the same.

We have also investigated the effect of 3AB on UV-induced repair replication in human lymphocytes. Because we were using unstimulated lymphocytes it was not possible to prelabel the DNA with $^{32}$P. We therefore modified the experimental protocol in such a way that we were using two neutral CsCl gradients instead of one neutral and one alkaline gradient. After the second neutral gradient the DNA in the

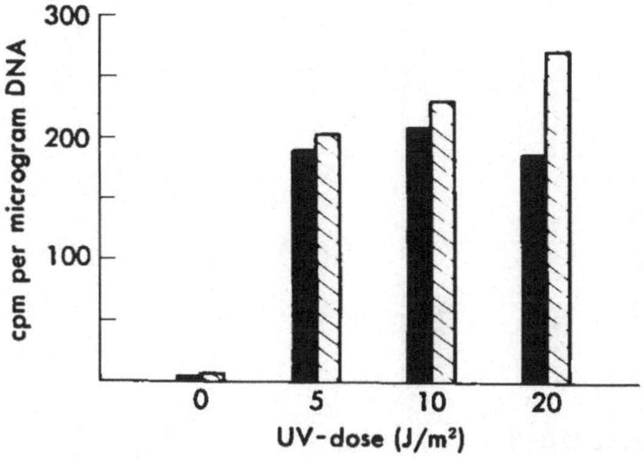


FIG. 5. Influence of 3 mM 3AB on UV-induced DNA repair replication in human lymphocytes. Lymphocytes were isolated from 100 ml human blood using a Ficoll-Histopaque gradient. The cells were resuspended in F10 medium containing 20% foetal calf serum, $10^{-5}$ M BrdUrd and $10^{-6}$ M FUdR. The cell suspension was split in 8 equal parts and incubated for 30 min at 37° C. 3 mM 3AB was added where indicated and incubation was continued for 30 min. The cells were centrifuged, resuspended in PBS and exposed to UV-light. Immediately following irradiation the cells were resuspended in F10 medium containing 20% foetal serum, 5 $\mu$Ci/ml [$^3$H]TdR (80 Ci/mmol), $10^{-5}$ M BrdUrd, $10^{-6}$ M FUdR and where indicated 3 mM 3AB. Cells were allowed to repair for a period of 4 hr and subsequently processed as described in the text. ■ 0 mM 3AB; ▨ 3 mM 3AB.

normal density region was collected and the amount determined by measuring the optical density at 260 nm. Then the amount of repair label was determined and expressed as cpm $^3$H/$\mu$g DNA. The repair label used in these experiments with lymphocytes was a mixture of 5 $\mu$Ci/ml [$^3$H]TdR (80 Ci/mmol) and $10^{-5}$ M unlabeled BrdUrd instead of $^3$H-labeled BrdUrd. The results shown in Fig. 5. indicate that 3AB does not have an effect on UV-induced repair replication in human lymphocytes. These results and those obtained with human fibroblasts treated with DMS are in contrast to the results published by several other investigators (*18*, *19*).

The difference between our protocol and that of other published results is that most other investigators are not using the CsCl density gradient technique but are measuring the incorporation of [$^3$H]TdR in cells outside the S-phase (unscheduled DNA synthesis, UDS) (*18*, *22*). The residual incorporation of [$^3$H]TdR due to semiconservative DNA synthesis in a few contaminating S-phase cells is then inhibited with hydroxyurea. We have carried out an experiment with human lymphocytes using the UDS technique. The results in Fig. 6 shows that there was a strong increase in the amount of UDS when measured in the presence of 3AB. Apparently there is a discrepancy between the results obtained with UDS and those obtained with the CsCl density labeling technique. We propose that these differences might be explained by possible effects of 3AB on the poolsize of deoxyribonucleotidetriphosphates in the cells. If, in the case of UDS 5 $\mu$Ci/ml, [$^3$H]TdR is used with a specific activity of 80 Ci/mmole, then the concentration of [$^3$H]TdR in the medium will be $6.25 \times 10^{-8}$ M. The concentration of [$^3$H]dTTP in the cell will be lower. This means that very small amounts of unlabeled dTTP made by the cell will considerably lower the specific activity of [$^3$H]dTTP and therefore the amount of incorporation in DNA of radio-

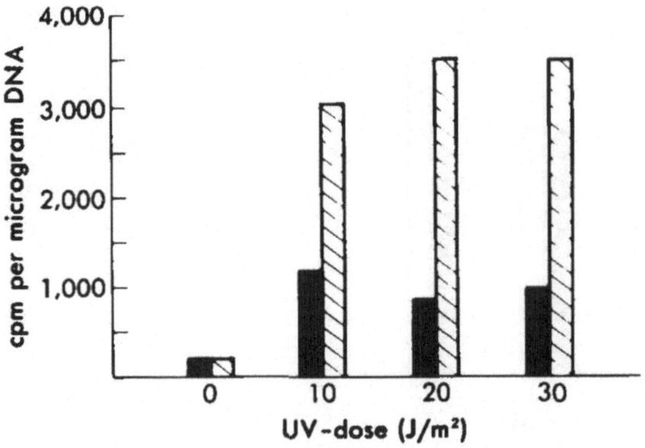


FIG. 6. Effect of 3AB on UV-induced UDS in human lymphocytes. Lymphocytes were isolated from 15 ml human blood using a Ficoll-Histopaque gradient. The cells were resuspended in F10 medium containing 20% foetal calf serum, and 10 mM hydroxyurea. The cell suspension was subsequently split in 8 equal parts and incubated at 37°C for 30 min. 3 mM 3AB was added where indicated and incubation was continued for another 30 min. The cells were centrifuged, resuspended in PBS, exposed to various doses of UV-light and resuspended in medium containing 10 mM hydroxyurea and 3 mM 3AB where indicated. The cells were incubated for 30 min at 37°C after which 5 μCi/ml [³H]TdR (80 Ci/mmol) was added and incubation was continued for 3 hr. Cells were centrifuged and washed 3 times with PBS. The amount of radioactivity incorporated was measured as described. ■ 0 mM 3AB; ▨ 3 mM 3AB.

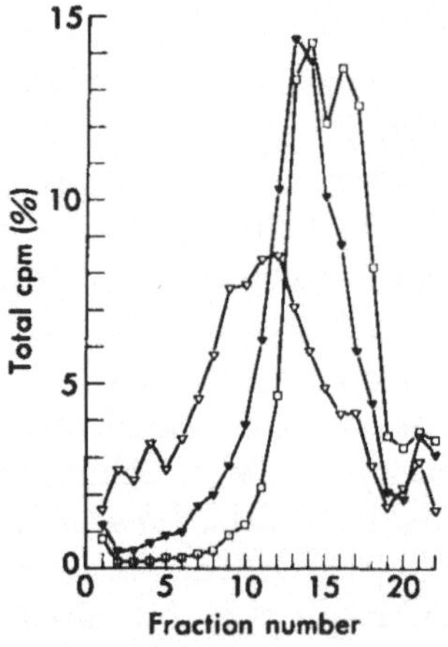


FIG. 7. Alkaline sucrose gradients containing DNA from human diploid fibroblasts (VH-12) treated with DMS. Cells were treated with 100 μM DMS for 20 min and either lysed directly atop the gradients (□) or incubated in medium for 40 min in the absence (▽) or presence (▼) of 3 mM 3AB before lysis on the gradients. Lysis conditions and centrifugation were exactly as described by Durkacz *et al.* (20). Centrifugation was from right to left. The cells were prelabelled with 2 μCi/ml [³H]TdR (6 Ci/mmol) for 24 hr.

active material. If 3AB inhibits the endogeneous synthesis of dTTP in the cell, an increase of radioactive precursor will be the result. In the CsCl density labelling technique a much higher concentration of repair label in the medium is used ($10^{-5}$ M). Possible effects of 3AB on the precursor pool do have much less effect in that case.

Furthermore FUdR was included in the medium which inhibits the endogeneous synthesis of dTTP by the cells.

The differences between our results and those in the literature are not caused by differences between the cell strains used. When we measured the effect of 3AB on the repair of DMS induced single strand breaks and alkali labile sites in human fibroblasts we observed a clear inhibition of repair in the presence of 3AB (Fig. 7).

In conclusion, our data show that 3AB does not influence the amount of DNA repair replication in human diploid fibroblasts treated with DMS or irradiated with UV-light and in UV-irradiated human lymphocytes. The amount of UDS in UV-irradiated lymphocytes was strongly enhanced by 3AB. The results can be explained if we assume that 3AB is able to lower the endogenous synthesis of dTTP in the cell.

ACKNOWLEDGMENT

Excellent technical assistance by Tonneke van Kesteren-van Leeuwen and Ineke Bussman and financial support from Koningin Wilhelminafonds (SG 81-91) and Euratom-Leiden University contract BIAN No. 52-64-1 are gratefully acknowledged.

REFERENCES

1. Griggs, H. G. and Bender, M. A. Photoreactivation of ultraviolet induced chromosomal aberrations. Science, *179*: 86–88, 1973.
2. van Zeeland, A. A., Natarajan, A. T., Verdegaal-Immerzeel, E.A.M., and Filon, A. R. Photoreactivation of UV-induced cell killing, chromosome aberrations, sister chromatid exchanges, mutations and pyrimidine dimers in *Xenopus laevis* fibroblasts. Mol. Gen. Genet., *180*: 495–500, 1980.
3. Natarajan, A. T., van Zeeland, A. A., and Degrassi, F. UV-induced SCEs and effects of photoreactivation. *In;* A. A. Sandberg (ed.), Sister Chromatid Exchanges, pp. 315–325, Alan R. Liss Inc. Ltd., New York, 1982.
4. Natarajan, A. T. and Obe, G. Molecular mechanisms involved in the production of chromosomal aberrations. I. Utilization of Neurospora endonuclease for the study of aberration production in G2 stage of the cell cycle. Mutat. Res., *52*: 137–149, 1978.
5. Natarajan, A. T. and Zwanenburg, T. S. B. Mechanisms for chromosomal aberrations in mammalian cells. Mutat. Res., *95*: 1–6, 1982.
6. Vogel, E. and Natarajan, A. T. The relation between reaction kinetics and mutagenic action of monofunctional alkylating agents in higher eukaryotic systems. III. Interspecific comparisons. *In;* F. J. de Serres and A. Hollaender (eds.), Chemical Mutagens, Principles and Methods for Their Detection, Vol. 7, pp. 295–336, Plenum Press, New York, 1982.
7. Benjamin, R. C. and Gill, D. M. Poly(ADP-ribose) synthesis *in vitro* programmed by damaged DNA. A comparison of DNA molecules containing different types of strand breaks. J. Biol. Chem., *255*: 10502–10508, 1980.
8. Natarajan, A. T., Csukas, I., Degrassi, F., van Zeeland, A. A., Palitti, F., Tanzarella, C., de Salvia, R., and Fiore, M. Influence of inhibition of repair enzymes on the induction of chromosomal aberrations by physical and chemical agents. *In;* A. T. Natarajan, G. Obe, and H. Altmann (eds.), DNA Repair, Chromosome Alterations and Chromatin Structure, pp. 47–59, Elsevier Biomedical Press, Amsterdam, 1982.
9. Natarajan, A. T., Csukas, I., and van Zeeland, A. A. Contribution of incorporated 5-bromodeoxyuridine in DNA to the frequencies of sister chromatid exchanges induced by inhibitors of poly(ADP-ribose) polymerase. Mutat. Res., *84*: 125–132, 1981.

10. Kihlman, B. A., Hansson, K., Palitti, F., Andersson, H. C., and Hartley, B. Potentiation of induced chromatid-type aberrations by hydroxyurea and caffeine in G2. *In;* A. T. Natarajan, G. Obe, and H. Altmann (eds.), DNA Repair, Chromosome Alterations and Chromatin Structure, pp. 11–24, Elsevier Biomedical Press, Amsterdam, 1982.
11. van Zeeland, A. A., Bussman, G.J.M., Degrassi, F., Filon, A. R., van Kesteren-van Leeuwen, A. C., Palitti, F., and Natarajan, A. T. Effects of aphidicolin on repair replication and induced chromosomal aberrations in mammalian cells. Mutat. Res., *92*: 379–392, 1982.
12. Perry, P. and Wolff, S. New Giemsa method for the differential staining of sister chromatids. Nature, *251*: 156–158, 1974.
13. Carrano, A. V., Thompson, L. H., Linde, P. A., and Minkler, J. L. Sister chromatid exchanges as indicator of mutagenicity. Nature, *271*: 551–553, 1978.
14. Oikawa, A., Thoda, H., Kanai, M., Miwa, M., and Sugimura, T. Inhibitors of poly-(adenosine diphosphate ribose) polymerase induced sister chromatid exchanges. Biochem. Biophys. Res. Commun., *97*: 1311–1316, 1980.
15. Natarajan, A. T., Csukas, I., and Degrassi, F. (unpublished results).
16. Natarajan, A. T. and Degrassi, F. (unpublished results).
17. Natarajan, A. T., Csukas, I., and Zwanenburg, T. S. B. (unpublished results).
18. Miwa, M., Kanai, M., Kondo, T., Hoshino, H., Ishihara, K., and Sugimura, T. Inhibitors of poly(ADP-ribose) polymerase enhance unscheduled DNA synthesis in human peripheral lymphocytes. Biochem. Biophys. Res. Commun., *100*: 463–470, 1981.
19. James, M. R. and Lehmann, A. R. The role of (ADP-ribose)$_n$ in DNA repair in human fibroblasts. *In;* A. T. Natarajan, G. Obe, and H. Altmann (eds.), DNA Repair, Chromosome Alterations and Chromatin Structure, pp. 177–182, Elsevier Biomedical Press, Amsterdam, 1982.
20. Durkacz, B. W., Omidiji, O., Gray, D., and Shall, S. (ADP-ribose)$_n$ participates in DNA excision repair. Nature, *283*: 593–596, 1980.
21. Smith, C. A., Cooper, P. K., and Hanawalt, P. C. Measurement of repair replication by equilibrium sedimentation. *In;* E. C. Friedberg and P. C. Hanawalt (eds.), DNA Repair, A Laboratory Manual of Research Procedures, pp. 289–305, Marcel Dekker, New York, 1982.
22. Lehmann, A. R. and Stevens, S. A rapid procedure for measurement of DNA repair in human fibroblasts and for complementation analysis of xeroderma pigmentosum cells. Mutat. Res., *69*: 117–190, 1980.

ADP-RIBOSYLATION, DNA REPAIR AND CANCER, M. MIWA ET AL. (EDS.),
JAPAN SCI. SOC. PRESS, TOKYO/VNU SCIENCE PRESS, UTRECHT, PP. 243-252, 1983

# Temperature Sensitivity of Poly(ADP-ribose) Synthetase in Intact Cells

William R. KIDWELL and Michael R. PURNELL
*Laboratory of Pathophysiology, National Cancer Institute, Bethesda, Maryland 20205, U.S.A.*

*Abstract:* The response of cells to various poly(ADP-ribose) synthetase inhibitors may vary with the cell type and inhibitor type. We have observed that most cells show an inhibition of growth, at least transiently, when exposed to synthetase inhibitors. For one cell type, there is, in contrast, a stimulation of proliferation. These results suggest that the synthetase plays different roles in different cell types or, the inhibitors also affect other cell processes that are variably represented in the different cell types. Because of these difficulties we have sought alternative methods of modulating poly(ADP-ribose) synthetase activity in intact cells. A method of inactivating the synthetase by heat shock is described and the consequences of this inactivation for the repair of single strand breaks in DNA is presented. Additionally, a method for elevating the synthetase in intact cells is given.

In previous attempts to define the role(s) of poly(ADP-ribose) synthetase in cells two major approaches have been followed. First, the levels of the synthetase and its products have been assessed *in vivo* and *in vitro* in cells in various physiological states (*1*, *2*). Second, the responses of cells to compounds that inhibit the synthetase have been examined (*1*, *2*). Based on such studies the synthetase has been proposed to play an important role in growth regulation (*3*, *4*), in altering or proof-reading chromatin structure (*5*, *6*), in promoting differentiation (*7*, *8*), and in DNA repair processes (*9*–*12*). At present it is not possible to state unequivocably that the synthetase is directly required for any of these functions, nor have any been conclusively ruled out. This state of uncertainty derives from the fact that cause and effect relationships are difficult to prove other than through mutation studies and also in part because of the differing responses of cells to synthetase inhibitors (*13*). In the present report we will describe some of the difficulties we have encountered and suggest some alternative methods for modulating (up and down) the synthetase activities of cells. In conjunction with the use of inhibitors and these alternative procedures, some limited conclusions about synthetase function can be drawn.

*Variable Responses of Cells to Poly(ADP-ribose) Synthetase Inhibitors*

In our attempts to understand the function of the synthetase we have utilized both major approaches as described earlier, the measurement of synthetase and its products in synchronously growing cells as well as assessing the effects of synthetase inhibitors on cell growth and survival. Both approaches have been consistent in suggesting that there is some involvement of the synthetase in the progression of cells beyond $G_2$ (*3*, *4*). Thus the synthetase and its products were found to be maximal at this time and some synthetase inhibitors, namely 5-methylnicotinamide and nicotinamide, caused synchronously growing cells to arrest at this stage of growth. Our hypothesis that the synthetase played some role in $G_2$ seems to be at variance with the recent view that the synthetase is involved in DNA repair processes (*1*). However, the two observations are easily reconciled because the large majority of cell cycle synchronizing agents cause some DNA chain breaks and because most drug treatments which cause DNA damage also produce an arrest of the cells in $G_2$, as we have pointed out (*13*). Although there is some controversy regarding the timing of the synthetase in the cell cycle, there is general agreement that the enzyme activity is maximal sometime between the end of S and the early stages of $G_1$ (*4*, *14–16*). These differences should not be surprising since growth restriction points are observed to vary depending on the cell type, the transformation state and even the type of agent used for cell transformation (*13*).

Some recent observations demonstrate, however, that generalizations about the role of the synthetase derived from studies with synthetase inhibitors are difficult to formulate. Although our findings with various synthetase inhibitors have usually been that such compounds arrest cell growth and usually do so in $G_2$, albeit transiently in some cases (*13*), one surprising exception has been found. In attempts to evaluate the synthetase inhibitor, 3-aminobenzamide on the growth of normal cells, we felt that freshly isolated cells would be most appropriate for analyses. Consequently, such cells were isolated from the mammary glands of mice and tested for their response to the inhibitor. To our surprise, there was no observed inhibition, but rather a stimulation of cell growth (Table 1). Other complications relating to the use of synthetase include the following. The responses of cells varies depending on the particular cell types, the particular inhibitor and even the mode of cell growth. As an example of the latter phenomenon consider the response of HeLa cells to 5-methylnicotinamide. At a concentration of 10 mM, the drug blocks the growth of these cells, causes them to arrest at $G_2$ and subsequently causes cell death. When the concentration of 5-methylnicotinamide is reduced to 2 mM, there is no discernible effect on HeLa cells grown in suspension culture, but the same cells in monolayer show a 50% decline in the rate of growth and an accumulation in $G_2$. The point about the variable response is further illustrated with observations utilizing 3-aminobenzamide. At a concentration of 2 mM, this agent only reduces the cloning efficiency of Chinese hamster ovary (CHO) cells by 5% after 24 hr exposure. However, it causes as much as a 50% stimulation in the rate of amino acid uptake in these cells and does so in the absence of any induced DNA damage (Fig. 1). In HeLa cells in suspension culture, the same concentration of inhibitor produces a transient growth arrest and then the cells recover in so far as growth rate is concerned but upon analysis of the cells one

TABLE 1. The Responses of Various Cells to Poly(ADP-ribose) Synthetase Inhibitors

| Cell | Growth condition | Synthetase inhibitor | Cell response |
|---|---|---|---|
| Mammary epithelium | Primary culture | 3 AB (3 mM) | 50% stimulation of growth, differential stimulation of collagen synthesis |
| CHO cells | Monolayer | 3 AB (3 mM) | 50% stimulation of amino acid uptake; 5% reduction in cloning efficiency |
| HeLa cells | Monolayer | 3 AB (3 mM) | 20% inhibition of growth rate |
| HeLa cells | Suspension culture | 3 AB (5 mM) | Transient growth inhibition. An increase in RNA and protein content after long-term exposure |
| CHO cells | Monolayer | 3 AAB (4 mM) | 50% decrease in cloning efficiency after 24 hr exposure |
| CHO cells | Monolayer | 3 AAB (2 mM) | 99% reduction in cloning efficiency after N-methylnitrosourea treatment |
| HeLa cells | Monolayer | 5 MN (3 mM) | 50% reduction in growth rate, $G_2$ accumulation |
| HeLa cells | Suspension culture | 5 MN (3 mM) | No effect on growth |
| HeLa cells | Suspension culture | 5 MN (10 mM) | Complete growth arrest in $G_2$ Followed by cell death |

AB, aminobenzamide; AAB, acetylaminobenzamide; MN, methylnicotinamide.

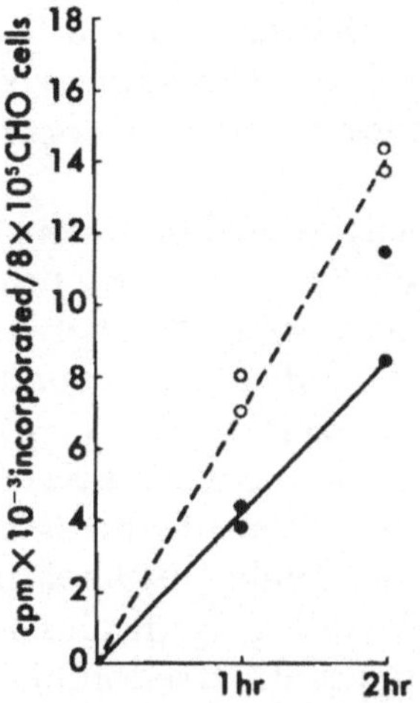


FIG. 1. Effect of 3-aminobenzamide on leucine incorporation into CHO cells. The inhibitor and labeled amino acid were added simultaneously and radioactivity incorporated determined thereafter. ○ 3-aminobenzamide; ● control.

finds that the inhibitor produces substantial increases in the average RNA and protein contents of the cells after 10 generations. One of the most potent inhibitors we have tested is 3-acetylaminobenzamide. This compound is very effective at potentiating the killing of CHO cells following their exposure to 0.1 mM methylnitrosourea. However it is also directly toxic to these cells and to HeLa cells even in the absence of induced DNA damage. The above results are summarized in Table 1.

It seems obvious then that the best experimental design for evaluating cell responses to synthetase inhibitors is one where comparisons are made on a particular cell line grown under fixed conditions. This we have done for a number of inhibitors in the benzamide series with CHO cells grown in monolayer. Analyses were made of the relative direct cytotoxicity of these inhibitors as well as their ability to reduce cell cloning efficiency after methylnitrosourea treatment. The experimental protocol followed was to plate out the cells under cloning conditions and allow them to attach for 24 hr, then subsequently expose them to various inhibitor concentrations for an additional 24 hr. Colonies were counted after about 5 days incubation in the absence of the inhibitors. Alkylating agent treatment consisted of exposure of the cells to 0.1 mM MNU with the subsequent treatment protocol as described above.

The results of these evaluations are as follows. The inhibitors were ranked according to their $K_i$, cytotoxicity potential ($I_D50$), and synergistic killing potency for cells treated with methylnitrosourea. In increasing order the $K_i$'s are: 3-acetylaminobenzamide = 3-methoxybenzamide < benzamide = 3-hydroxybenzamide ≪ 3-aminobenzamide. $I_D50$ was attained at approximately $10^4$ times the $K_i$ for all except 3-methoxybenzamide which required a 2-fold higher concentration indicating a rough correlation between $K_i$ and $I_D50$. In methylnitrosourea treated cells, however, the inhibitors show more diversity in their ability to synergistically enhance cell killing. 3-aminobenzamide is equally as potent as 3-hydroxybenzamide and benzamide at a given concentration although its $K_i$ is 3-fold higher. Additionally, 3-acetylaminobenzamide is much more potent than 3-methoxybenzamide although they are equally good inhibitors of poly(ADP-ribose) synthetase. $I_D50$ values for the benzamides were determined for HeLa cells. The values obtained were qualitatively similar to those obtained in CHO cells except 3-hydroxybenzamide became disproportionately cytotoxic at higher concentrations.

Thus even with the same series of structurally closely related inhibitors and with the same cell type, it is not possible to predict absolutely how a cell will respond to a synthetase inhibitor. What is clear is that there is as good a correlation between the $K_i$ of a compound and its $I_D50$ as there is between the $K_i$ and the compound's ability to kill cells with DNA damage produced by methylnitrosourea. The question arises as to whether cell killing in either case is *via* direct effects on the synthetase or by some other cell process, also affected. The probability is that the synthetase is the target but that there are differences in the degree of inhibitor uptake, metabolism, *etc.* (It is even possible that the synergistic killing effect of alkylating agent plus synthetase inhibitor is simply due to permeability changes for the synthetase inhibitor caused by alkylating agent treatment). There are some published reports which suggest that some synthetase inhibitors may have effects on cells other than blocking synthetase. For example, nicotinamide is an inducer of nicotinamide deamidase (*17*). Benzamide and 3-aminobenzamide are probably inhibitors of nicotinamide deamidase since both compounds are deamidated by the enzyme (*18*). Additionally, the deamidase is induced by DNA damaging treatments such as by radiation (*19*). Benzamide is also an inhibitor of certain proteases (*20*) and the compound has been reported to disrupt the microfibrillar network in the slime mold (*21*).

### *Heat Shock Effects on Poly(ADP-ribose) Synthetase*

Because of some uncertainties regarding results obtained with synthetase inhibitors, we felt that other methods might also be of value in assessing synthetase

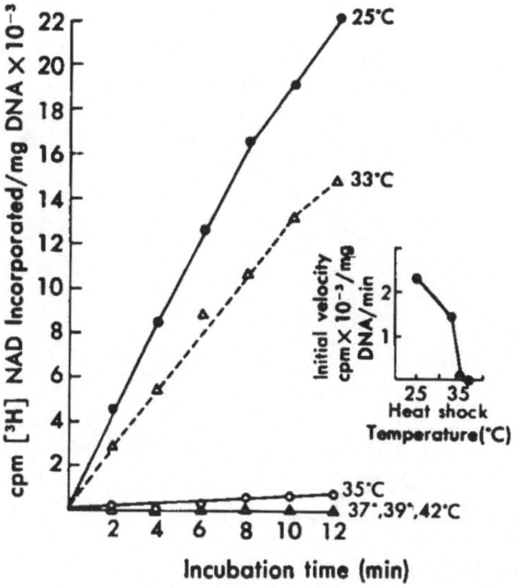


FIG. 2. Effect of heat shock temperature on synthetase activity in *D. melanogaster* cells. Cells were heat shocked for 5 min at the indicated temperature and synthetase measurements made on isolated nuclei. (Reproduced from Radiat. Res., *90*: 187, 1982).

function. One such method was suggested during our attempts to characterize poly-(ADP-ribose) synthetase activity in *Drosophila melanogaster* cells. The growth optimum for these cells is about 29°C. Measurements of the synthetase activity either in permeabilized cells or in isolated nuclei indicated that synthetase activity was heat labile. Not only was the enzyme labile in the subcellular state, but it was also found to be inactivated when intact cells were briefly heat-shocked (*22*). This is depicted in Fig. 2. Pretreatment of the cells for 5 min before assaying the synthetase *in vitro* under optimal temperature conditions caused a total loss of assayable synthetase when the heat shock temperature was 35–37°C. At a heat shock temperature of 33°C, the enzyme was reduced by 35%. To determine whether the synthetase levels measured *in vitro* really reflected changes in the synthetase *in vivo*, radioimmunoassays were made of polymer levels before and after heat shocking and were found to be reduced about 5-fold by a 5 min heat shock at 35°C. Most interestingly, the heat shock at this temperature had no detectable effect on the subsequent viability of *Drosophila* cells (*22*). Comparisons of the heat shock lability of the synthetase enzyme with two other chromatin bound enzymes, RNA and DNA polymerases, also showed that the synthetase was uniquely temperature sensitive.

Having found a condition under which the synthetase could be rendered inactive in intact cells, we then asked whether the repair of single strand DNA breaks could take place in the absence of detectable synthetase. *Drosophila* cells were irradiated after prelabeling their DNA and the strand rejoining process was analyzed during the interval before synthetase activity was regenerated. This analysis indicated that the repair process progressed to the same extent in heat shocked and control cells (*22*). Using an alkaline elution assay (*23*) which is very sensitive, we have confirmed reports that synthetase inhibitors do marginally slow the rate of single strand break repair (*24*, *25*) but our results show unequivocably that this type of repair is

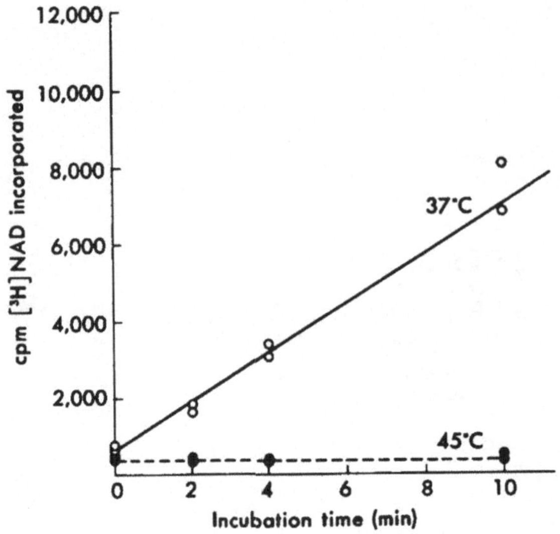


FIG. 3. Heat shock effect on synthetase activity in HeLa cells. Cells were rapidly adjusted to 45°C and maintained at this temperature for 30 min, then rapidly cooled. Nuclei were isolated and assayed for residual activity.

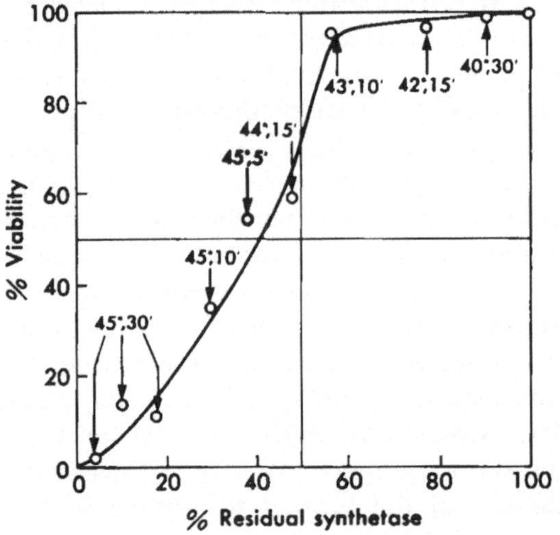


FIG. 4. Comparison of heat shock effects on synthetase activity and HeLa cell viability. Cells were heat shocked as described and cell aliquots analyzed for nuclear synthetase activity or viability by cloning efficiency measurements.

fully accomplished in heat shocked cells with no detectable synthetase, albeit at a slightly slowed rate. It is therefore, not surprising that James and Lehmann failed to observe any effects of 3-aminobenzamide on the viability of irradiated diploid fibroblasts (*26*).

*D. melanogaster* cells are not unique in so far as the differential effects of heat shock on poly(ADP-ribose) synthetase are concerned. Heat shock reduces the synthetase in HeLa cells and CHO cells. The effects of a 30 min heat shock at 45°C

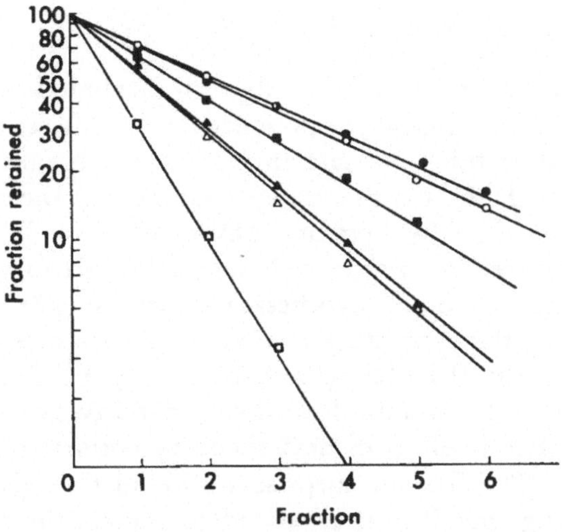


FIG. 5. Ability of heat-shocked HeLa cells to repair $\gamma$-ray induced DNA chain breaks. Cells were heat-shocked for 15 min at 45°C, rapidly cooled and irradiated (600 rads). Repair was permitted by incubation at 37°C as indicated. Chain break repair was assessed using the alkaline elution technique. ● 10′; ○ H 10′; ■ C 5′; ▲ H 5′; △ HO′; □ CO′.

on the synthetase, measured in isolated HeLa cell nuclei, is depicted in Fig. 3. Heat shock before isolating the cell nuclei results in a near complete inactivation of the synthetase. Both HeLa and CHO cells differ from the *Drosophila* cells in that the synthetase cannot be completely inactivated without considerable loss of cell viability. This is shown in Fig. 4. About 50% of the synthetase activity can be inactivated by heat shocking the HeLa cells for 10 min at 43°C without much loss in cell viability. At increased temperatures or longer times of heat shock, the cell viability is rapidly lost along with further reductions of synthetase activity.

In spite of the limitations in interpreting the capability of heat shocked HeLa cells to repair single strand breaks in DNA, we attempted such an experiment. Cells were heat shocked for 15 min, then cooled in ice and irradiated in parallel with non-heat shocked controls. The cells were then incubated to permit repair and analyses of the DNA elution rates under alkaline conditions was performed (*23*). As demonstrated in Fig. 5, the interpretation of these experiments was not clear-cut. The apparent number of DNA chain breaks in the heat shocked cells, at zero repair time, was considerably less than the controls. The controls had repaired DNA breaks extensively after 5 min incubation at 37°C while the heat shocked cells appeared not to accomplish any repair. By 10 min of repair time, the DNA elution rates were the same in heat shocked and control cells. We interpret these results to mean that artefacts in the elution rate are produced by the heat shock process, possibly by denatured proteins which slows down the rate of DNA unwinding under alkaline conditions. What is important in the experiments is that there is considerable strand rejoining in HeLa cells following a heat shock treatment that inactivates about 80% of the synthetase in these cells. Therefore strand rejoining in HeLa as well as *Drosophila* cells is probably not absolutely dependent on synthetase activity.

*Stimulation of Poly(ADP-ribose) Synthetase Activity in Intact Cells*

Previously we reported that treatment of NRK cells with picolinic acid resulted in an accumulation of the cells in $G_2$ and elevated polymer levels (*27*). These results were seen in the absence of any detectable increase in DNA chain breaks as determined with the alkaline elution method. It is therefore apparent that the synthetase may be activated by means other than by breaking DNA, although DNA chain breaks clearly activate the synthetase. We postulated that some sort of conformational changes in the chromatin might promote synthetase activation (*27*).

A second method of activating the synthetase in intact cells has now been discovered. Because heat shock treatment of HeLa cells that not only inactivated poly-(ADP-ribose) synthetase also caused a dramatic loss of cell viability, we attempted to protect HeLa cells against the killing effect of heat shock by performing the heat shock in the presence of glycerol (*28*). To our surprise we found that glycerol not only blocked the heat shock killing, but it also appeared to protect the synthetase against heat shock inactivation. On closer examination, however, the glycerol treatment was found to activate the synthetase in non-heat shocked controls. This activation was rapid, occurring after 1/2 hr treatment, and it was apparently produced in the absence of DNA chain breaks. Radioimmunoassays were performed on control and glycerol treated cells and the results confirmed that synthetase activity was elevated in intact cells treated with glycerol (*29*). While the significance of the glycerol stimulation of the synthetase remains to be elucidated, the technique may be useful in permitting an analysis of mechanisms involved in enzyme activation. Our guess is that some type of metabolic pertubation of the cell leads to a stimulation of synthetase, possibly *via* altering the chromatin structure with which the synthetase is associated, or by a direct modification of the enzyme itself.

## REFERENCES

1. Purnell, M. R., Stone, P. R., and Whish, W.J.D. ADP-ribosylation of nuclear proteins. Biochem. Soc. Trans., *8*: 215–227, 1980.
2. Hayaishi, O. and Ueda, K. Poly(ADP-ribose) and ADP-ribosylation of proteins. Annu. Rev. Biochem., *46*: 95–116, 1977.
3. Kidwell, W. R. and Burdett, K. E. Poly(ADP-ribose) synthesis and cell division. Biochem. Biophys. Res. Commun., *61*: 766–773, 1974.
4. Kidwell, W. R. and Mage, M. Changes in poly(ADP-ribose) and poly(ADP-ribose) synthetase in synchronous HeLa cells. Biochemistry, *15*: 1213–1217, 1976.
5. Stone, P. R., Lorimer III, W. S., and Kidwell, W. R. Properties of the complex between histone H1 and poly(ADP-ribose) synthesized in HeLa cell nuclei. Eur. J. Biochem., *81*: 9–18, 1977.
6. Giri, C. P., West, M.H.P., and Smulson, M. Nuclear protein modification and chromatin substructure. Biochemistry, *17*: 3495–3504, 1978.
7. Terada, T., Yamada, M., Fujiki, H., Nakayasu, M., Umezawa, K., Matsushima, T., and Sugimura, T. Differentiation of leukemic cells and poly(ADP-ribose). *In;* M. Smulson and T. Sugimura (eds.), Novel ADP-ribosylations of Regulatory Enzymes and Proteins, pp. 251–260, Elsevier/North-Holland Publ. Co., New York, 1980.
8. Kanai, M., Miwa, M., Kondo, T., Tanaka, Y., Nakayasu, M., and Sugimura, T.

Involvement of poly(ADP-ribose) metabolism in induction of differentiation of HL-60 promyelocytic leukemia cells. Biochem. Biophys. Res. Commun., *105*: 404–411, 1982.
9. Juarez-Salinas, H., Sims, J. L., and Jacobson, M. K. Poly(ADP-ribose) levels in carcinogen treated cells. Nature, *282*: 740–741, 1979.
10. Althaus, F. R., Lawrence, S. D., Sattler, G. L., and Pitot, H. C. The effect of nicotinamide on unscheduled DNA synthesis in cultured hepatocytes. Biochem. Biophys. Res. Commun., *95*: 1063–1070, 1980.
11. Durkacz, B. W., Omidiji, O., Gray, D. A., and Shall, S. (ADP-ribose)$_n$ participates in DNA excision repair. Nature, *283*: 593–596, 1980.
12. Miwa, M., Kanai, M., Kondo, T., Hoshino, H. O., Ishihara, K., and Sugimura, T. Inhibitors of poly(ADP-ribose) polymerase enhance unscheduled DNA synthesis in human peripheral lymphocytes. Biochem. Biophys. Res. Commun., *100*: 463–470, 1981.
13. Kidwell, W. R., Nolan, N., and Stone, P. R. Variations in poly(ADP-ribose) synthetase and poly(ADP-ribose) in synchronously dividing HeLa cells. *In;* K. Ueda and O. Hayaishi, (eds.), ADP-ribosylation Reactions, pp. 373–388, Academic Press, New York, 1982.
14. Berger, N. A., Weber, G., and Kaichi, A. S. Characterization and Comparison of poly(ADP-ribose) synthesis and DNA synthesis in nucleotide-permeable cells. Biochem. Biophys. Acta, *519*: 87–104, 1978.
15. Kanai, Y., Tanuma, S., and Sugimura, T. Immunofluorescent staining of poly(ADP-ribose) *in situ* in HeLa cell chromosomes in the M phase. Proc. Natl. Acad. Sci. U.S., *78*: 2801–2804, 1981.
16. Miwa, M., Sugimura, T., Inui, N., and Takayama, S. Poly(ADP-ribose) synthesis during the cell cycle of transformed hamster lung cells. Cancer Res., *33*: 1306–1309, 1973.
17. Curt, B. and Porcellati, G. The nicotinamidase activity extracted from pigeon liver. Ital. J. Biochem., *12*: 12–26, 1963.
18. Bray, H. G., James, S. P., Ryman, B. E., and Thorpe, W. V. The fate of certain organic acids and amides in the rabbit. Biochem. J., *42*: 274–281, 1948.
19. Zicha, B., Buric, L., and Benes, J. Effect of irradiation on the activity of nicotinamide deamidase of rat liver. Stralentherapie, *137*: 183–186, 1969.
20. Bender, M. L., Gibian, M. J., and Whelan, D. J. The alkaline pH dependency of chymotrypsin reactions; Postulation of a pH-dependent intramolecular competitive inhibition. Proc. Natl. Acad. Sci. U.S., *56*: 833–839, 1966.
21. Korohoda, W. and Wohlfarth-Botterman, K. E. Effects of relaxation and contraction stimulating solutions on the structure of cytoplasmic fibrils in plasmodia of *Physarium polycephalum*. Acta Protozool., *15*: 195–202, 1976.
22. Nolan, N. and Kidwell, W. R. Effect of heat shock on poly(ADP-ribose) synthetase and DNA repair in *Drosophila* cells. Radiat. Res., *90*: 187–203, 1982.
23. Blakly, W. F., Ward, J. F., and Joner, T. I. A quantitative assay of DNA strand breaks and their repair in mammalian cells. Anal. Biochem., *124*: 125–133, 1982.
24. Durkacz, B. W., Irwin, J., and Shall, S. The effect of inhibition of (ADP-ribose)$_n$ biosynthesis on DNA repair assayed by the nucleoid technique. Eur. J. Biochem., *121*: 65–69, 1981.
25. Zwilling, L. A., Kerrigan, D., and Pommier, Y. Inhibitors of poly(ADP-ribose) synthesis slow the resealing rate of X-ray induced DNA strand breaks. Biochem. Biophys. Res. Commun., *104*: 897–902, 1982.
26. James, M. R. and Lehmann, A. R. Role of poly(ADP-ribose) in DNA repair in human fibroblasts. Biochemistry, *21*: 4007–4013, 1982.

27. Kidwell, W. R., Stone, P. R., and Johnson, G. S. Poly(ADP-ribose) synthesis and chromatin structure. *In;* M. Smulson and T. Sugimura (eds.), Novel ADP-Ribosylations of Regulatory Enzymes and Proteins, pp. 73–82, Elsevier/North-Holland Publ. Co., New York, 1980.
28. Henle, K. J. and Warters, R. L. Heat protection by glycerol *in vitro*. Cancer Res., *42*: 2171-2176, 1982.

ADP-RIBOSYLATION, DNA REPAIR AND CANCER, M. MIWA ET AL. (EDS.),
JAPAN SCI. SOC. PRESS, TOKYO/VNU SCIENCE PRESS, UTRECHT, PP. 253-265, 1983

# DNA Polymerase Function in Repair Synthesis in Human Fibroblasts

Steven L. DRESLER and Michael W. LIEBERMAN
*Department of Pathology, Washington University School of Medicine, St. Louis, Missouri 63110, U.S.A.*

*Abstract:* We have analyzed DNA polymerase function in repair synthesis in human diploid fibroblasts. Previous studies from many laboratories have resulted in confusing and apparently contradictory findings; our studies suggest a unifying hypothesis and indicate that previous results may not actually be in conflict. Nongrowing (confluent) cells were damaged with N-methyl-N-nitrosourea (MNU), bleomycin, X-ray, UV radiation, or N-acetoxy-2-acetylaminofluorene (NA-AAF) over a wide range of doses, and repair synthesis was studied in the presence of one of three polymerase inhibitor, aphidicolin, dideoxythymidine triphosphate (ddTTP), or N-ethyl maleimide. We find that both polymerase $\alpha$ and a non-$\alpha$ polymerase, probably polymerase $\beta$, are involved in repair synthesis. Furthermore, there is a dose dependence for polymerase function: at low doses of damage repair synthesis is mediated primarily by the non-$\alpha$ polymerase; as the amount of damage is increased, polymerase $\alpha$ participates to an increasing extent and at high levels of damage is responsible for ~50–80% of the repair synthesis. In a study in rapidly growing cells, repair synthesis following a high dose of UV was refractory to aphidicolin; these findings indicate that in growing and quiescent cells DNA polymerases may function differently in repair synthesis. All of the above results may be analyzed in terms of a simple model in which two repair synthesis systems, each involving one of the polymerases, compete to put repair patches into damaged DNA.

As in so many other areas of eukaryotic molecular biology, the analysis of the enzymology of DNA repair lags far behind our understanding of similar processes in prokaryotes. In fact, the enzymology of DNA excision repair in mammalian cells has proved to be surprisingly intractable when standard biochemical methods are used. In spite of the intensive efforts of more than a dozen laboratories around the world we know relatively little of the molecular details of this process. Because of an extensive literature characterizing the three known eukaryotic DNA polymerases ($\alpha$, $\beta$, and $\gamma$; $\delta$ for present purposes is considered "$\alpha$-like") (*1*) and the availability of some relatively specific inhibitors of each, one may begin to probe the function of these polymerases in mammalian DNA excision repair. In this paper we summarize previous attempts to analyze DNA polymerase function in repair and present some novel findings of our own.

### *Early Attempts to Analyze DNA Polymerase Function in DNA Excision Repair*

In the mid 1970s an Italian group and our laboratory independently began to examine human cell lines with known or presumed repair deficiencies for absent or reduced levels of DNA polymerases (*2–4*). Implicit in these studies was the hypothesis that one or a number of these lines might contain a polymerase mutation similar to *Escherichia coli pol A* mutations. It was found, however, that all the lines analyzed, including representatives of xeroderma pigmentosum complementation groups A, C, D, and E, Bloom's syndrome, Fanconi's anemia, and ataxia telangiectasia, had normal or near normal levels of polymerases $\alpha$, $\beta$, and $\gamma$.

Another approach, used chiefly by Spadari and his collaborators, involved physiologic manipulations of cells to analyze polymerase function. For example, in phytohemagglutinin-stimulated lymphocytes, levels of DNA polymerases $\alpha$ and $\beta$ were followed and compared to repair synthesis of following damage with UV radiation (*5*). On the basis of correlation between levels of polymerase $\beta$ and repair synthesis, the investigators concluded that polymerase $\beta$ was the repair polymerase. A similar, more recent paper from this group analyzed repair in rat brain neurons following UV damage. Since these cells contain only polymerase $\beta$ and perform excision repair of UV damage, the group again concluded that this polymerase was involved in repair (*6*). Ingenious as these experiments are, they analyze systems which are very specialized and probably do not provide a complete description of polymerase function in excision repair.

### *Studies of Repair Synthesis with Inhibitors of DNA Polymerases*

Recently investigators have used inhibitors of eukaryotic DNA polymerases to study repair synthesis in mammalian cells. Aphidicolin is a highly specific inhibitor of polymerase $\alpha$ (*7–11*) while dideoxythymidine triphosphate (ddTTP) inhibits polymerase $\beta$ more strongly than polymerase $\alpha$ (*11–16*), and N-ethyl maleimide inhibits polymerases $\alpha$ and $\gamma$ but has little effect on polymerase $\beta$ (*17–20*). Many studies have used aphidicolin to examine the role of polymerase $\alpha$ in the repair of UV damage in both intact and permeable cells (*21–30*). In general, these studies have implicated polymerase $\alpha$ in the repair of UV damage, and in some reports 90% or more of the repair synthesis was thought to be mediated by polymerase $\alpha$ (*22, 23*). In contrast, recent studies using aphidicolin from both Spadari's laboratory (*24, 25*) and Regan's laboratory (*30*) suggest an important role for polymerase $\beta$ or some other non-$\alpha$ polymerase in the repair of UV damage. Investigations of the role of polymerase function in the repair of "short patch and/or X-ray mimetic" agents such as N-methyl-N-nitrosourea (MNU), N-methyl-N-nitro-N-nitrosoguanidine (MNNG), bleomycin, neocarzinostatin, and X-ray, itself, have led to another seemingly incompatible set of conclusions (*21, 22, 29–34*). Repair synthesis after MNNG and MNU was only partially inhibited by aphidicolin or ddTTP and thus seemed to involve both $\alpha$ and $\beta$ polymerases (or additional polymerases or complexes not sensitive to these agents) (*21, 22, 29, 34*). In contrast, repair of X-ray damage is reported to be largely independent of DNA polymerase $\alpha$ (*33*). Most of the data on bleomycin suggest that polymerase $\beta$ is involved in its repair; however, a recent paper indicates that repair synthesis following bleomycin is about 50% inhibitable by aphidicolin (*29*). The above results suggest that DNA polymerase function in DNA repair synthesis is a complex phenomenon which can only be studied using a comprehensive approach.

### *Initial Studies from Our Laboratory*

Several years ago we began to develop a permeable cell system to study the enzymology of DNA repair in human cells (*35*). Subsequent improvements have made the system rapid and convenient for studying repair synthesis (*36*). Confluent (nondividing) human diploid fibroblasts are made permeable by suspending the cells in a Tris/$MgCl_2$/EDTA/sucrose solution at 4°C in a dounce homogenizer, subjecting them to five strokes with a loose pestle, and maintaining them at 4°C for 30 min (*36*). Such preparations are permeable to both low molecular weight compounds (*e.g.*, dNTPs, ATP, ddTTP) as well as at least some proteins (*e.g.*, *Micrococcus luteus* UV endonuclease) and allow the study of all aspects of excision repair from incision to ligation and rearrangement. Figure 1 presents the time course of repair synthesis in permeable human cells damaged with UV. Under these conditions incorporation is linear or nearly linear for about 30 min, and the degree of stimulation is large enough to be measured directly by trichloroacetic acid precipitation thus obviating the need for density shift experiments. Repair synthesis following UV damage in this system shows the expected dependence on dNTPs and ATP (Table 2), and other studies have characterized its salt dependencies and pH optimum (*35, 36*).

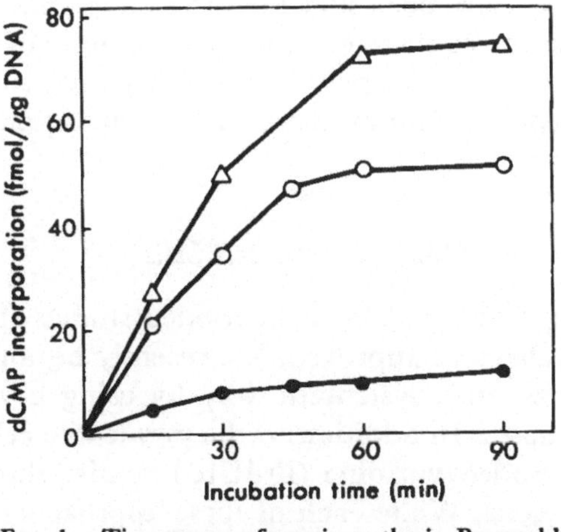


FIG. 1. Time course of repair synthesis. Permeable fibroblasts (AG1518), irradiated with 0 (●), 40 (○), or 400 J/m² (△) UV, were incubated with reaction mix for the times indicated. The mix contained 40 mM Tris (pH 7.6 at 37 C), 8 mM $MgCl_2$, 15 mM KCl, 5 mM ATP, 3 μM TTP, 3 μM dCTP, 3 μM dGTP, 3 μM dATP, 20 μCi/ml [$\alpha$-$^{32}$P]dCTP (410 Ci/mmol), 167 mM sucrose, 2 mM dithiothreitol, and 0.67 mM EDTA. Each point is the average of two determination. (Reproduced with permission from ref. *36*).

TABLE 1. Effect of Repair Synthesis of Various Modifications of the Reaction Mixture[a)]

| Modification | Repair synthesis (%) |
|---|---|
| None (control) | 100 |
| −TTP | 16 |
| −Unlabeled dCTP[b)] | 56 |
| −dGTP | 18 |
| −dATP | 95 |
| −TTP, dGTP, dATP, unlabeled dCTP[b)] | 7 |
| −ATP | 0 |
| −ATP, +GTP | 0 |
| +GTP | 107 |
| +1 mM spermidine | 53 |
| +53 μM novobiocin | 95 |
| +530 μM novobiocin | 49 |
| +1,060 μM novobiocin | 0 |
| +2.6 mM nalidixic acid | 130 |
| +150 μM dideoxy-TTP | 89 |
| +25 μM aphidicolin | 10 |

a) Permeable AG1518 cells, unirradiated and irradiated with 20 or 40 J/m² UV, were incubated for 15 min with reaction mixes modified as indicated from that described in the legend to Fig. 1. Repair synthesis was determined as described in ref. *36* from which the table is reproduced. Each value is the average of two determinations. (Reproduced with permission from ref. *36*). b) Unlabeled dCTP was omitted, but 20 μCi/ml [α-$^{32}$P]-dCTP was present, giving a final dCTP concentration of 57 nM.

We used this system to study synthesis following UV damage and found after relatively high doses of UV (20–40 J/m²) repair synthesis was about 90% inhibited by aphidicolin but only about 10% inhibited by ddTTP (Table 1). Taken at face value these data suggest that DNA polymerase $\alpha$ is responsible for most of the repair synthesis in nondividing human fibroblasts. We also found that high doses of novobiocin inhibited repair synthesis. While it has been suggested that such data represent evidence for the role of topoisomerases in repair (*e.g.*, *37*), the high doses of inhibitor required and the known effects of novobiocin on polymerase $\alpha$ (*e.g.*, *38*) suggest the possibility that novobiocin inhibits repair synthesis primarily by inhibition of polymerase $\alpha$.

### *Studies of DNA Repair Synthesis Function with Aphidicolin in Intact Cells*

Much of the confusion stemming from the above-mentioned studies (including our own) results from lack of a comprehensive approach. We recently began to study DNA polymerase involvement in repair in a systematic way by using a variety of different agents over a wide range of doses. In addition to the permeable cell experiments, we have also performed 5-bromodeoxyuridine (BrdUrd) density shift experiments and autoradiography in intact cells. While each of these approaches has certain drawbacks, together they help provide an understanding of many apparently incongruous results.

Our first experiments were designed to find a dose of aphidicolin which would produce maximal inhibition of damage-stimulated repair synthesis. Confluent fibroblasts were damaged with relatively high levels of four agents [12 J/m² UV;

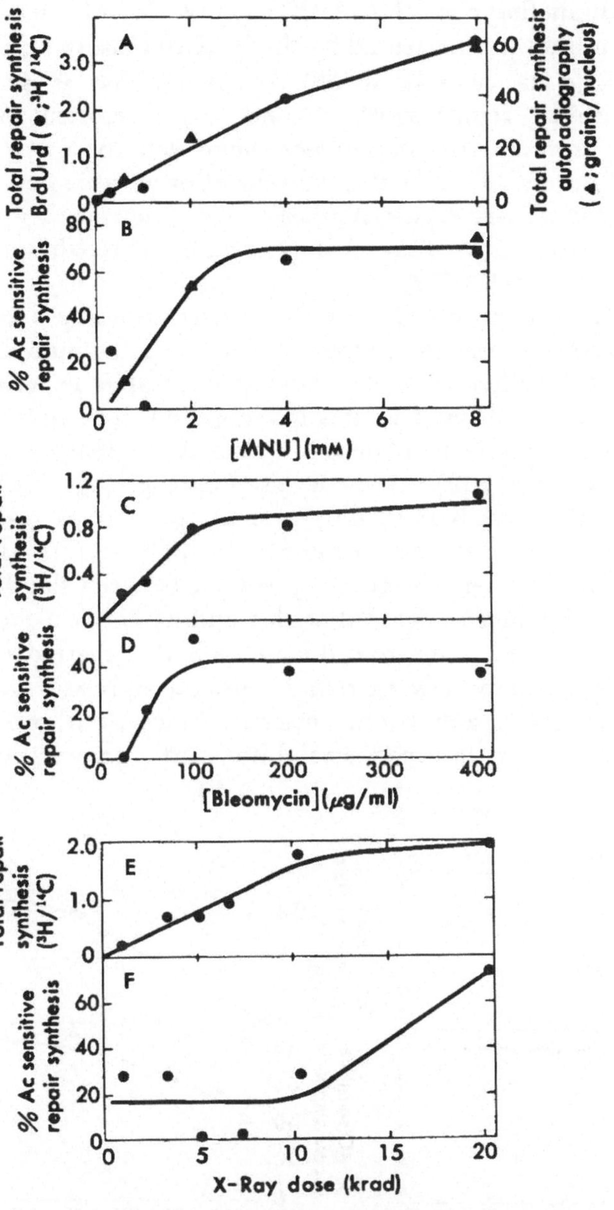


FIG. 2. Inhibition by aphidicolin of DNA repair synthesis induced by various doses of MNU, bleomycin and X-ray. Confluent fibroblasts were damaged with the indicated doses of either MNU (A, B), bleomycin (C, D), or X-ray (E, F) and repair synthesis was determined either without aphidicolin or in the presence of 65 μM aphidicolin by the BrdUrd density shift technique (●) or by autoradiography (▲); see ref. *39*. In the autoradiography experiment, undamaged cells incorporated 1.5 grains/nucleus in the absence of aphidicolin and 1.1 grains/nucleus in the presence of 65 μM aphidicolin. Total repair synthesis is the amount of repair synthesis observed in the absence of aphidicolin. Percent aphidicolin (Ac) resistant repair synthesis was calculated as: 100 × (1 − (repair synthesis in presence of aphidicolin/repair synthesis in absence of aphidicolin)). (Reproduced from ref. *39*).

40 μM N-acetoxy-2-acetylminofluorene (NA-AAF); 6 mM MNU; or 100 μg/ml bleomycin], and repair synthesis was measured by the BrdUrd density shift technique in alkaline CsCl with [$^3$H]dThd as a label (*39*). We found that as little as 1 μM aphidicolin partially inhibited repair synthesis and 16 μM resulted in maximal inhibition. To be certain, however, that studies were done well above the $K_i$, a dose of 65 μM was chosen for future studies. At this concentration of aphidicolin and the levels of damage mentioned above, repair synthesis in response to UV, NA-AAF, and MNU was inhibited ~60–70%, while bleomycin-induced repair synthesis was inhibited to a lesser extent (20–25%) (*39*).

A series of experiments was performed to analyze the relationship between levels of damage (and repair synthesis) and the degree to which repair synthesis was aphidicolin sensitive in intact, nondividing cells. Both density shift experiments and autoradiographic experiments were used in this investigation. The data for MNU, bleomycin and X-rays are striking in that there is a clear dependence of aphidicolin sensitivity on degree of damage (Fig. 2). At low levels of damage (*e.g.*, 0.5 mM MNU, 25 μg/ml bleomycin, and <5 krad X-ray), only a small percentage (~0–20%) of the repair synthesis is aphidicolin sensitive. In contrast, at high levels of damage, 40–75% is aphidicolin sensitive. Further, the good agreement between the density shift data and the autoradiography data indicate that this finding is not a measurement artifact (Fig. 2A, B). Another conclusion from these data is that there does not seem to be a straightforward relationship between relative amount of repair synthesis and the degree to which the process is aphidicolin sensitive. Since aphidicolin is a very specific inhibitor of polymerase α, the most straightforward interpretation of these

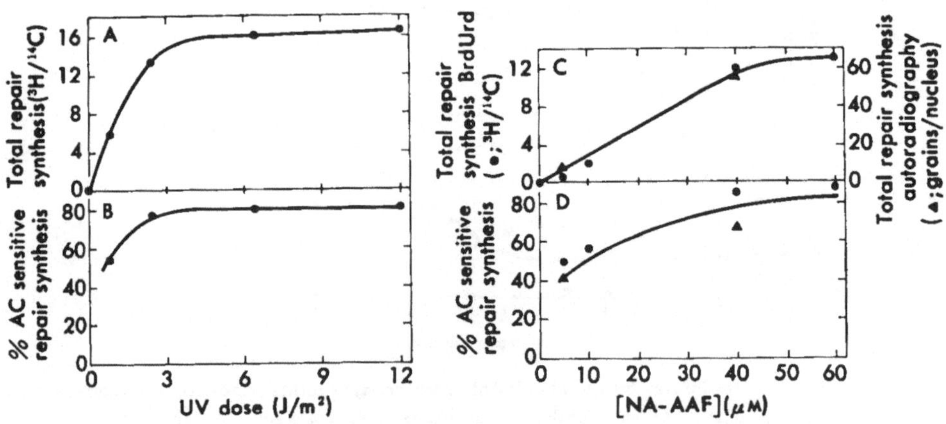


FIG. 3. Inhibition by aphidicolin of repair synthesis induced by various doses of UV and NA-AAF. Confluent fibroblasts were damaged with the indicated dose of either UV (A, B) or NA-AAF (C, D) and repair synthesis was determined either without aphidicolin or in the presence of 65 μM aphidicolin by the BrdUrd density shift technique (●) or by autoradiography (▲); see ref. *39*. In the autoradiography experiment, undamaged cells incorporated 1.5 grains/nucleus in the absence of aphidicolin and 1.1 grains/nucleus in the presence of 65 μM aphidicolin. Total repair synthesis is the amount of repair synthesis observed in the absence of aphidicolin. Percent aphidicolin (Ac) sensitive repair synthesis was calculated as described in the legend to Fig. 2. (Reproduced from ref. *39*).

data is that at low levels of damage repair synthesis is largely mediated by a non-$\alpha$ polymerase while at higher levels of damage polymerase $\alpha$ plays a major role.

We performed similar studies in which UV and NA-AAF were the damaging agents (Fig. 3). Once again, the dose dependent relation between amount of damage and percent aphidicolin sensitive repair synthesis is seen. With these agents, however, it is difficult to use levels of damage sufficiently low to reduce the percent aphidicolin sensitive to the low levels seen with the other agents. At high levels of damage, however, repair synthesis is more than 80% aphidicolin sensitive. With NA-AAF damage as with MNU damage, density shift data, and autoradiography data are in good agreement. Thus it is clear that with five different agents the amount of damage is closely related to the aphidicolin sensitivity of repair synthesis. Further, although quantitative differences may exist, this relationship appears to hold both for agents which are repaired principally by a "short patch" mechanism (incorporation of only a few nucleotides; MNU, X-ray) and for those repaired by a "long patch" mechanism (30–70 nucleotides incorporated; UV, NA-AAF) (*40*).

*Studies of DNA Repair Synthesis with Aphidicolin, ddTTP, and N-Ethyl Maleimide in Permeable Cells*

We have extended our studies of repair synthesis in intact nondividing cells by using our permeable cell system for additional inhibitor studies. Intact cells were damaged with 12 J/m² UV, 40 $\mu$M NA-AAF, 6 $\mu$M MNU, and 100 $\mu$g/ml bleomycin, made permeable, and assayed for repair synthesis at increasing concentrations of aphidicolin or ddTTP (Fig. 4). Because these cells are damaged prior to being made

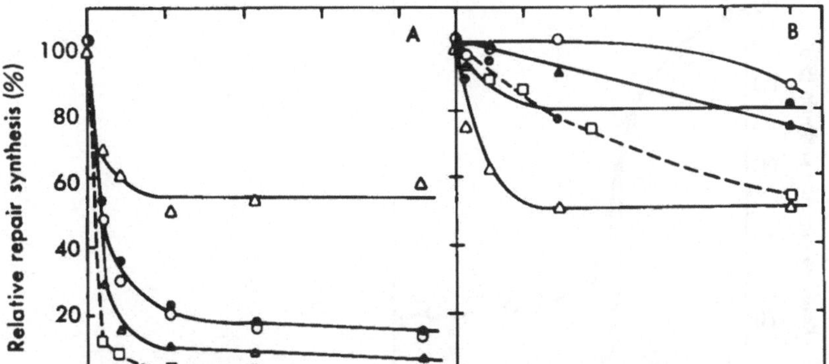


FIG. 4. Inhibition by aphidicolin or ddTTP of DNA repair synthesis in permeable human fibroblasts. Intact confluent fibroblasts were damaged with either 12 J/m² UV (●), 40 $\mu$M NA-AAF (○), 6 mM MNU (▲), or 100 $\mu$g/ml bleomycin (△) as described in ref. *39*. The cells were then made permeable and repair synthesis was assayed in the presence of the indicated concentrations of either (A) aphidicolin or (B) ddTTP. See legend to Table 1 and Fig. 1 for assay conditions. Inhibition of replicative synthesis (---□---) in growing cells, made permeable and assayed under the same conditions as were used for assaying repair, was also determined. (Reproduced from ref. *39*).

permeable, data from permeable and intact cells are directly comparable. Inhibition of replicative synthesis in permeable growing cells was also studied. In permeable cells, repair synthesis following a high dose of UV, NA-AAF, or MNU is 80–90% inhibitable by aphidicolin (Fig. 4A), a finding in good agreement with data from intact cells (Figs. 2, 3). Likewise in both systems, repair synthesis in response to 100 μg/ml bleomycin is about 50% inhibited by aphidicolin.

It is known that high levels of dNTPs can antagonize the effects of aphidicolin on replicative synthesis (*41*, *42*). Conceivably in intact cells damage might alter dNTP pools: by alkylating nucleotides or nucleosides or by damaging enzymes involved in their production, high levels of damaging agents could lower dNTP pools and increase the effectiveness of aphidicolin in blocking repair synthesis. Our permeable cell data rule out this possibility as the cause of the observed relationship between level of damage and the aphidicolin sensitivity of repair synthesis since the four dNTPs are supplied at 3 μM in the reaction mix. (The data in Table 1 indicate that the level of residual cellular dNTPs in permeable cells is negligible.).

When ddTTP is used to inhibit repair synthesis induced by these agents, a complementary pattern emerges. Even at high levels of ddTTP (at 100 μM ddTTP, ddTTP/dTTP=33) repair synthesis induced by UV, NA-AAF, and MNU was reduced only 15–25%, while following bleomycin damage inhibition was about 50%. Since ddTTP is known to inhibit polymerase β more readily than polymerase α, these data are additional evidence that at high doses of UV, NA-AAF, and MNU damage, DNA repair synthesis proceeds largely *via* polymerase α. In addition the bleomycin data in Fig. 2D are in good agreement with the data in Fig. 4: at a concentra-

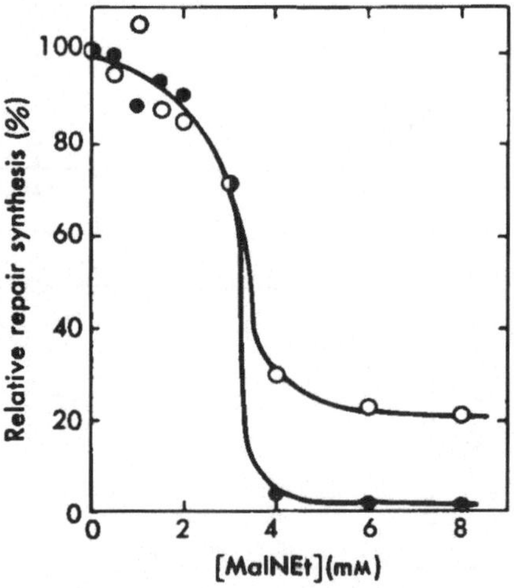


FIG. 5. Inhibition by N-ethyl maleimide of DNA repair synthesis in permeable human fibroblasts. Intact confluent fibroblasts were damaged with either 12 J/m² UV (●) or with 25 μg/ml bleomycin (○) as described in ref. *39*. The cells were then made permeable and repair synthesis was assayed in the presence of the indicated concentration of N-ethyl maleimide. Assay conditions are described in the legend to Table 1 and Fig. 1. Note that the levels of N-ethyl maleimide which inhibit repair synthesis in this system cannot be compared directly with values from the literature since 2 mM dithiothreitol (equivalent to 4 mM-SH) is present in the reaction. (Reproduced from ref. *39*).

tion of 100 μg/ml bleomycin-induced repair synthesis is about 50% inhibited by aphidicolin and 50% by ddTTP. Hence under these conditions it appears that repair synthesis is about 50% mediated by polymerase α and about 50% mediated by a non-α polymerase, probably polymerase β.

We have taken advantage of the sensitivity of polymerases α and γ to N-ethyl maleimide and the resistance of polymerase β to this agent in other studies of polymerase specificity. Confluent fibroblasts were damaged with 25 μg/ml bleomycin or 12 J/m² UV, and repair synthesis was measured in permeable cells in the presence of increasing levels of N-ethyl maleimide (Fig. 5). UV-induced repair synthesis can be completely inhibited by this agent while bleomycin-induced repair synthesis is only about 80% inhibited. These data are consistent with polymerase α involvement in the repair synthesis seen at high levels of UV damage. The bleomycin data suggest that at this level of damage some of the repair synthesis is mediated by polymerase β and thus are in agreement with the data presented in Fig. 2D. These N-ethyl maleimide data as well as the low concentration of polymerase γ in the nucleus (*19, 43, 44*) suggest that polymerase γ does not play a major role in nuclear excision repair synthesis.

### *Repair Synthesis in Dividing Cells*

To determine whether these findings were unique to nondividing cells or were representative of all cells, we performed an autoradiography experiment in which nondividing (confluent) fibroblasts and growing fibroblasts were damaged with 12 J/m² UV and repair synthesis was measured in the presence or absence of 65 μM aphidicolin (Table 2). As expected, in confluent cells repair synthesis was about 80% inhibited by aphidicolin (compare with Figs. 3B, 4A). However, in growing cells no inhibition of repair synthesis was seen even though the concentration of aphidicolin used is many times higher than that needed to inhibit replicative synthesis completely in growing cells (Fig. 4A). These findings, in agreement with those from other laboratories (*e.g.*, *30*), demonstrate that the physiologic state of the cell plays a major role in how repair synthesis occurs. It may be that polymerase α in these cells is con-

TABLE 2. Aphidicolin Inhibition of UV-induced Repair Synthesis in Confluent and Growing Fibroblasts

| State of cells | Repair synthesis (grains/nucleus) | | % inhibition by aphidicolin |
|---|---|---|---|
| | Control | +Aphidicolin | |
| Confluent | 48.1 (±1.0) | 9.1 (±0.6) | 81 |
| Growing | 26.5 (±0.8) | 29.2 (±1.3) | 0 |

Confluent or growing fibroblasts, grown on glass slides, were damaged with 12 J/m² UV and incubated for 2 hr in medium containing 5 μCi/ml [³H]dThd, either without aphidicolin or in the presence of 65 μM aphidicolin. The slides were processed for autoradiography and grains were counted. Undamaged confluent cells incorporated a mean of 0.6 grains/non S-phase nucleus when incubated with or without aphidicolin. Undamaged growing cells incorporated a mean of 1.2 grains/non S-phase nucleus without aphidicolin and 1.1 grains/non S-phase nucleus with aphidicolin. Repair synthesis data are mean grains/nucleus (±standard error of the mean). Aphidicolin sensitive repair synthesis was calculated as described in the legend to Fig. 2. The dose of aphidicolin used totally suppressed S-phase synthesis in both irradiated and unirradiated cultures of growing cells. (Reproduced from ref. *39*).

cerned chiefly with DNA replication and that some other polymerase (?$\beta$) carries out repair synthesis even at high levels of damage. Another possibility is that the pool of polymerase $\alpha$ molecules which perform repair in growing cells is somehow modified or sequestered as a complex so that it is insensitive to aphidicolin. Thus the meaning of this unusual finding must await additional experimentation.

*Towards a Comprehensive Model of Polymerase Function in DNA Repair Synthesis*

Our data provide the beginnings of a detailed model of polymerase function in repair synthesis. It is clear that at least two DNA polymerases—polymerase $\alpha$ and a non-$\alpha$ polymerase, probably polymerase $\beta$—are involved in repair synthesis. Further, the extent to which each of these polymerases is involved is damage dependent: at low levels of damage the non-$\alpha$ polymerase functions in repair synthesis, while at higher levels of damage polymerase $\alpha$ participates more extensively.

It is tempting to suggest, as others have (*45*), that "polymerase $\beta$" is involved in short patch repair while polymerase $\alpha$ is involved in long patch repair. It is clear from the data in Figs. 2, 3, and 4 that repair synthesis following X-ray damage, a short patch agent, is relatively insensitive to aphidicolin except at very high doses of X-ray, while repair synthesis following long patch agents (UV and NA-AAF) is very sensitive to aphidicolin except at very low doses. In addition, other investigators using naked DNA templates have found that polymerase $\beta$ can utilize templates with nicks or gaps of only a few nucleotides, while polymerase $\alpha$ prefers templates with gaps of 20 nucleotides or more (*46–48*).

One may think of the two different polymerase functions as competing for the same substrate(s) in the cell. While it is not clear, for example, whether the substrates represent incised damaged sites or gaps left following excision and how the structure of chromatin may affect substrates for repair synthesis, one can imagine competition between two polymerase systems with differing "affinities" for substrates. If one assumes that these affinities are related to an apparent $K_m$ for each process, then it can be shown that simple Michaelis-Menten kinetics can be used to explain apparent "switching" from polymerase $\beta$ to polymerase $\alpha$ at increased levels of damage. Although we have not modeled more complex situations, obviously different types of damage may result in different substrates for repair synthesis for which the apparent $K_m$s of the two polymerases may differ. Thus a model must take into account not only amount of damage but type of damage. Further, most damaging agents produce a spectrum of products, the relative distribution of which might be concentration dependent. Different affinities for different types of damage sites might well explain why it is difficult experimentally to find a level of UV or NA-AAF low enough to produce low levels of polymerase $\alpha$ involvement (see Figs. 2, 3). It is not unreasonable, however, to think that in the near future some simple experimental systems may be developed in which such parameters may be measured and related to polymerase function.

ACKNOWLEDGMENTS

This work was supported by NIH Grant CA 20513 and the following companies: Brown & Williamson Tobacco Corporation, Philip Morris Incorporated, R. J. Reynolds

Tobacco Company and United States Tobacco Company. S. L. Dresler was supported by an NIEHS postdoctoral fellowship (ES 07066).

## REFERENCES


1. Weissbach, A. The functional roles of mammalian DNA polymerase. Arch. Biochem. Biophys., *198*: 386–396, 1979.
2. Bertazzoni, U., Stefanini, M., Pedrali-Noy, G., Nuzzo, F., and Falaschi, A. Levels of DNA polymerase $\alpha$ and $\beta$ in normal and xeroderma pigmentosum fibroblasts. Nucleic Acids Res., *4*: 141–148, 1977.
3. Bertazzoni, U., Scovassi, A. I., Stefanini, M., Giulotto, E., Spadari, S., and Pedrini, A. M. DNA polymerases $\alpha$, $\beta$, and $\gamma$ in inherited diseases affecting DNA repair. Nucleic Acids Res., *5*: 2189–2196, 1978.
4. Parker, V. P. and Lieberman, M. W. Levels of DNA polymerases $\alpha$, $\beta$, and $\gamma$ in control and repair-deficient human diploid fibroblasts. Nucleic Acids Res., *4*: 2029–2037, 1977.
5. Bertazzoni, U., Stefanini, M., Pedrali-Noy, G., Giulotto, E., Fiorella, N., Falaschi, A., and Spadari, S. Variations of DNA polymerases-$\alpha$ and -$\beta$ during prolonged stimulation of human lymphocytes. Proc. Natl. Acad. Sci. U.S., *73*: 785–789, 1976.
6. Waser, J., Hubscher, U., Kuenzle, C. C., and Spadari, S. DNA polymerase $\beta$ from brain neurons is a repair enzyme. Eur. J. Biochem., *97*: 361–368, 1979.
7. Huberman, J. A. New views of the biochemistry of eucaryotic DNA replication revealed by aphidicolin, an unusual inhibitor of DNA polymerase $\alpha$. Cell, *23*: 647–648, 1981.
8. Wist, E. and Prydz, H. The effect of aphidicolin on DNA synthesis in isolated HeLa cell nuclei. Nucleic Acids Res., *6*: 1583–1590, 1979.
9. Ikegami, S., Taguchi, T., Ohashi, M., Oguro, M., Nagano, H., and Mano, Y. Selective inhbition by aphidicolin of the activity of DNA polymerase alpha leads to blockade of DNA synthesis and cell division in sea urchin embryos. Dev. Growth Differ., *21*: 119–127, 1979.
10. Pedrali-Noy, G. and Spadari, S. Effect of aphidicolin on viral and human DNA polymerases. Biochem. Biophys. Res. Commun., *88*: 1194–1202, 1979.
11. Krokan, H., Schaffer, P., and DePamphilis, M. L. Involvement of eucaryotic deoxyribonucleic acid polymerases $\alpha$ and $\gamma$ in the replication of cellular and viral deoxyribonucleic acid. Biochemistry, *18*: 4431–4443, 1979.
12. Edenberg, H. J., Anderson, S., and DePamphilis, M. L. Involvement of DNA polymerase $\alpha$ in simian virus 40 DNA replication. J. Biol. Chem., *253*: 3273–3280, 1978.
13. Waqar, M. A., Evans, M. J., and Huberman, M. A. Effect of 2′,3′-dideoxythymidine-5′-triphosphate on HeLa cell *in vitro* DNA synthesis: evidence that DNA polymerase $\alpha$ is the only polymerase required for cellular DNA replication. Nucleic Acids Res., *5*: 1933–1946, 1978.
14. Abboud, M. M. and Horwitz, M. S. The DNA polymerases associated with the adenovirus type 2 replication complex: effect of 2′3′-dideoxythymidine-5′-triphosphate on viral DNA synthesis. Nucleic Acids Res., *6*: 1026–1039, 1979.
15. Allaudeen, H. S. Inhibition of deoxyribonucleic acid polymerases of human leukemic leukocytes by 2′,3′-dideoxythymidine triphosphate. Biochem. Pharmacol., *29*: 1149–1153, 1980.
16. van der Vliet, P. C. and Kwant, M. M. Role of DNA polymerase $\gamma$ in adenovirus DNA replication. Mechanism of inhibition by 2′,3′-dideoxynucleoside 5′-triphosphates. Biochemistry, *20*: 2628–2632, 1981.

17. Knopf, K.-W., Yamada, M., and Weissbach, A. HeLa cell DNA polymerase $\gamma$: Further purification and properties of the enzyme. Biochemistry, *15*: 4540–4548, 1976.
18. Chang, L.M.S. Phylogeny of DNA polymerase-$\beta$. Science, *191*: 1183–1185, 1976.
19. Hubscher, U., Kuenzle, C. C., and Spadari, S. Identity of DNA polymerase $\gamma$ from synaptosomal mitochondria and rat-brain nuclei. Eur. J. Biochem., *81*: 249–258, 1977.
20. Yamaguchi, M., Tanabe, K., Taguchi, Y. N., Nishizawa, M., Takahashi, T., and Matsukage, A. Chick embryo DNA polymerase $\beta$. J. Biol. Chem., *255*: 9942–9948, 1980.
21. Berger, N. S., Kurohara, K. K., Petzold, S. J., and Sikorski, G. W. Aphidicolin inhibits eukaryotic DNA replication and repair — Implications for involvement of DNA polymerase $\alpha$ in both processes. Biochem. Biophys. Res. Commun., *89*: 218–225, 1979.
22. Ciarrocchi, G., Jose, J. G., and Linn, S. Further characterization of a cell-free system for measuring replicative and repair DNA synthesis with cultured human fibroblasts and evidence for involvement of DNA polymerase $\alpha$ in DNA repair. Nucleic Acids Res., *7*: 1205–1219, 1979.
23. Hanaoka, F., Kato, H., Ikegami, S., Ohashi, M., and Yamada, M. Aphidicolin does inhibit repair replication in HeLa cells. Biochem. Biophys. Res. Commun., *87*: 575–580, 1979.
24. Hardt, N., Pedrali-Noy, G., Focher, F., and Spadari, S. Aphidicolin does not inhibit DNA repair synthesis in ultraviolet-irradiated HeLa cells. Biochem. J., *199*: 453–455, 1981.
25. Pedrali-Noy, G. and Spadari, S. Aphidicolin allows a rapid and simple evaluation of DNA-repair synthesis in damaged human cells. Mutat. Res., *70*: 389–394, 1980.
26. Snyder, R. D. and Regan, J. D. Aphidicolin inhibits repair of DNA in UV-irradiated human fibroblasts. Biochem. Biophys. Res. Commun., *99*: 1088–1094, 1981.
27. Waters, R. Aphidicolin: an inhibitor of DNA repair in human fibroblasts. Carcinogenesis, *2*: 795–797, 1981.
28. Collins, A.R.S., Squires, S., and Johnson, R. T. Inhibitors of repair DNA synthesis. Nucleic Acids Res., *10*: 1230–1213, 1982.
29. Miller, M. R. and Chinault, D. N. The roles of DNA polymerases $\alpha$, $\beta$, and $\gamma$ in DNA repair synthesis induced in hamster and human cells by different DNA damaging agents. J. Biol. Chem., *257*: 10204–10209, 1982.
30. Snyder, R. D. and Regan, J. D. Differential responses of log and stationary phase human fibroblasts to inhibition of DNA repair by aphidicolin. Biochim. Biophys. Acta, *697*: 229–234, 1982.
31. Castellot, J. J., Miller, M. R., Lehtomaki, D. M., and Pardee, A. B. Comparison of DNA replication and repair enzymology using permeabilized baby hamster kidney cells. J. Biol. Chem., *254*: 6904–6908, 1979.
32. Seki, S. and Oda, T. Effects of 2′,3′-dideoxythymidine triphosphate on replicative DNA synthesis and unscheduled DNA synthesis in permeable mouse sarcoma cells. Biochim. Biophys. Acta, *606*: 246–250, 1980.
33. Waters, R., Crocombe, K., and Mirzayans, R. The inhibition of DNA repair by aphidicolin or cytosine arabinoside in X-irradiated normal and xeroderma pigmentosum fibroblasts. Mutat. Res., *94*: 229–234, 1981.
34. Miller, M. R. and Chinault, D. N. Evidence that DNA polymerases $\alpha$ and $\beta$ participate differentially in DNA repair synthesis induced by different agents. J. Biol. Chem., *257*: 46–49, 1982.
35. Roberts, J. D. and Lieberman, M. W. Deoxyribonucleic acid repair synthesis in

permeabilized human fibroblasts exposed to ultraviolet radiation or *N*-acetoxy-*2*-acetylaminofluorene. Biochemistry, *18*: 4499–4505, 1979.
36. Dresler, S. L., Roberts, J. D., and Lieberman, M. W. Characterization of deoxyribonucleic acid repair synthesis in permeable human fibroblasts. Biochemistry, *21*: 2557–2564, 1982.
37. Mattern, M. R. and Scudiero, D. A. Dependence of mammalian DNA synthesis on DNA supercoiling. III. Characterization of the inhibition of replication and repair-type DNA synthesis by novobiocin and nalidixic acid. Biochim. Biophys. Acta, *653*: 248–253, 1981.
38. Edenberg, H. J. Novobiocin inhibition of simian virus 40 DNA replication. Nature, *286*: 529–531, 1980.
39. Dresler, S. L. and Lieberman, M. W. Identification of DNA polymerases involved in DNA excision repair in diploid human fibroblasts. J. Biol. Chem., 1983, in press.
40. Francis, A. A., Snyder, R. D., Dunn, W. L., and Regan, J. D. Classification of chemical agents as to their ability to induce long- or short-patch DNA repair in human cells. Mutat. Res., *83*: 159–169, 1981.
41. Oguro, M., Shioda, M., Nagano, H., and Mano, Y. The mode of action of aphidicolin on DNA synthesis in isolated nuclei. Biochem. Biophys. Res. Commun., *92*: 13–19, 1980.
42. Krokan, H., Wist, E., and Krokan, R. H. Aphidicolin inhibits DNA synthesis by DNA polymerase $\alpha$ and isolated nuclei by a similar mechanism. Nucleic Acids Res., *5*: 4709–4719, 1981.
43. Kornberg, A. DNA Replication, 742 pp., W. H. Freeman and Company, San Francisco, 1980.
44. Bertazzoni, U., Scovassi, A. I., and Brun, G. M. Chick-embryo DNA polymerase $\gamma$. Identity of $\gamma$-polymerases purified from nuclei and mitochondria. Eur. J. Biochem., *81*: 237–248, 1977.
45. Grossman, L. Enzymes involved in the repair of damaged DNA. Arch. Biochem. Biophys., *211*: 511–522, 1981.
46. Wang, T. S.-F. and Korn, D. Reactivity of KB cell deoxyribonucleic acid polymerases $\alpha$ and $\beta$ with nicked and gapped deoxyribonucleic acid. Biochemistry, *19*: 1782–1790, 1980.
47. Fisher, P. A. and Korn, D. Properties of the primer-binding site and the role of magnesium ion in primer-template recognition by KB cell DNA polymerase $\alpha$. Biochemisty, *20*: 4570–4578, 1981.
48. Wang, T. S.-F. and Korn, D. Specificity of the catalytic interaction of human DNA polymerase $\beta$ with nucleic acid substrates. Biochemistry, *21*: 1597–1608, 1982.

ADP-RIBOSYLATION, DNA REPAIR AND CANCER, M. MIWA ET AL. (EDS.),
JAPAN SCI. SOC. PRESS, TOKYO/VNU SCIENCE PRESS, UTRECHT, PP. 267-276, 1983

# Infidelity of DNA Synthesis as a Cause of Mutagenesis

Lawrence A. Loeb, Philip K. Liu, Shishir K. Das, and John R. Silber
*The Joseph Gottstein Memorial Cancer Research Laboratory, Department of Pathology SM-30, University of Washington, Seattle, Washington 98195, U.S.A.*

*Abstract:* The concept underlying these studies is that a major determinant of mutagenesis involves perturbations in the fidelity of DNA replication. *i.e.*, the accuracy by which DNA polymerases copy DNA templates. To investigate this relationship, we have designed *in vitro* assays to measure the accuracy of DNA replication and used these systems to screen for and to quantitate factors that promote errors in DNA synthesis. Using DNA polymerase from bacteria, the frequency of mistakes with $\phi$X174 DNA as a template approaches $10^{-7}$ and is similar to the spontaneous mutation rates in bacterial cells. In contrast, DNA polymerases from animal cells are more error-prone.

The differences in fidelity among mammalian DNA polymerases which lack error-correcting mechanisms suggest that these enzymes enhance accuracy by improving base-selection. Thus, mutants in DNA polymerase-$\alpha$ might be altered in base-selection. Chinese hamster V79 cell mutants selected by resistance to aphidicolin, a specific inhibitor of DNA polymerase-$\alpha$, have been reported (Somatic Cell Genet., 7: 235–253, 1981). DNA polymerase-$\alpha$ was purified from mitochondria-free crude extracts of these mutants by sequential column chromatography using DEAE-cellulose and phosphocellulose. DNA polymerase-$\alpha$ purified from one of the mutants is 10-fold more resistant to aphidicolin than the same enzyme purified from the parental cells. Moreover, the apparent $K_m$ for dCTP is 1.0±0.4 $\mu$M for the mutant polymerase and 10±4 $\mu$M for the parental enzyme. These observed differences are in accord with the known competition between aphidicolin and dCTP, and provide a mechanism for the aphidicolin resistance of the mutant, *i.e.*, the decrease in $K_m$ for dCTP. The elevated spontaneous and induced mutation rate exhibited by this mutant could be mediated by the alteration in DNA polymerase-$\alpha$.

With DNA replicating enzymes from a variety of sources, enhancement of mutagenesis has been demonstrated by alteration in precursor pools, damage to DNA templates, loss of nucleotide bases on DNA, metal ions that interact with nucleotide bases, and organic compounds that intercalate into DNA. The alterations of deoxynucleoside triphosphate pools also occur after treatment of animal cells with known mutagens. This observation may provide a new mechanism for mutagenesis by these agents independent of alterations in DNA.

In this presentation, we have summarized our recent studies on the accuracy of DNA replication. We have focused on the question of how animal cells can replicate the $10^9$ nucleotides in their DNA genomes without apparent error during each division cycle. The corollary question is whether or not perturbations in this mechanism are causally associated with mutagenesis and carcinogenesis. Our approach to this problem has been to design model *in vitro* systems that accurately copy DNA using purified components. We have then altered these systems with mutagens and carcinogens and asked whether these perturbations alter the frequency and specificity of misincorporations during DNA replication.

### *Assay Systems to Measure the Fidelity of DNA Synthesis In Vitro*

Until recently, most studies on the fidelity of DNA synthesis utilized synthetic polynucleotides as templates. Errors were quantitated by measuring the frequency of incorporation of nucleotides that are noncomplementary to the base(s) in the template (*1*). Natural DNA is not a suitable template for this assay since it contains all four nucleotides. With natural DNA, one cannot simply measure the incorporation of a noncomplementary radioactive base and equate this with mistakes in DNA synthesis. This difficulty was overcome by our use of a genetic assay in which a biologically active DNA was copied by purified DNA polymerases *in vitro*. Errors were quantitated *in vivo* by measuring the frequency with which a single base change reverts the mutant to wild type. To measure the fidelity of copying natural DNA, we chose $\phi$X174 single-stranded DNA as a template (*2*), since: 1) the nucleotide sequence of the DNA is established, 2) a catalog of single-base mutations is available, 3) when provided with a primer, DNA polymerases from all known sources can copy this DNA template, 4) the biological activity of the DNA can be measured by infection of spheroplasts, and 5) $\phi$X174 DNA serves as a template for the replicating protein complexes delineated in *Escherichia coli* (*3*). The $\phi$X174 fidelity assay is diagrammed in Fig. 1. The template to be copied is a single-strand circular DNA molecule from an amber mutant containing a single nucleotide substitution. Synthesis is started at a specified position on the template by using as a primer a DNA restriction fragment that is hybridized to the template. After copying past the amber mutation *in vitro* with any purified DNA polymerase, the frequency of errors in the newly synthesized daughter strand is determined from the reversion frequency of the amber mutation to wild type. This is assessed by transfecting the copied DNA into *E. coli* spheroplasts and measuring the number of mutant and wild type phage by plating on permissive and nonpermissive *E. coli* indicator strains. It should be noted that this assay measures only single nucleotide substitutions. The identity of the noncomplementary nucleotide that is substituted can be surmised by measuring the effects of separately varying the concentration of each of the nucleotides in the reaction mixture, or established by sequencing the DNA from revertant phage (*4*).

Two other methods have been developed subsequently to measure the fidelity of DNA synthesis. In the method developed in Alberts' laboratory, double-stranded

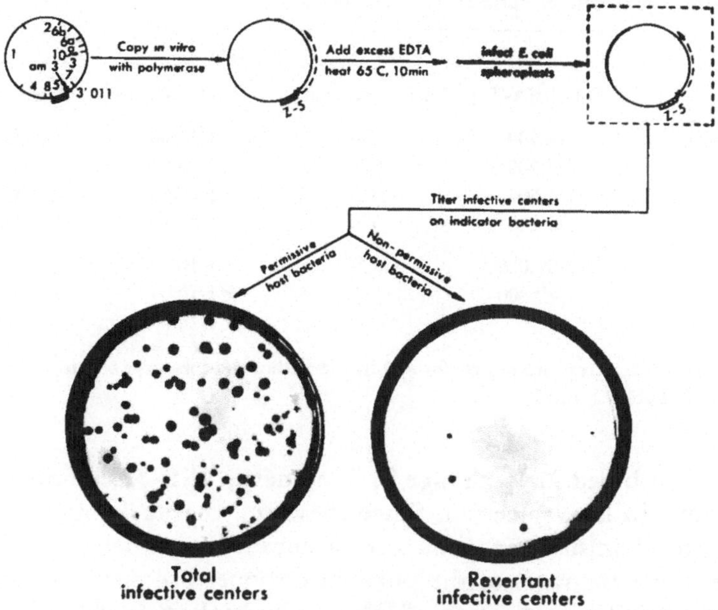


FIG. 1. $\phi$X174 fidelity assay. In the example shown, the template is *am*3 $\phi$X174 DNA, and the primer is the Hae III Z-5 restriction enzyme fraction whose 3′-OH terminus is 83 nucleotides away from the *am*3 codon. All mutations observed are at the middle position of the amber codon 587 (*3*). All three mismatches can be observed and indicate A: C, A: G, and A: A mispairings.

$\phi$X174 replicative-form DNA is used as a template (*5*). Synthesis is catalyzed by a seven-protein replicating complex and produces, by a rolling-circle mechanism, 25 to 30 product molecules per input RF II molecule. The method offers the advantage of enhanced sensitivity. However, the requirement for the entire replicating complex prevents one from easily assigning functions to each of the components in the complex. Also, rigorous interpretation on the nature of mismatches is more difficult when copying two complementary DNA strands. In the method developed by Fersht (*6*), $\phi$X174 single-stranded DNA serves as a template. DNA synthesis is catalyzed by the purified *E. coli* replication apparatus consisting of the holoenzyme complex and four other added proteins. Since single-stranded molecules are being copied, a direct analysis of error rates and nucleotide specificity is obtainable. However, one cannot measure accuracy of single purified DNA polymerases. In this paper, we will consider the results on the initial $\phi$X174 fidelity assay (Fig. 1).

### *Fidelity of Purified DNA Polymerases*

The error rates of purified DNA polymerases using poly(d(A-T)) and $\phi$X174 *am*3 DNA as templates are tabulated in Table 1. Of the enzymes tested, the DNA polymerases (also termed reverse transcriptase) from RNA tumor viruses are the least accurate with both templates (*7*). Eukaryotic DNA are also highly inaccurate, the lowest error rates observed being approximately 1 in 40,000 for DNA polymerase-$\alpha$ (*8*), the enzyme believed to be involved in DNA replication. The fact that the error rate is much less than that achieved on the basis of difference in free energies of base-

TABLE 1. Representative Error Rate of Purified DNA Polymerases

| Source | Error rate | | | |
|---|---|---|---|---|
| | Poly [d(A-T)] | Misincorporation | am3 $\phi$X174 | Misincorporation |
| Avian myeloblastosis | 1/1,400 | (G) | 1/1,200 | (C, A, G) |
| Calf thymus-$\alpha$ | 1/4,500 | (G) | 1/30,500 | (C, A, G) |
| Rat hepatoma-$\beta$ | 1/4,500 | (G) | 1/6,600 | (C, A, G) |
| HeLa cell-$\gamma$ | 1/3,120 | (G) | 1/8,070 | (C, A, G) |
| *E. coli* Pol I | 1/80,000 | (G) | 1/$10^6$ | (G) |
| Bacteriophage T4 | 1/42,300 | (G) | <1/$10^7$ | (C) |
| *E. coli* holoenzyme | | | <1/$10^7$ | (C) |

The values in this table were taken from studies performed in the author's laboratory, and from the individual references detailed in ref. *1*, Tables 1 and 2.

pairings (1/100–1/1,000 based on a change in free energy of 2–3 kcal/mol) indicates a role of these enzymes in base selection. Since these enzymes lack any proofreading exonuclease, other mechanisms for enhanced accuracy have been proposed (*9*). Whatever the case, the error rates of the purified enzymes are orders of magnitude higher than that required for replication of DNA in animal cells, which is estimated to be $10^{-7}$ to $10^{-11}$, based on known spontaneous mutation rates (*10*). Thus, other and still undefined cellular components must enhance accuracy. The simple expectation of a separate proofreading exonuclease that works in concert with a eukaryotic DNA polymerase is not yet supported by experimental studies (*11*).

Prokaryotic DNA polymerases are highly accurate in the $\phi$X174 assay; error rates range from 1/$10^6$ to <1/$10^7$. With the addition of single-strand DNA binding protein, there is a further 10-fold increase in accuracy (*12*). In fact, we find that the accuracy achieved in the test tube approaches that achieved in cells (*13*). To a large extent this accuracy is due to the associated $3' \rightarrow 5'$ proofreading activity of these enzymes (*14*). Our ability to mimic the accuracy of DNA replication *in vivo* indicates that in prokaryotes, DNA polymerase is the predominant factor in achieving accuracy, or that the conditions in cells are not optimal for fidelity. The discrepancy between error rates with poly(d(A-T)) and natural DNA suggests either that the poly-(d(A-T)) measurements are not sufficiently sensitive or that proofreading is not fully operational on polynucleotide templates (*15*).

### *Mutator DNA Polymerases*

Mutant DNA polymerases from bacteriophage T4 have altered fidelity *in vitro* and exhibit a mutator phenotype *in vivo* (*16*). These polymerases have been important in establishing the role of the proofreading exonuclease in fidelity. In animal cells mutant DNA polymerase has only recently become obtainable. Aphidicolin, a tetracyclic diterpenoid mycotoxin, has been shown to specifically inhibit DNA polymerase-$\alpha$, but not polymerase-$\beta$ or -$\gamma$ (*17, 18*). This specific inhibitor has been used to further establish a pivotal role for DNA polymerase-$\alpha$ in DNA replication and also an additional role for this enzyme in DNA repair (*19–21*). A large number of mutant cells resistant to aphidicolin in culture have been obtained (*22–24*). Until now, none of the

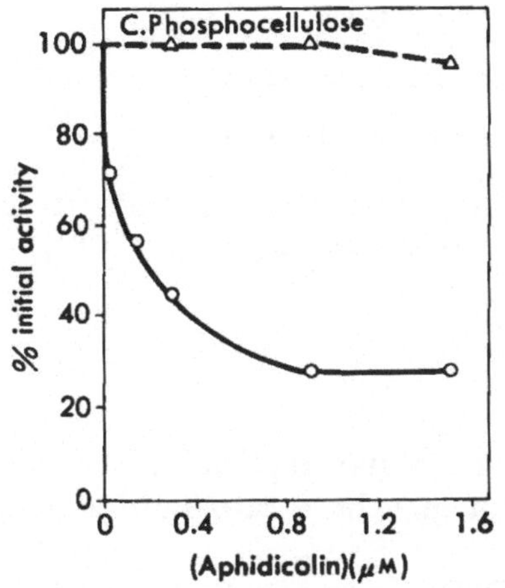


FIG. 2. Aphidicolin sensitivity of DNA polymerase-α from wild type (○) and *aph*^r-4-2 (△) cells. DNA polymerase was purified from a mitochondria-free crude extract by sequential chromatograph on DEAE-cellulose and phosphocellulose. Assays were carried out with activated DNA as a template, $Mg^{2+}$ as a metal activator, 100 μM of dATP, dTTP, dGTP, and 5 μM dCTP.

aphidicolin-resistant animal cells have been shown to contain a mutant DNA polymerase α. Resistance has been based on amplification of DNA polymerase, diminished transport of aphidicolin by altered membranes, or elevation in pyrimidine nucleotides (*i.e.*, dCTP or dTTP) which are competitive with aphidicolin (*25–27*). In 1981, Chang *et al.* (*28*) described a Chinese hamster V79 cell mutant (*aph*^r-4-2) obtained by 2-step mutagenesis using BrdU/black light and UV irradiation, and then selected for resistance to aphidicolin. In order to study the properties of DNA polymerase-α, the mutant cells were grown sequentially for 1-week intervals in the presence and absence of 1 μM aphidicolin to obtain $10^9$ cells (*29*). Partially purified DNA polymerase-α from the mutant and parental lines were assayed in the presence of increasing concentrations of aphidicolin (Fig. 2). Under conditions in which DNA polymerase-α from the parental line was inhibited by 75%, the enzyme from the mutant was refractory to inhibition. The identification of *aph*^r-4-2 as the first mutant in DNA polymerase-α is reinforced by kinetic studies that varied substrate concentration (*29*). The apparent $K_m$ for dCTP is 1.0±0.4, and 10±4 for the mutant and wild type polymerase-α, respectively. In contrast, no significant difference was noted in the $K_m$ for dATP. The decrease in apparent $K_m$ for dCTP provides a logical mechanism for resistance to aphidicolin, a competitive inhibitor of dCTP. Moreover, the alteration in the $K_m$ for dCTP should result in more frequent errors by the mutant polymerase, *i.e.*, substitutions of dCTP for the complementary nucleotide. Studies to test this hypothesis are in progress. This selective nucleotide polymerase interaction could account for the mutator phenotype observed in these cells (*29, 30*).

### *Factors That Affect the Fidelity of DNA Synthesis In Vitro*

Fidelity assays have provided a convenient method to assess the possible mutagenic potential of a variety of agents. The guiding assumption is that misincor-

TABLE 2. Effect of Exogenous Agents on the Fidelity of DNA Synthesis

| Little or no effect on fidelity | Enhanced misincorporation |
|---|---|
| Template (length) | Ratio of incorrect to correct nucleotide substrates |
| Type of primer | Alkylation of template (small adducts) |
| pH (6–8) | Different metal activators for DNA polymerase ($Mn^{2+}$, $Co^{2+}$, $Ni^{2+}$) |
| Urea (<4 M) | Nonactivating metal mutagens or carcinogens ($Be^{2+}$, $Cr^{3+}$) |
| Anions | Intercalators |
| Caffeine | Depurination of template |
| Ethanol | Steroids |

porations made by DNA polymerases during cellular replication are not always corrected. The fidelity of DNA polymerization may be a rate-limiting step in mutagenesis by certain agents.

A list of agents that alter the fidelity of DNA is presented in Table 2. It should first be stressed that not very many changes in reaction conditions diminish fidelity, and so far enhancement in accuracy has been observed only with single-strand binding proteins. In general, changes that alter fidelity for one DNA polymerase invariably do so for all. As a general rule, those agents that increase infidelity are also mutagenic. This is most firmly documented with metals where there is a strong correlation between alterations in fidelity by divalent metal ions and the reported mutagenicity and carcinogenicity by the same metal (*31*). Of particular interest is the recent elucidation of the mutagenic effects of nucleotide pool imbalances and depurination, since these two situations might reflect common intermediates in mutagenesis by diverse agents.

*Alterations in Nucleotide Pools*

A relationship between mutagenesis and deoxynucleotide pools in cells was initially suggested by studies of Bernstein *et al.* (*32*), indicating that alterations in deoxynucleoside triphosphate pools were mutagenic in T4 infected cells. Studies on the fidelity of DNA synthesis show a strong error-rate dependency on nucleotide concentration. Increasing the concentration of the incorrect nucleotide results in an increased frequency of misincorporation; conversely, increasing the concentration of the correct nucleotide reduces the frequency of misincorporation (*33, 34*). This proportionality has been observed with all eukaryotic DNA polymerases tested (*8*). In addition, with DNA polymerases containing an associated $3' \rightarrow 5'$ exonuclease, increasing the concentration of the next nucleotide enhances misincorporation at the previous site, presumably by increasing the rate of the reaction and allowing less time for excision (*6*).

This mutagenicity *in vitro* by purified DNA polymerases is in accord with recent data on increased mutagenesis *in vivo* observed in cells with altered nucleotide pools due to mutations in ribonucleotide reductase and other key enzymes in nucleotide metabolism (*35*). In order to further explore this interdependent relationship, we studied the effects of known mutagens on nucleotide metabolism. The rapid generation of imbalances in endogenous dNTP pools that we observed in mammalian cells is an unexpected finding (Table 3) (*36*). Treatment of V79 cells with either UV ir-

TABLE 3. Alteration in Deoxynucleoside Triphosphate Concentrations by Mutagen Treatment of V79 Cells

| Treatment | dATP (pmol/$10^6$) | dTTP (pmol/$10^6$) | dCTP (pmol/$10^6$) | dGTP (pmol/$10^6$) |
|---|---|---|---|---|
| None | 40.0 | 43.0 | 78.0 | 4.0 |
| MNNG, 5.0 μM for 1 hr | 120.0 (3×) | 135.0 (3×) | 82.0 | 4.0 |
| UV (5 J/$m^2$) | 91.0 (2×) | 140.0 (3.5×) | 72.0 | 3.8 |

Exponentially growing cultures were used. dNTP pools were measured 4 hr after treatment with the mutagen. Values reported are pmol/$10^6$ cells. No cell-cycle perturbations were detected by the time of pool measurement.

radiation or N-methyl-N′-nitro-N-nitrosoguanidine (MNNG) results within 4 hr in a 2- to 4-fold expansion of dATP and dTTP pools, but not dGTP and dCTP. This selective pool expansion might provide a significant contribution to mutagenesis by these agents, particularly with respect to untargeted mutagenesis (mutations at a site distant from the DNA lesion) (*37*). However, the presence of a DNA-replicating complex which kinetically funnels early nucleotide precursor into DNA (*38*, *39*) suggests that a direct relationship between nucleotide-mediated errors in DNA synthesis *in vitro* and effects of pool sizes and mutagenesis *in vivo* is likely to be an oversimplification. In fact, the alteration in nucleotide pools at the site of DNA replication may be much larger than those shown by pool measurements.

### *Depurination and Mutagenesis*

Depurination is a frequent spontaneous event (*40*) that results from the inherent chemical instability of the glycosylic bond, the bond between the purine base and the sugar. The rate of depurination is increased markedly by chemicals that form stable adducts at the $^3$N and $^7$N positions of purines, and by glycosylases that excise altered bases. For these reasons, we have measured the frequency and specificity of copying depurinated sites by DNA polymerases. Enhanced infidelity is observed during DNA synthesis *in vitro* (*41*) with all DNA polymerases tested. In *E. coli*, mutagenicity is dependent on the induction of an error-prone SOS-dependent pathway (*42*). In both situations, misincorporation is not random, since the predominant substitution is deoxyadenosine opposite a putative apurinic site ((*43*, *44*), and unpublished results). The mechanism for the resultant transversions is under investigation.

### *Speculation on Future Studies*

Studies on the fidelity of DNA replication have not yet been considered with respect to poly(ADP-ribose). From the results outlined, there are at least four important questions to be targeted for investigation: could DNA polymerases serve as acceptor molecules, and if so would ADP-ribosylation alter their fidelity? Does ribosylation account for the marked heterogeneity exhibited by purified DNA polymerase-α? As a DNA-binding protein, does poly(ADP-ribose) enhance accuracy during DNA repair or replication? Does reduction of NAD during poly(ADP-ribose) polymerization contribute to alteration in deoxynucleoside pools and mutagenicity? Methods are now at hand to answer these questions.

ACKNOWLEDGMENT

This work was supported by grants from the National Institutes of Health (CA-24845, CA-24988), the National Science Foundation (PCM-7680439), the Environmental Protection Agency (R809623010), and the Department of Energy (76RLO2225).

REFERENCES

1. Loeb, L. A. and Kunkel, T. A. Fidelity of DNA synthesis. Annu. Rev. Biochem., *52*: 429–457,1982.
2. Weymouth, L. A. and Loeb, L. A. Mutagenesis during *in vitro* DNA synthesis. Proc. Natl. Acad. Sci. U.S., *75*: 1924–1928, 1978.
3. Kornberg, A. DNA Replication, W. H. Freeman and Company, San Francisco, 1980.
4. Kunkel, T. A. and Loeb, L. A. On the fidelity of DNA replication: The accuracy of *Escherichia coli* DNA polymerase I in copying natural DNA *in vitro*. J. Biol. Chem., *255*: 9961–9966, 1980.
5. Liu, C. C., Burke, R. L., Hibner, U., Barry, J., and Alberts, B. M. Probing DNA replication mechanisms with the T4 bacteriophage *in vitro* system. Cold Spring Harbor Symp. Quant. Biol., *43*: 469–487, 1979.
6. Fersht, A. R. Fidelity of replication of phage $\phi$X174 DNA by DNA polymerase III holoenzyme: Spontaneous mutation by misincorporation. Proc. Natl. Acad. Sci. U.S., *76*: 4946–4950, 1979.
7. Gopinathan, K. P., Weymouth, L. A., Kunkel, T. A., and Loeb, L. A. On the fidelity of DNA replication: Mutagenesis *in vitro* by DNA polymerases from an RNA tumor virus. Nature, *278*: 857–859, 1979.
8. Kunkel, T. A. and Loeb, L. A. The fidelity of mammalian DNA polymerases. Science, *213*: 765–767, 1981.
9. Hopfield, J. J. The energy relay: A proofreading scheme based on dynamic cooperativity and lacking all characteristic symptoms of kinetic proofreading in DNA replication and protein synthesis. Proc. Natl. Acad. Sci. U.S., *77*: 5248–5252, 1980.
10. Drake, J. W. Comparative rates of spontaneous mutation. Nature, *221*: 1132, 1969.
11. Wang, M-L. J., Stellwagen, R. H., and Goodman, M. F. Evidence for the absence of DNA proofreading in HeLa cell nuclei. J. Biol. Chem., *256*: 7097–7100, 1981.
12. Kunkel, T. A., Meyer, R., and Loeb, L. A. Single-strand binding protein enhances the fidelity of DNA synthesis *in vitro*. Proc. Natl. Acad. Sci. U.S., *76*: 6331–6335, 1979.
13. Loeb, L. A., Kunkel, T. A., and Schaaper, R. M. Fidelity of copying natural DNA templates. *In;* Mechanistic Studies on DNA Replication and Genetic Recombination, ICN-UCLA Symposia on Molecular and Cellular Biology, XIX, pp. 735–751, Academic Press, New York, 1980.
14. Brutlag, D. and Kornberg, A. Enzymatic synthesis of deoxyribonucleic acid XXXVI. A proofreading function for the $3' \rightarrow 5'$ exonuclease activity in deoxyribonucleic acid polymerases. J. Biol. Chem., *247*: 241–248, 1972.
15. Loeb, L. A., Dube, D. K., Beckman, R. A., and Gopinathan, K. P. On the fidelity of DNA replication: Nucleoside monophosphate generation during polymerization. J. Biol. Chem., *256*: 3978–3987, 1981.
16. Muzyczka, N., Poland, R. L., and Bessman, M. J. Studies on the biochemical basis of spontaneous mutation. I. A comparison of the deoxyribonucleic acid polymerases of mutator, antimutator, and wild type strains of bacteriophage T4. J. Biol. Chem., *247*: 7116–7122, 1972.
17. Ikegami, S., Taguchi, T., Ohashi, M., Oguro, M., Nagano, H., and Mano, Y.

Aphidicolin prevents mitotic cell division by interfering with the activity of DNA polymerase-α. Nature, *275*: 458–460, 1978.
18. Ohashi, M., Taguchi, T., and Ikegami, S. Aphidicolin: A specific inhibitor of DNA polymerases in the cytosol of rat liver. Biochem. Biophys. Res. Commun., *82*: 1084–1090, 1978.
19. Ciarrocchi, G., Jose, J. G., and Linn, S. Further characterization of a cell-free system for measuring replicative and repair DNA synthesis with cultured human fibroblasts and evidence for the involvement of DNA polymerase α in DNA repair. Nucleic Acids Res., *7*: 1205–1219, 1979.
20. Dresler, S. L., Roberts, J. D., and Lieberman, M. W. Characterization of deoxyribonucleic acid repair synthesis in permeable human fibroblasts. Biochemistry, *21*: 2257–2664, 1982.
21. Berger, N. A., Kurohara, K. K., Petzold, S. J., and Sikorski, G. W. Aphidicolin inhibits eukaryotic DNA replication and repair — implications for involvement of DNA polymerase α in both processes. Biochem. Biophys. Res. Commun., *89*: 218–225, 1979.
22. Sugino, A. and Nakayama, K. DNA polymerase α mutants from a *Drosophila melanogaster* cell line. Proc. Natl. Acad. Sci. U.S., *77*: 7049–7053, 1980.
23. Ayusawa, D., Iwata, K., Kozu, T., Ikegami, S., and Seno, T. Increase in dATP pool in aphidicolin-resistant mutants of mouse FM3A cells. Biochem. Biophys. Res. Commun., *91*: 946–954, 1979.
24. Aizawa, S., Loeb, L. A., and Martin, G. M. Isolation and preliminary characterization of mouse teratocarcinoma cells resistant to aphidicolin and arabinofuranosyl cytosine. J. Cell. Physiol., *115*: 9–14, 1983.
25. Holmes, A. M. Studies on the inhibition of highly purified calf thymus 8S and 7.3S DNA polymerase-α by aphidicolin. Nucleic Acids Res., *9*: 161–168, 1981.
26. Ayusawa, D., Iwata, K., Ikegami, S., and Seno, T. Reversal of aphidicolin-directed inhibition of DNA synthesis *in vivo* and *in vitro* by deoxyribonucleosides or their 5'-triphosphates. Cell Struct. Funct., *5*: 147–154, 1980.
27. Sugino, A., Kojo, H., Greenberg, B. D., Brown, P. O., and Kim, K. C. *In vitro* replication of yeast 2 μm plasmid DNA. ICN-UCLA Symp. Mol. Cell. Biol., *22*: 529–553, 1981.
28. Chang, C-c., Boezi, J. A., Warren, S. T., Sabourin, C.L.K., Liu, P. K., Glatzer, L., and Trosko, J. E. Isolation and characterization of a UV-sensitive hypermutable aphidicolin-resistant Chinese hamster cell line. Somatic Cell Genet., *7*: 235–253, 1981.
29. Liu, P. K., Chang, C-c., Trosko, J. E., Dube, D. K., Martin, G. M., and Loeb, L. A. Mammalian mutator mutant with an aphidicolin-resistant DNA polymerase-α. Proc. Natl. Acad. Sci. U.S., *80*: 797–801, 1983.
30. Liu, P. K., Chang, C-c., and Trosko, J. E. Association of mutator activity with UV-sensitivity in an aphidicolin-resistant mutant of Chinese hamster V79 cells. Mutat. Res., *106*: 317–332, 1982.
31. Sirover, M. A. and Loeb, L. A. Infidelity of DNA synthesis *in vitro*: Screening for potential metal mutagens or carcinogens. Science, *190*: 1434–1436, 1979.
32. Bernstein, C., Bernstein, H., Mufti, S., and Strom, B. Stimulation of mutation in phage T4 by lesions in gene 32 and by thymidine imbalance. Mutat. Res., *16*: 113–119, 1972.
33. Battula, N. and Loeb, L. A. The infidelity of avian myeloblastosis virus deoxyribonucleic acid polymerase in polynucleotide replication. J. Biol. Chem., *249*: 4086–4093, 1974.
34. Kunkel, T. A., Silber, J. R., and Loeb, L. A. The mutagenic effect of deoxynucleotide substrate imbalances during DNA synthesis with mammalian DNA polymerases. Mutat. Res., *94*: 413–419, 1982.

35. Meuth, M., Trudel, M., and Siminovitch, L. Selection of Chinese hamster cells auxotrophic for thymidine by 1-β-D-arabinofuranosyl cytosine. Somatic Cell Genet., *5*: 303–318, 1979.
36. Das, S. K., Silber, J. R., Kunkel, T. A., and Loeb, L. A. Rapid generation of dNTP pool imbalances in mammalian cells. Proceedings of Pacific Slope Biochemical Conference, p. 24, 1982.
37. Schaaper, R. M. and Glickman, B. W. Mutability of bacteriophage M13 by ultraviolet light: Role of pyrimidine dimers. Mol. Gen. Genet., *185*: 404–407, 1982.
38. Mathews, C. K., North, T. H., and Reddy, G.P.V. Multienzyme complexes in DNA precursor biosynthesis. *In;* G. Weber (ed.), Advances in Enzyme Regulation, Vol. 17, pp. 133–156, Pergamon Press, New York, 1979.
39. Reddy, G.P.V. and Pardee, A. B. Multienzyme complex for metabolic channeling in mammalian DNA replication. Proc. Natl. Acad. Sci. U.S., *77*: 3312–3316, 1980.
40. Lindahl, T. and Nyburg, B. Rate of depurination of native deoxyribonucleic acid. Biochemistry, *11*: 3610–3618, 1972.
41. Shearman, C. W. and Loeb, L. A. Depurination decreases fidelity of DNA synthesis *in vitro*. Nature, *270*: 537–538, 1977.
42. Schaaper, R. M. and Loeb, L. A. Depurination causes mutations in SOS-induced cells. Proc. Natl. Acad. Sci. U.S., *78*: 1773–1777, 1981.
43. Schaaper, R. M., Kunkel, T. A., and Loeb, L. A. Infidelity of DNA synthesis associated with bypass of apurinic sites. Proc. Natl. Acad. Sci. U.S., *80*: 487–491, 1983.
44. Strauss, B., Rabkin, S., Sagher, D., and Moore, P. The role of DNA polymerase in base substitution mutagenesis on non-instructional templates. Biochimie, 1983, in press.

ADP-RIBOSYLATION, DNA REPAIR AND CANCER, M. MIWA ET AL. (EDS.),
JAPAN SCI. SOC. PRESS, TOKYO/VNU SCIENCE PRESS, UTRECHT, PP. 277-286, 1983

# Induction of Diphtheria Toxin-resistant Cells by Mutagen-carcinogens

Masaaki TERADA, Michie NAKAYASU, Hiromi SAKAMOTO, Fumiko NAKASATO, and Takashi SUGIMURA

*National Cancer Center Research Institute, Tokyo 104, Japan*

*Abstract:* A mutation assay system with Chinese hamster lung (CHL) cells in culture has been established using diphtheria toxin resistance as a phenotypic selection marker. The results of series of studies on this mutation assay system are summarized and presented. Dose-dependent increase in the number of diphtheria toxin-resistant ($DT^r$) cells was observed when the cells were exposed to a mutagen-carcinogen and then incubated in fresh medium for an expression period of 7 to 8 days. After exposure to ethyl methanesulfonate, the number of $DT^r$ cells was much higher than the number of thioguanine-or ouabain-resistant cells. ADP-ribosylation of elongation factor 2 (EF-2) catalyzed by diphtheria toxin was measured in cell-free extracts from the parent cells and 17 $DT^r$ cells, including 6 spontaneous $DT^r$ cells and 11 $DT^r$ cells induced by mutagens; the numbers of ADP-ribose molecules transfer to EF-2 in extracts of mutant cells were less than 1% of that in extract of the parent cells.

Various mutagen-carcinogens, including heterocyclic amines isolated from pyrolysates of amino acids, proteins, and broiled fish, have been assayed with this mutation assay system in the presence or absence of a metabolic activation system and the results of these studies are also presented.

In addition, a method to detect $DT^r$ cells *in situ* has been developed using autoradiographic method. The potential usefulness of this method for detection of $DT^r$ cells with no proliferative capacity to form colony and for analysis of cellular events occurring after exposure of the cells to mutagen-carcinogens is discussed.

There is a high correlation between mutagenicity and carcinogenicity and the mutation assay method using *Salmonella typhimurium* has been widely and successfully used in screening for carcinogens (*1–5*). Mammalian cells, however, differ from bacteria in their organization of genetic materials, metabolism, DNA repair processes, and transport systems.

We have established a mutation assay system with Chinese hamster lung (CHL) cells in culture, using diphtheria toxin resistance as a selective marker (*6*). Use of this marker has proved effective in assays of mutagenic activities on human fibroblasts (*7*) and Chinese hamster ovary cells (*8*). Diphtheria toxin enters the cells *via* receptors and catalyzes ADP-ribosylation of elongation factor 2 (EF-2) of the toxin-sensitive cells. Once EF-2 is ADP-ribosylated, it loses activity, and so protein synthesis stops and the cells die (*9*, *10*). ADP-ribosylation has been shown to occur on a single unique amino acid, diphthamide, in EF-2 (*11*). With this assay system using diphtheria toxin resistance as a marker, mutagen-carcinogens, including heterocyclic amines isolated from amino acid and protein pyrolysates (*12*, *13*) and nitropyrenes (*6*) have been assayed. Diphtheria toxin-resistant ($DT^r$) cells obtained in this assay system have also been characterized.

Analysis of cellular events after exposure to a mutagen-carcinogen has been hampered by the lack of a proper assay method to detect a mutant cell which does not have proliferative capacity. In the presence of diphtheria toxin, CHL cells stop synthesizing protein, while $DT^r$ cells do not. Based on these results, a method to detect $DT^r$ cells *in situ* has been developed. This method will also be useful for detection of somatic mutation *in vivo*.

### *Mutation Assay*

For characterization of the mutation assay system with diphtheria toxin resistance as a phenotypic selection marker, the mutagenicities of alkylating agents, such as ethyl methanesulfonate (EMS), methyl methanesulfonate (MMS), methylnitrosourea (MNU), and N-methyl-N′-nitro-N-nitrosoguanidine (MNNG) in this assay have been determined (*6*). The mutagenicities of benzo(*a*)pyrene (BP) and dimethylnitrosamine (DMNA) have also been determined (*12*). When CHL cells were exposed to different concentrations of alkylating agents for 3 hr, and cultured in fresh medium for an expression time of 7 days, there was a dose-dependent increase in the

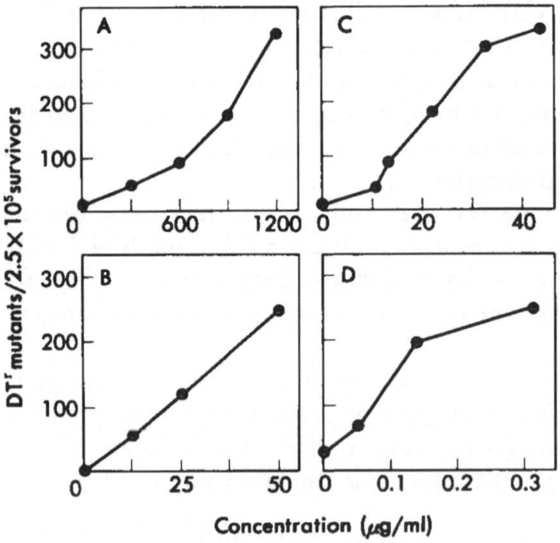


FIG. 1. Dose-dependency of the mutagenicities of alkylating agents on CHL cells. The cells were exposed to various concentrations of alkylating agents for 3 hr, and the numbers of $DT^r$ cells were determined after an expression time of 7 days (*6*). A, EMS; B, MNU; C, MMS; D, MNNG.

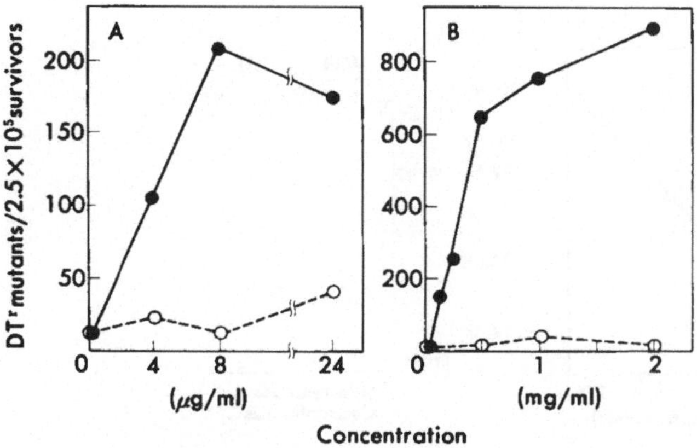


FIG. 2. Dose-dependency of the mutagenicities of BP (A) and DMNA (B). The cells were exposed to various concentrations of BP or DMNA in the presence or absence of S9 mix for 3 hr, and the numbers of DT$^r$ cells were determined (*12*). ● +S9 mix; ○ −S9 mix.

FIG. 3. Induction of DT$^r$ and 6-thioguanine-and ouabain-resistant cells by EMS. CHL cells were incubated with or without 1.32 mg/ml of EMS for 3 hr. After expression periods, the cells were cultured for 7 days in the presence of 0.1 Lf/ml of diphtheria toxin, 50 μM thioguanine, or 3 mM ouabain.

number of DT$^r$ cells, which were identified as cells capable of forming colonies in the presence of 0.1 Lf/ml of the toxin (Fig. 1). Since concentrations of 0.01 to 1 Lf/ml diphtheria toxin were cytotoxic to CHL cells, 0.1 Lf/ml of the toxin was used in the mutation assay. This concentration of diphtheria toxin resulted in almost complete brock in protein synthesis; 20 hr after addition of toxin, the incorporation of [$^{14}$C] L-leucine into the trichloroacetic acid-insoluble fraction was less than 0.05% of that of cells cultured in the absence of the toxin. Under the conditions used, an expression time of 7 to 9 days was required for induction of the maximum

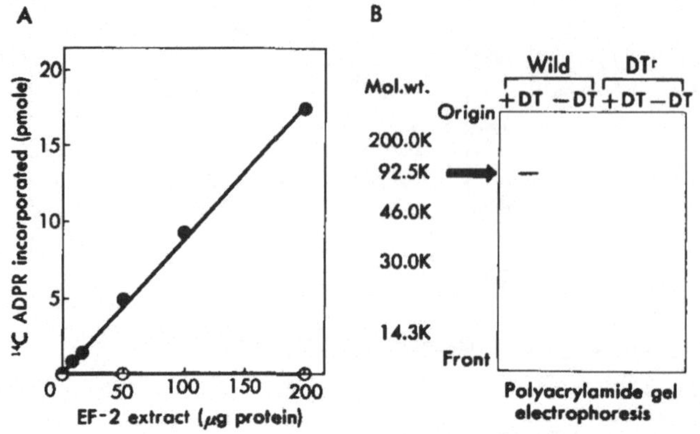


FIG. 4. ADP-ribosylation of EF-2 catalyzed by diphtheria toxin. Cytoplasmic extracts were prepared from parent CHL cells (●), and DT$^r$ cells (○). ADP-ribosylation of EF-2 catalyzed by the toxin was measured (*6*) (A). Samples from parent CHL cells or DT$^r$ cells were analyzed by SDS-polyacrylamide gel electrophoresis followed by autoradiography (B).

number of DT$^r$ cells after exposure of the cells to a mutagen. An activation system containing the postmitochondrial fraction (S9) from the liver of rats treated with polychlorinated biphenyls was required for BP and DMNA to show mutagenic activity on CHL cells. The optimum concentration of S9 differed, depending on the test compounds; it was 17.5 μl/ml for BP and 100 μl/ml for DMNA, when the protein concentration of S9 was 35 mg/ml. At the optimum concentration of S9, and with an expression time of 7 to 9 days, BP and DMNA induced dose-dependent increase in the number of DT$^r$ cells (Fig. 2).

In mammalian mutation assay systems, 6-thioguanine resistance or ouabain resistance has been widely used as a phenotypic selection marker. Therefore, these assay systems have been compared with our method using diphtheria toxin resistance as a marker. CHL cells were exposed to different concentrations of EMS for 3 hr, and mutant cells were selected for resistance to 0.1. Lf/ml of diphtheria toxin, 50 μM 6-thioguanine, or 3 mM ouabain. At the same concentration of EMS, the number of DT$^r$ cells induced was 3–5 times that of 6-thioguanine-resistant cells and 20–30 times that of ouabain-resistant cells (Fig. 3). Unlike in mutation assays with 6-thioguanine or ouabain-resistance as a marker, in this assay with diphtheria toxin resistance as a marker, the number of induced DT$^r$ cells did not change, when the concentration of diphtheria toxin used for selection was increased to 1 Lf/ml.

### *Characterization of DT$^r$ Cells*

Seventeen DT$^r$ clones, including six spontaneous, five EMS-induced and six 2-amino-3-methylimidazo [4,5-*f*]quinoline (IQ)-induced mutant colonies resistant to diphtheria toxin were isolated. EF-2 extracts were prepared from the parent CHL cells and these 17 mutant cells, and the diphtheria toxin-catalyzed ADP-ribosylation of EF-2 were measured with nicotinamide-[U-$^{14}$C]adenine dinucleotide as a substrate (*14*). With the parent CHL cells, ADP-ribosylation increased linearly with increase in the amount of EF-2 from the cells, and the amount of ADP-ribose covalent-

TABLE 1. Specific Mutagenic Activities of Various Chemicals in CHL Cells

| DT[r] mutants/$10^6$ survivors/$\mu g$[a)] | | | |
|---|---|---|---|
| Compounds not requiring S9 mix | | Compounds requiring S9 mix[b)] | |
| MNNG | 4,000 | Trp-P-2 | 160 |
| 4NQO | 952 | BP | 90 |
| 1,6-Dinitropyrene | 720 | IQ | 40 |
| 1,8-Dinitropyrene | 520 | MeIQ | 38 |
| 1,3-Dinitropyrene | 256 | Trp-P-1 | 33 |
| 1,3,6-Trinitropyrene | 192 | AαC | 20 |
| MMS | 25 | MeIQx | 5.7 |
| MNU | 18 | DMNA | 4.0 |
| Harman | 4.7 | Glu-P-1 | 1.2 |
| AF-2 | 4.4 | Glu-P-2 | 0.3 |
| Quercetin | 2.0 | | |
| Norharman | 1.7 | | |
| EMS | 0.6 | | |
| Coffee | 0.2 | | |

[a)] The specific mutagenic activity of each compound to CHL cells was calculated from the dose-response curve, and is expressed as DT[r] cells per $10^6$ survivors induced by culture with each compound at the concentration of 1 $\mu g$/ml for 3 hr. [b)] The activity was measured at optimum concentration of S9 for each compound.

ly linked to EF-2 was 113 pmol/mg of protein of the postribosomal supernatant fraction of the cells (Fig. 4). In contrast, there was no measurable ADP-ribosylation of EF-2 from any of the 17 mutants resistant to 0.1 Lf/ml of diphtheria toxin. The amounts of ADP-ribose transferred to EF-2 in extracts from these mutants were less than 1% of that transferred in the extract of the parent diphtheria toxin-sensitive cells. It has been known that there are two classes of DT[r] cells: one involves a membrane alteration leading to impermeability to the toxin, and the other involves an alteration of EF-2 such that ADP-ribosylation by diphtheria toxin does not occur, and so protein synthesis is not inhibited by the toxin (*14*). All 17 mutants isolated under the selective conditions used were of the latter type, because their EF-2 molecules could not be ADP-ribosylated in the cell-free system by the toxin. The high concentration of diphtheria toxin present in the selection medium probably did not allow survival of variants with an altered membrane.

### *Mutagenic Activities of Various Environmental Mutagen-carcinogens*

Since there is a good correlation between mutagenicity and carcinogenicity, the mutation assay method using *S. typhimurium* has been widely and successfully used in screening for mutagen-carcinogens in our daily environment. Bacteria, however, differ from mammalian cells in their organization of genetic material, cell permeability, metabolism, and possibly also DNA repair processes. Moreover, mutation assay with CHL cells using diphtheria toxin resistance as a marker measures forward mutation, while the assay with *Salmonella* measures reverse mutation. Accordingly, the mutagenicities of a number of compounds on CHL cells were tested with diphtheria toxin resistance as a marker.

Nitropyrenes are the most potent mutagenic compounds on *S. typhimurium* TA98 found so far (*15*). They and other nitroarenes are probably ubiquitous in the environment, since polycyclic aromatic hydrocarbons produced by internal combustion engines or coal-fired power plants are readily nitrated. Incubation of CHL cells with 1,3-,1,6-, and 1,8-dinitropyrenes and 1,3,6-trinitropyrene at various concentrations for 3 hr without a metabolic activation system induced significant increases in the number of $DT^r$ cells over those of untreated control cells. The number of $DT^r$ cells after an expression time of 7 days increased with increasing concentration of the compounds. 1,6- and 1,8-Dinitropyrenes were the most mutagenic and 1,3-dinitropyrene and 1,3,6-trinitropyrene were less mutagenic. 1-Nitropyrene and 1,3,6,8-tetranitropyrene did not induced $DT^r$ cells, although both these compounds were mutagenic to *Salmonella*. Comparison of the specific mutagenic activities of nitropyrenes with those of known mutagen-carcinogens showed that the activities of 1,3-, 1,6-, and 1,8-dinitropyrenes and 1,3,6-trinitropyrene were much higher than that of MMS or MNU; 1,6- and 1,8-dinitropyrenes were comparable that of to 4-nitroquinoline 1-oxide (4NQO) (Table 1). It has been reported that reduction of the nitro groups is required for the mutagenic activity of nitropyrenes on *S. typhinurium*, and that this reduction can be achieved with a bacterial enzyme. Thus, the finding that nitropyrenes are mutagenic to CHL cells indicates that the cells have a nitroreductase.

Several heterocyclic amines that have potent mutagenic activity to *S. typhimurium* in the presence of a metabolic activation system have been isolated from pyrolysis products of amino acids and proteins, and cooked foods (*16, 17*). These compounds include 3-amino-1,4-dimethyl-5*H*-pyrido[4,3-*b*]indole (Trp-P-1), 3-amino-1-methyl-5*H*-pyrido-[4,3-*b*]indole (Trp-P-2) (*18*), 2-amino-6-methyldipyrido[1,2-*a*:3′,2′-*d*]imidazole(Glu-P-1), 2-amino-dipyrido[1,2-*a*:3′,2′-*d*]imidazole(Glu-P-2) (*19*), 2-amino-9*H*-pyrido[2,3-*b*]indole (AαC) (*20*), 2-amino-3-methylimidazo[4,5-*f*]quinoline (IQ) (*21*), 2-amino-3,4-dimethylimidazo[4,5-*f*]quinoline (MeIQ) (*22*), and 2-amino-3,8-dimethylimidazo[4,5-*f*]quinoxaline (MeIQx) (*23*). The carcinogenicities of Trp-P-1, Trp-P-2, Glu-P-1, Glu-P-2, and AαC have already been demonstrated (*16*). The mutagenicities of these heterocyclic amines have been assayed in CHL cells in the presence and absence of a metabolic activation system, with diphtheria toxin resistance as a selective marker. Trp-P-2 was the most mutagenic of the heterocyclic amines tested, followed in order by IQ, MeIQ, Trp-P-1, AαC, MeIQx, Glu-P-1, and Glu-P-2. All these compounds except Trp-P-1 (*12*) required a metabolic activation system for mutagenicity on CHL cells, the optimum concentration of S9 differing for different compounds and ranging from 12 μl/ml to 50 μl/ml. The order of the specific mutagenic activities of the compounds toward CHL cells differed from that of their activities toward *S. typhimurium*.

Norharman and harman have called attention of scientists because they do not have mutagenic activity to *S. typhimurium*, but their additions to certain non-mutagenic carcinogens results in marked mutagenicity to *Salmonella* (*24, 25*). However, these compounds were found to be mutagenic to CHL cells in the absence of S9 mix (*13*). The reasons for the difference in response between *Salmonella* and CHL cells to these compounds have remained unknown. Regardless of the mechanisms involved in the mutagenic activity of these compounds to CHL cells, it is necessary to test carcinogenicity of these compounds: these compounds were both found at high

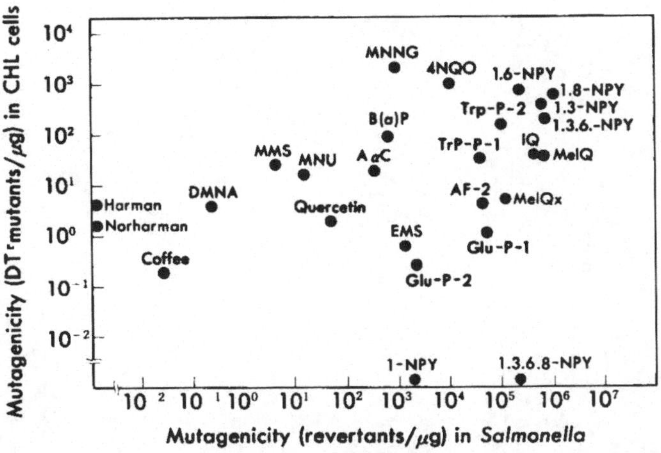


FIG. 5. Comparison of mutagenic activities to CHL cells with those to *S. typhimurium*. Specific mutagenic activity to CHL cells was expressed as number of DT$^r$ cells per $10^6$ survivors induced by culture for 3 hr with each compound at 1 μg/ml. The activity to *Salmonella* was expressed as the number of revertants obtained per μg of each substance. NPY, nitropyrene.

concentrations in a tryptophan pyrolysate (*26*), and were also found in pyrolysis products of proteins and foods and in tobacco tar (*27*).

The mutagenic activities of the compounds tested in the CHL assay system are summarized in Table 1. It is noteworthy that the mutagenicities of crude materials such as coffee could be measured with this assay system. The CHL assay system gives relatively reproducible results and can be recommended for use as a mammalian mutation method. The compounds show a wide range of specific mutagenic activities, and the order of their specific mutagenic activities to CHL cells is different from that to *S. typhimurium* (Fig. 5). Comparison of the mutagenic activities to *Salmonella* and CHL cells with the carcinogenic potential of each compound should give information on not only the qualitative but also the quantitative correlation between mutagenicity and carcinogenicity.

### *Mutation Assays In Situ*

In mammalian mutation assays, mutant cells have been usually detected by their capability to proliferate and form colonies in the presence of the selective agents. But, with these methods, it is difficult to follow the process in which a wild-type parent cell becomes a mutant cell. With these colony methods, it is also difficult to detect a mutant cell *in vivo*, since most of the somatic cells can not proliferate to form colonies. We have developed a method to detect a DT$^r$ cell by autoradiographic techniques. Parent CHL cells or DT$^r$ cells were exposed to diphtheria toxin at 1 Lf/ml for 16 hr and pulse-labeled with [$^3$H] L-leucine at 10 μCi/ml for 1 hr. The cells were fixed, washed and examined by autoradiography. More than 99.95% of parent CHL cells did not contain grains, since protein synthesis was inhibited. On the other hand, all the DT$^r$ cells were heavily labeled. Increasing numbers of DT$^r$ cells were mixed with a fixed number of parent CHL cells, incubated in the presence of diph-

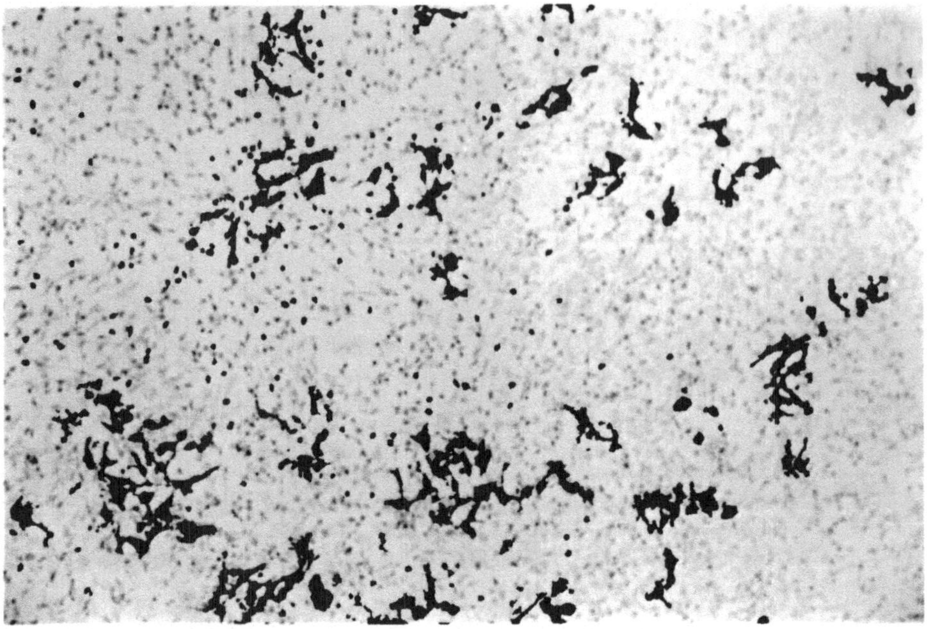

FIG. 6. Autoradiographic detection of DT$^r$ cells. Parent CHL cells were mixed with DT$^r$ cells, incubated for 16 hr, and then labeled with [$^3$H]leucine for 1 hr. The cells were washed and subjected to autoradiography. The cells were stained with Giemsa.

theria toxin at 1 Lf/ml for 16 to 20 hr and then pulse-labeled with [$^3$H]leucine for 1 hr. On autoradiography, the expected numbers of DT$^r$ cells were recovered as heavily labeled cells (Fig. 6).

Next, parent CHL cells were exposed to different concentrations of EMS, and after a expression time of 7 days, the numbers of DT$^r$ cells were measured by the conventional colony formation method and autoradiographical analysis. With both methods, dose-dependent increase in the number of DT$^r$ cells were observed. However, the number of DT$^r$ cells detected by autoradiography was higher than that obtained by the colony formation method. This difference was probably due to the fact that autoradiographic analysis can detect DT$^r$ cells with little or no proliferative capacity while the colony formation method can detect only DT$^r$ cells with proliferative capacity.

This method for detecting DT$^r$ cells *in situ* will be useful in analyzing mutational events occurring at the cellular level after exposure of the cells to a mutagen. It will also be possible to use this autoradiographic method for analyzing DT$^r$ cells *in vivo*, and a method to detect DT$^r$ cells *in vivo* are being developed in our labolatory.

## CONCLUSION

Mono-ADP-ribosylation as well as poly-ADP-ribosylation plays various important biological functions. It is known that diphtheria toxin inhibits protein synthesis and kills the sensitive cells by ADP-ribosylation of EF-2. We have introduced a mammalian forward mutation assay method with CHL cells, using diphtheria toxin resistance as a selective marker. This method is relatively simple and gives reproducible results, and can be used for studies on the environmental mutagen-carcinogens. The mutant cells can easily be characterized biochemically and could also be used for understanding further on the biological function of diphtheria toxin-catalyzed ADP-ribosylation of EF-2 at molecular levels. Development of a method to detect a single $DT^r$ cell *in situ* might open up new fields to analyze mutational events at cellular level and more importantly to detect somatic mutation *in vivo*.

## ACKNOWLEDGMENT

The work described in this paper was supported by grants from the Ministry of Health and Welfare of Japan, the Ministry of Education, Science and Culture of Japan, and the Princess Takamatsu Cancer Research Fund.

## REFERENCES


1. McCann, J., Choi, E., Yamasaki, E., and Ames, B. N. Detection of carcinogens as mutagens in the *Salmonella*/microsome test: Assay of 300 chemicals. Proc. Natl. Acad. Sci. U.S., *72*: 5135–5139, 1975.
2. McCann, J., Spingarn, N. E., Kobori, J., and Ames, B. N. Detection of carcinogens as mutagens: Bacterial tester strains with R factor plasmid. Proc. Natl. Acad. Sci. U.S., *72*: 979–983, 1975.
3. Sugimura, T., Sato, S., Nagao, M., Yahagi, T., Matsushima, T., Seino, Y., Takeuchi, M., and Kawachi, T. Overlapping of carcinogens and mutagens. *In;* P. N. Magee *et al.* (eds.), Fundamentals in Cancer Prevention, pp. 191–215, Japan Scientific Society Press, Tokyo/University Park Press, Baltimore, 1976.
4. Purchase, I.F.M., Longstaff, E., Ashby, J., Styles, J. A., Anderson, D., Lefevere, P. A., Westwood, F. R. Evaluation of six short term tests for detecting organic chemical carcinogens and recommendation for their use. Nature, *264*: 624–627, 1976.
5. Nagao, M., Sugimura, T., and Matsushima, T. Environmental mutagens and carcinogens. Annu. Rev. Genet., *12*: 117–159, 1978.
6. Nakayasu, M., Sakamoto, H., Wakabayashi, K., Terada, M., Sugimura, T., and Rosenkranz, H. S. Potent mutagenic activity of nitropyrenes on Chinese hamster lung cells with diphtheria toxin resistance as a selective marker. Carcinogenesis, *3*: 917–922, 1982.
7. Glover, T. W., Chang, C.-C., Trosko, J. E., and Li, S. S.-L. Ultraviolet light induction of diphtheria toxin-resistant mutants in normal and xeroderma pigmentosum human fibrobrasts. Proc. Natl. Acad. Sci. U.S., *76*: 3982–3986, 1979.
8. Gupta, R. S. and Siminovitch, L. Genetic markers for quantitative mutagenesis studies in Chinese hamster cells: Characteristics of some recently developed selective systems. Mutat. Res., *69*: 113–126, 1980.
9. Honjo, T., Nishizuka, Y., and Hayaishi, O. Diphtheria toxin-dependent adenosine diphosphate ribosylation of aminoacyl transferase II and inhibition of protein synthesis. J. Biol. Chem., *243*: 3553–3555, 1968.
10. Pappenheimer, A. M. Jr. Diphtheria toxin. Annu. Rev. Biochem., *46*: 69–94, 1977.
11. Van Ness, B. G., Howard, J. B., and Bodley, J. W. ADP-ribosylation of elongation factor 2 by diphtheria toxin. J. Biol. Chem., *255*: 10710–10716, 1980.
12. Nakayasu, M., Nakasato, F., Sakamoto, H., Terada, M., and Sugimura, T. Mutagenic activity of heterocyclic amines in Chinese hamster lung cells with diphtheria toxin resistance as a marker. Mutat. Res., *118*: 91–102, 1983.

13. Nakayasu, M., Nakasato, F., Sakamoto, H., Terada, M., and Sugimura, T. Mutagenic activity of norharman and harman in Chinese hamster lung cells in assays with diphtheria toxin resistance as a marker. Cancer Lett., *17*: 249–255, 1983.
14. Moehring, T. J. and Moehring, M. J. Selection and characterization of cells resistant to diphtheria toxin and pseudomonas exotoxin A: presumptive translation mutants. Cell, *11*: 447–454, 1977.
15. Rosenkranz, H. S., McCoy, E. C., Saunders, D. R., Butler, M., Kiriazides, D. K., and Mermelstein, R. Nitropyrenes: isolation, identification and reduction of mutagenic impurities in a carbon black in toners. Science, *209*: 1039–1043, 1980.
16. Sugimura, T. Mutagens, carcinogens and tumor promoters in our daily food. Cancer, *49*: 1970–1984, 1982.
17. Sugimura, T. and Nagao, M. Modification of mutagenic activity. *In*: F. J. de Serres *et al.* (eds.), Chemical Mutagens, Vol. 6, pp. 41–60, Plenum Press, New York, 1980.
18. Sugimura, T., Kawachi, T., Nagao, M., Yahagi, T., Seino, Y., Okamoto, T., Shudo, K., Kosuge, T., Tsuji, K., Wakabayashi, K., Iitaka, Y., and Itai, A. Mutagenic principle(s) in tryptophan and phenylalanine pyrolysis products. Proc. Jpn. Acad., *53*: 58–61, 1977.
19. Yamamoto, T., Tsuji, K., Kosuge, T., Okamoto, T., Shudo, K., Takeda, K., Iitaka, Y., Yamaguchi, K., Seino, Y., Yahagi, T., Nagao, M., and Sugimura, T. Isolation and structure determination of mutagenic substances in L-glutamic acid pyrolysate. Proc. Jpn. Acad., *54B*: 248–250, 1978.
20. Yoshida, D., Matsumoto, T., Yoshimura, R., and Matsuzaki, T. Mutagenicity of amino-$\alpha$-carbolines in pyrolysis products of soybean globulin. Biochem. Biophys. Res. Commun., *83*: 951–920, 1978.
21. Kasai, H., Nishimura, S., Wakabayashi, K., Nagao, M., and Sugimura, T. Chemical synthesis of 2-amino-3-methylimidazo[4,5-*f*]quinoline (IQ), a potent mutagen isolated from broiled fish. Proc. Jpn. Acad., *56B*: 382–384, 1980.
22. Kasai, H., Yamaizumi, Z., Wakabayashi, K., Nagao, M., Sugimura, T., Yokoyama, S., Miyagawa, T., and Nishimura, S. Structure and chemical synthesis of Me-IQ, a potent mutagen isolated from broiled fish. Chem. Lett., 1391–1394, 1980.
23. Kasai, H., Yamaizumi, Z., Shiomi, T., Yokoyama, S., Miyazawa, T., Wakabayashi, K., Nagao, M., Sugimura, T., and Nishimura, S. Structure of a potent mutagen isolated from fried beef. Chem. Lett., 485–488, 1981.
24. Nagao, M., Yahagi, T., Honda, M., Seino, Y., Matsushima, T., and Sugimura, T. Demonstration of mutagenicity of aniline and *o*-toluidine by norharman. Proc. Jpn. Acad., *53B*: 34–37, 1977.
25. Nagao, M., Yahagi, T., Kawachi, T., Sugimura, T., Kosuge, T., Tsuji, K., Wakabayashi, K., Mizusaki, S., and Matsumoto, T. Comutagenic action of norharman and harman. Proc. Jpn. Acad., *53*: 95–98, 1977.
26. Yasuda, T., Yamaizumi, Z., Nishimura, S., Nagao, M., Takahashi, Y., Fujiki, H., Sugimura, T., and Tsuji, K. Detection of comutagenic compounds, harman and norharman in pyrolysis products of proteins and food by gas chromatography-mass spectrometry. Proc. 3rd Annu. Meet. Med. GCMS Soc., 97–100, 1978.
27. Poindexter, E. H. Jr. and Carpenter, R. O. The isolation of harman and norharman from tobacco and cigarette smoke. Phytochemistry, *1*: 215–221, 1962.

ADP-RIBOSYLATION, DNA REPAIR AND CANCER, M. MIWA ET AL. (EDS.),
JAPAN SCI. SOC. PRESS, TOKYO/VNU SCIENCE PRESS, UTRECHT, PP. 287-294, 1983

# Interference with DNA Repair Mechanisms of Mammalian Cells: Cell Cycle Dependence

Arthur B. PARDEE, Robert J. BOORSTEIN, and Ching C. LAU

*Dana-Farber Cancer Institute and Department of Pharmacology, Harvard Medical School, Boston, Massachusetts 02115, U.S.A.*

*Abstract:* Agents that prevent repair of lesions in DNA increase cell death. Caffeine enhances lethality of agents such as HN2. We propose that damaged cells stop in $G_2$ phase until repair is completed, and caffeine prevents this $G_2$ arrest. Therefore caffeine lets damaged cells go through mitosis; their chromosomes are shattered and they die. This action of caffeine requires rapid synthesis of protein(s).

Some classes of damage by alkylating agent (methyl methane sulfonate (MMS)) are removed slowly in human cells, followed by rapid repair synthesis. Lethality results if gaps created during repair of DNA are present during semiconservative replication. Nicotinamide analogs prevent ligation, the gaps persist, and lethality is greater. This effect is S phase specific. Further experiments may specify conditions and agents giving better therapeutic results than can be obtained with anticancer drugs used alone.

X-rays and many drugs (*1*, *2*) used in chemotherapy kill cells and damage DNA, by alkylation or by interfering with DNA synthesis (*3*) (for reviews see *4*–*6*). Inability to repair DNA lesions increases lethality by orders of magnitude, as shown with xeroderma pigmentosum (XP) cells (*7*, *8*), and with repair inhibitory compounds (*4*, *9*). Also, many carcinogens damage DNA. The importance of DNA repair in carcinogenesis is indicated by cancer proneness of individuals who are genetically repair-defective (*8*, *10*).

Two major DNA repair mechanisms are called excision repair and postreplication repair (see reviews cited). In excision repair defective DNA is excised and new DNA is synthesized to replace it. This process is relatively specific and error-free. Inhibitors of poly(ADP-ribose) synthesis such as 3-aminobenzamide(3-AB) block excision repair (*11*).

In postreplication repair, DNA lesions are bypassed as DNA is newly synthesized, leaving gaps (*6*, *12*). Later these gaps are somehow filled. Caffeine, by far the best known proposed inhibitor of this process (*13*, *14*), enhances lethality (*15*, *16*), and chromosomes are broken (*13*, *15*).

Changes in mutation and transformation frequencies under the influence of repair modifiers such as caffeine (*17, 18*) or genetic defects (*7, 8*) suggest that excision repair is more error-free than postreplication repair (*19, 20*). Altered mutagenicity or carcinogenicity of DNA damaging agents was seen in cultured cells and animals when caffeine was also provided (*21, 22*).

The outcome of DNA damage may depend upon a race between repair and lethal events (*19, 23*). Cells that receive much DNA damage generally stop in S phase. Damage prevents completion of their DNA synthesis, and they cannot initiate subsequent mitosis or cell division (*24*). Cells receiving sublethal DNA damage are arrested in $G_2$ phase; mitosis occurs only after a time that depends on the degree of damage. DNA synthesis continues in these cells, although modifications remain as shown with radioactive alkylating agents or cis-platinum (*4, 25, 26*).

Caffeine is reported to interfere with postreplication gap production and filling, thereby enhancing lethality (*12, 13*); it can increase the rate of DNA synthesis, and may stimulate replicon initiation in damaged DNA (*23*).

Our work is designed to provide fundamental information regarding mechanisms by which defective repair enhances lethality. Our objective is to improve therapeutic procedures and diminish carcinogenicity, based upon this fundamental understanding. The cells of interest in both chemotherapy and carcinogenesis are the minority that survive damage, and grow into tumors. These are cells whose DNA lesions have been repaired. This DNA repair is a biochemical process, and therefore can be modified by pharmacological agents.

### *Caffeine and Post-replication Repair*

Caffeine greatly changes lethality, mutation, and transformation of mammalian cells treated with DNA damaging agents (*13, 14*). Literature data conflict because

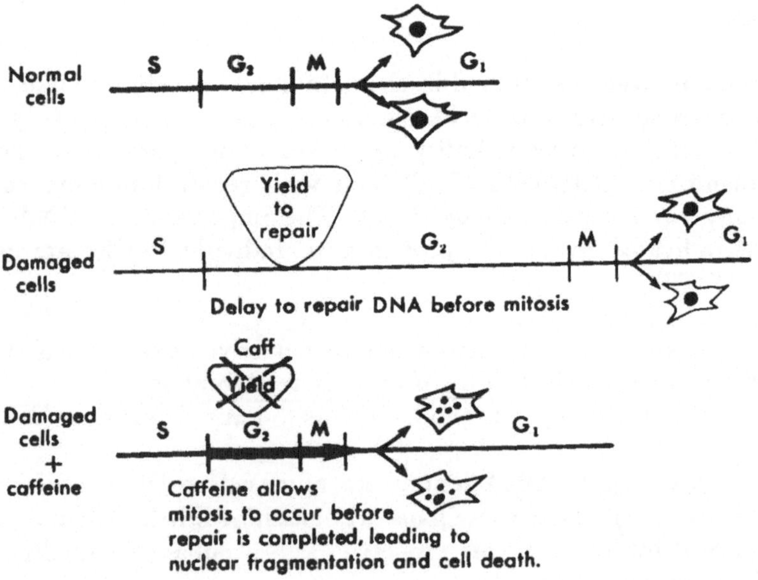


FIG. 1. Schematic representation of the proposed mechanism by which caffeine enhances lethality of alkylating agents.

various cells (rodent and human) are used, several repair systems are acting simultaneously to different degrees, and their roles are not readily separated. Mechanisms have been unclear. We have studied lethal effects of caffeine on baby syrian hamster kidney (BHK) cells treated with bifunctional alkylating agents such as nitrogen mustard (HN2). The lethal effects are large, rapidly quantitated, and relate to one major type of repair.

Damaged cells need sufficient time, in $G_2$, to repair DNA lesions before they divide (Fig. 1). Caffeine prevents this $G_2$ arrest, so that damaged cells go through mitosis, shatter their chromosomes, and daughter cells emerge with fragmented nuclei (Fig. 2). These events are lethal for cells that otherwise would have been repaired. Painter and Young (*27*) proposed that the genetic defect in ataxia telangiectasia cells causes chromosome damage and lethality by this kind of mechanism, and hypothesized that caffeine also might have this effect.

Supporting data (*28, 29*) are briefly as follows: treatment of exponential BHK cells with a low concentration of HN2 (0.5 $\mu$M for 1 hr) had little effect on survival. Cells were seen, using a flow microfluorimeter, to accumulate with a $G_2$ DNA content for several hours, and then to pass on through the cycle. This $G_2$ arrest—mitosis delay—has been seen often (*4, 24*). HN2 had little effect on S phase of cells synchronized at the $G_1$/S interface with hydroxyurea. It caused a 6 hr $G_2$ arrest, during which a period of unscheduled DNA synthesis ([$^3$H]thymidine uptake) was observed by autoradiography. These results suggest that these cells cannot enter mitosis until damage is repaired.

With caffeine present, these cells were not blocked in $G_2$, and entered mitosis

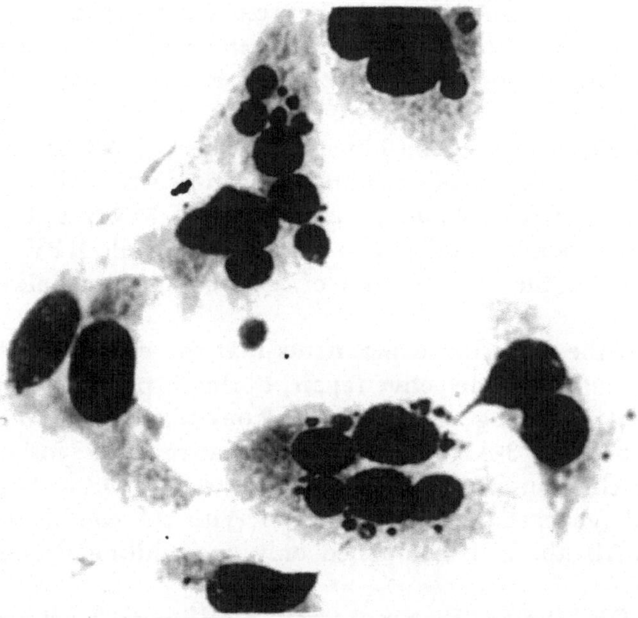

FIG. 2. Nuclear fragmentation in HN2-treated BHK cells after caffeine post-treatment. Exponentially growing cells were treated with HN2 (0.5 $\mu$M) for 1 hr followed by 12 hr of incubation with caffeine (2 mM). Cells were fixed and stained with Giemsa. Pairs of both intact and fragmented post-mitotic nuclei are shown. ×1,200.

at the normal time (*30*). Extensive pulverization of chromosomes resulted.* Highly fragmented nuclei were seen, whose numbers correlated very well with lethality. Both effects were equally inhibited by low doses (0.2 μg/ml) of cycloheximide. Synchronized HN2-treated S phase cells were sensitive to caffeine added during $G_2$, but not during $G_1$ or S, nor were quiescent cells sensitive.

The molecular basis for caffeine's lethality enhancement is not understood (*4, 13, 14*). Protein synthesis is needed to trigger mitosis. All effects of caffeine described above were abolished by 0.2 μg/ml or less of cycloheximide, a concentration that partly inhibits protein synthesis. We propose that cycloheximide blocks synthesis of a protein needed for mitosis, thereby making time available for DNA repair in caffeine-treated, damaged cells. Caffeine may permit mitotic initiation through permitting specific $G_2$ protein synthesis.

### *3-AB and Excision Repair*

Shall and his coworkers (*31*) have shown that 3-AB inhibits excision repair. 3-AB and caffeine act on different repair mechanisms. Enhanced lethality with caffeine did not appear to involve excision repair, since under our conditions 3-AB did not enhance lethality or cause nuclear fragmentation due to HN2, MNNG, and UV. Post-treatment with 2 mM 3-AB for 24 hr increased the lethality of MMS 3-fold. By contrast, caffeine increased the lethality of all of these agents 5 to 10-fold. Lethality was maximally increased by 3-AB during the first 6 hr after MMS addition, in contrast to a 12 hr delayed effect of caffeine. Cycloheximide did not affect the lethality of 3-AB on MMS treated cells, in contrast to its neutralizing effect with caffeine. When caffeine and 3-AB were both present post-treatment, their effects were synergistic. Thus, these drugs will provide useful tools for dissecting different DNA repair mechanisms, permitting us selectively to inhibit one while studying others.

We are studying effects of 3-AB on human fibroblasts. 3-AB, in contrast to caffeine, had no effect on the lethality of cross-linking agents such as HN2, cis-platinum, mitomycin C, or BCNU. 3-AB enhanced the lethality of MMS and MNNG more than 2-fold. XPA cells responded similarly to normal cells, while XPV cells showed a greater lethal response to the combination of a monofunctional alkylating agent and 3-AB.

We propose from the following experiments that the effect of 3-AB on MMS treated human cells results from defective repair, during S phase, of damage caused by this monofunctional alkylating agent (MMS). Non-toxic damage that is repaired or diluted is made lethal with 3-AB only during replicative DNA synthesis.

Asynchronous, cycling human fibroblasts treated with MMS or 3-AB continued growth uninterrupted (generation time 20–24 hr). With the combination of MMS (1 hr) followed by 3-AB, the cells completed only one additional doubling in the next 96 hr.

In rodent cells, DNA lesions that make cells sensitive to 3-AB are rapidly removed (*11*). In human fibroblasts, the DNA damage that makes cells sensitive to

* We thank Dr. I. Gadi from Dr. Sager's laboratory for karyotypes.

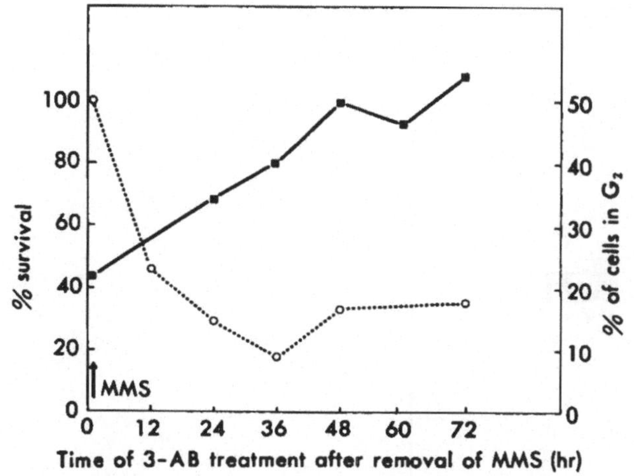


FIG. 3. Slow responses to 3-AB. Normal human fibroblasts GM 2936 (Institute for Medical Research, Camden, NJ) were plated at a density of $5 \times 10^4$ per 35 mm tissue culture plate in DME plus 10% FCS. After 48 hr, cells were treated with 0.5 mM MMS for 1 hr. The cells were washed, and 4 mM 3-AB added. Colony formation assays and flow microfluorimetry were done at times indicated. Survival (O) is expressed as % of control colonies. Percent of cells in $G_2$ (■) was determined from integration of $G_2$ peaks of FMF histograms.

3-AB is much more slowly repaired. Growth of cells treated with MMS for 1 hr continued to be sensitive to 3-AB for 48 hr (Fig. 3). Thus, some class of MMS-induced lesions persisted, even after cell replication. These lesions were not toxic by themselves, but 3-AB made them lethal.

3-AB caused MMS treated human cells to arrest irreversibly in $G_2$. The number of cells in $G_2$ reached a maximum after 48 hr (Fig. 3). 3-AB or MMS alone caused little $G_2$ accumulation.

To learn whether this prolonged effect of 3-AB is slowly continuous or depends on reaching a stage in the cell cycle, cells stopped in $G_0$ by serum starvation or density arrest were treated with MMS and immediately replated in the presence of 3-AB (Fig. 4). Synchronous DNA synthesis occurred 24–36 hr after replating. A colony formation assay showed no lethality was due to 3-AB given for 24 hr after MMS treatment, but 48 hr incubation with 3-AB caused 70% lethality (during S phase).

Cells synchronized with hydroxyurea at the $G_1$/S interface and treated with MMS showed 60% lethality when 3-AB was present during the first 12 hr after growth restarted, during S phase, and no more lethality with longer 3-AB incubations (Fig. 4). Furthermore, the lesions responsible for lethality were not removed from $G_0$ or hydroxyurea blocked cells, that is, in the absence of replicative DNA synthesis. They were still present in the arrested cells 24 hr after MMS addition, as demonstrated by lethality of 3-AB added after growth initiation. Other kinetic data support the conclusion that to increase lethality 3-AB must be present during S phase.

These cell biological studies were complemented by directly measuring lesions, using the alkaline elution technique (*32*). MMS damage could no longer be detected in cycling control cells 48 hr after treatment, when cells treated with 3-AB still showed considerable damage. Thus 3-AB inhibited removal of some lesions. Inhibition

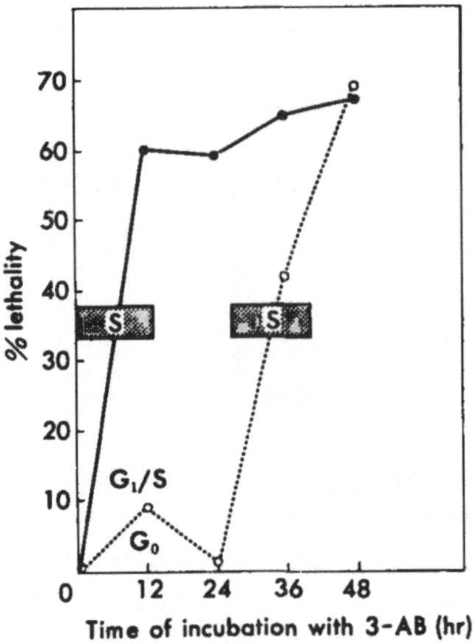


FIG. 4. Correlations of lethality and S phase. Normal human fibroblasts ($5 \times 10^4$ cells per 35 mm dish) were synchronized in $G_0$ by serum starvation (○) or at $G_1/S$ by HU blockade (●). Cells were released with complete medium, treated with 0.5 mM MMS for 1 hr, washed and replated at low density in medium containing 4 mM 3-AB, then fresh medium was added and colonies allowed to grow for 2 weeks. Lethality is expressed as percent of colonies from control cultures. The interval of maximum S phase DNA synthesis for each condition is indicated by the shaded boxes.

by 3-AB of repair, as measured by alkaline elution, correlated with enhanced lethality. To study the relationship of replicative DNA synthesis to 3-AB inhibition of DNA repair and cell lethality, we used cells density arrested in $G_0$, or blocked at $G_1/S$ with hydroxyurea. These cells showed much less repair of MMS induced damage as measured by alkaline elution than did cells released from $G_0$ or $G_1/S$. 3-AB blocked repair as detected by alkaline elution, during S phase, the phase in which 3-AB was lethal.

We conclude from these and other experiments that inhibition of DNA ligation by 3-AB (*33*) is not necessarily lethal. It becomes lethal only if cells attempt to replicate DNA having incompletely or inappropriately ligated DNA. Such an attempt can cause strand breaks, $G_2$ arrest, and lethality.

## CONCLUSIONS

Different repair systems exist for monofunctional and cross-linking alkylating agents, in both human and rodent cells. Repair of bifunctional agents' damage in rodent cells is strongly prevented by caffeine; the mechanism involves stimulation of $G_2$ transit. Damaged human cells are much less sensitive to caffeine. The repair of monofunctional alkylating agents' damage in rodent cells is sensitive to both caffeine and 3-AB, drugs that act by different mechanisms. Repair of damage caused by monofunctional alkylating agents in human cells is sensitive only to 3-AB. 3-AB in rodent cells inhibits a rapid repair process, while in human cells its inhibition of repair and enhancement of lethality is dependent on DNA synthesis in S phase. The effects on repair of both caffeine and 3-AB are cell cycle dependent, though through completely different mechanisms.

ACKNOWLEDGMENT

This work was aided by Grants NIGMS 5 T32 GM 07753 and NCI PO1-22427.

REFERENCES

1. Sieber, S. M. and Adamson, R. H. The clastogenic, mutagenic, teratogenic and carcinogenic effects of various antineoplastic agents. *In;* Pharmacological Basis of Cancer Chemotherapy, pp. 401–467, Williams and Wilkins, Baltimore, 1975.
2. Frindel, E. and Tubiana, M. Radiobiology and the cell cycle. *In;* R. Baserga (ed.), The Cell Cycle and Cancer, Marcel Dekker, New York, pp. 391–447, 1971.
3. Kunz, B. A., Barclay, B. J., Game, J. C., Little, J. G., and Haynes, R. H. Induction of mitotic recombination in yeast by starvation for thymine nucleotides. Proc. Natl. Acad. Sci. U.S., *77*: 6057–6061, 1980.
4. Roberts, J. J. The repair of DNA modified by cytotoxic, mutagenic, and carcinogenic chemicals. Adv. Radiat. Biol., *7*: 211–436, 1978.
5. Hanawalt, P. C., Cooper, P. K, Ganesan, A. K., and Smith, C. A. DNA repair in bacteria and mammalian cells. Annu. Rev. Biochem., *48*: 783–836, 1979.
6. Lehmann, A. R. and Karran, P. DNA repair. Int. Rev. Cytol., *72*: 101–146, 1981.
7. Maher, V. M., Yang, L. L., and McCormick, J. J. The role of DNA repair in preventing the cytotoxic and mutagenic effects of carcinogens in human cells. *In;* B. Pullman *et al.* (eds.), Carcinogenesis: Fundamental Mechanisms and Environmental Effects, pp. 479–490, D. Reidel, Dordrecht, 1980.
8. Setlow, R. B. Repair deficient human disorders and cancer. Nature, *271*: 713–717, 1978.
9. Walker, J. G. and Reid, B. R. Caffeine potentiation of the lethal action of alkylating agents on L-cells. Mutat. Res., *12*: 101–104, 1971.
10. German, J. L. III. Chromosome Breakage and Neoplasia, Academic Press, New York, 1982.
11. Durkacz, B. W., Omidiji, O., Gray, D. A., and Shall, S. (ADP-ribose)$_n$ participates in DNA excision repair. Nature, *283*: 593–596, 1980.
12. Lehmann, A. R., Kirk-Bell, S., Arlett, C. F., Harcourt, S. A., de Weers-Kastelein, E. A., Keijzer, W., and Hall-Smith, P. Repair of ultraviolet light damage in a variety of human fibroblast cell strains. Cancer Res., *37*: 904–910, 1977.
13. Kihlman, B. A. Caffeine and Chromosomes, Elsevier, Amsterdam, 1977.
14. Timson, J. Caffeine. Mutat. Res., *47*: 1–52, 1977.
15. Nilsson, K. and Lehmann, A. R. The effect of methylated oxypurines on the size of newly synthesized DNA and on the production of chromosome aberrations after UV irradiation in Chinese hamster cells. Mutat. Res., *30*: 255–266, 1975.
16. Busse, P. M., Bose, S. K., Jones, R. W., and Tolmach, L. J. The action of caffeine on X-irradiated HeLa cells. II. Synergistic lethality. Radiat. Res., *71*: 666–677, 1977.
17. Kakunaga, T. Caffeine inhibits cell transformation by 4-nitroquinoline-1-oxide. Nature, *258*: 248–250, 1975.
18. Arlett, C. F. and Harcourt, S. A. The induction of 8-azaguanine-resistant mutants in cultured Chinese hamster cells by ultra-violet light. The effect of changes in post-irradiation conditions. Mutat. Res., *14*: 431–437, 1972.

19. Phillips, R. A. and Tolmach, L. J. Repair of potentially lethal damage in X-irradiated HeLa cells. Radiat. Res., *29*: 413–432, 1966.
20. Trosko, J. E. and Chu, E.H.Y. The role of DNA repair and somatic mutation in carcinogenesis. Adv. Cancer Res., *21*: 391–425, 1975.
21. Rothwell, K. Dose-related inhibition of chemical carcinogenesis in mouse skin by caffeine. Nature, *252*: 69–70, 1974.
22. Donovan, P. J. and Dipaolo, J. A. Caffeine enhancement of chemical carcinogen-induced transformation of cultured Syrian hamster cells. Cancer Res., *34*: 2720–2727, 1974.
23. Cleaver, J. E. DNA repair and its coupling to DNA replication in eukaryotic cells. Biochim. Biophys. Acta, *516*: 489–516, 1978.
24. Canti, R. G. and Spear, F. G. The effect of gamma irradiation on cell division in tissue culture *in vitro*. Proc. Roy. Soc., *B102*: 92–101, 1927.
25. Roberts, J. J. and Fraval, H.N.A. Repair of cis-platinum (II) diammine dichloride-induced DNA damage and cell sensitivity. *In;* A. W. Prestayko, S. T. Crooke, and S. K. Carter (eds.), Cisplatin, pp. 57–77, Academic Press, New York, 1980.
26. Johnson, N. P., Hoeschele, J. D., Rahn, R. O., O'Neill, J. P., and Hsie, A. W. Mutagenicity, cytotoxicity, and DNA binding of platinum (II) — chloroammines in Chinese hamster ovary cells. Cancer Res., *40*: 1463–1468, 1980.
27. Painter, R. B. and Young, B. R. Radiosensitivity in ataxia telangiectasia: A new explanation. Proc. Natl. Acad. Sci. U.S., *77*: 7315–7317, 1980.
28. Lau, C. C. and Pardee, A. B. Mechanism by which caffeine enhances lethality of nitrogen mustard. Proc. Natl. Acad. Sci. U.S., *79*: 2942–2946, 1982.
29. Das, S. K., Lau, C. C., and Pardee, A. B. Abolition by cycloheximide of caffeine enhanced lethality of alkylating agents in hamster cells. Cancer Res., *42*: 4499–4504, 1982.
30. Walters, R. A., Gurley, L. R., and Tobey, R. A. Effects of caffeine on radiation-induced phenomena associated with cell cycle traverse of mammalian cells. Biophys. J., *14*: 99–118, 1974.
31. Shall, S. ADP-ribose in DNA repair. *In;* O. Hayaishi and K. Ueda (eds.), ADP-ribosylation Reactions, pp. 477–520, Academic Press, New York, 1982.
32. Kohn, K. W., Ewig, R.A.G., Erickson, L. C., and Zwelling, L. A. Measurement of strand breaks and cross-links by alkaline elution. *In;* Friedberg, E. C. and Hanawalt, P. C. (eds.), DNA Repair, A Laboratory Manual of Research Procedures, Vol. 1, Part B, pp. 379–401, Marcel Dekker, New York, 1981.
33. Creissen, D. and Shall, S. Regulation of DNA ligase activity by poly(ADP-ribose). Nature, *296*: 271–272, 1982.

# ADP-RIBOSYLATION AND CANCER

ADP-RIBOSYLATION, DNA REPAIR AND CANCER, M. MIWA ET AL. (EDS.),
JAPAN SCI. SOC. PRESS, TOKYO/VNU SCIENCE PRESS, UTRECHT, PP. 297-308, 1983

# DNA Strand Breaks and Poly(ADP-ribose) Synthetase Activation in Pancreatic Islets—A New Aspect to Development of Insulin-dependent Diabetes and Pancreatic B-cell Tumors

Hiroshi OKAMOTO and Hiroshi YAMAMOTO

*Department of Biochemistry, Toyama Medical and Pharmaceutical University School of Medicine, Toyama 930-01, Japan*

*Abstract:* Alloxan and streptozotocin, which produce diabetes mellitus in experimental animals, have been known to inhibit various functions of pancreatic islets including proinsulin synthesis. However, little is known about the mechanisms underlying the action of these agents in pancreatic islets. Our recent *in vivo* and *in vitro* study using rats and isolated islets showed that one of the primary targets of the diabetogenic agents is the DNA of pancreatic islets. The first step is the generation of hydroxyl radical by alloxan which attacks DNA to produce strand breaks. In the case of streptozotocin, the alkylating activity of this compound may be causally related to its ability to induce DNA strand breaks. Subsequently, the fragmented DNA activates poly(ADP-ribose) synthetase which depletes cellular NAD. Since NAD is the most abundant of cellular coenzymes and participates in many biological reactions in mammalian cells, the reduction in intracellular NAD to such a nonphysiological level may severely affect islet cell functions including proinsulin synthesis. These results, in turn, raise the possibility that insulin-dependent diabetes may be preventable by inhibiting the occurrence of DNA strand breaks or the poly(ADP-ribose) synthetase. In fact, by poly(ADP-ribose) synthetase inhibitors such as nicotinamide and picolinamide, alloxan- and streptozotocin-induced NAD depletion was completely prevented, and B-cell functions including proinsulin synthesis proceeded normally. However, poly(ADP-ribose) synthetase inhibitors did not prevent the DNA strand breaks at all. Therefore, B-cells may survive with the residual DNA damage within their genome. About one year after the combined administration to rats of alloxan or streptozotocin with poly(ADP-ribose) synthetase inhibitors, diabetes did not develop but islet B-cell tumors were found frequently. This suggests that insulin-dependent diabetes and B-cell tumors are closely related with respect to their developmental processes. In other words, DNA breaks initiate two kinds of pathological state in B-cells, one is degeneratively and the other is oncogenically expressed.

Several years ago, we succeeded in developing a method for isolation of a large number of "intact" islets from rat pancreas; since then we have pursued *in vitro* investigations on insulin biosynthesis and its regulation using the isolated islets (*1–11*).

Our recent study also has been focused on the mechanism by which insulin biosynthesis is inhibited in various pathological conditions. Alloxan and streptozotocin exhibit the most potent diabetogenicity, and have been widely used for induction of experimental diabetes (*12, 13*). Understanding the mechanisms of action of the typical diabetogenic agents seems to be of great importance for elucidating not only the causes of insulin-dependent diabetes but also its prevention. Alloxan and streptozotocin have been known to inhibit various functions of pancreatic islets including proinsulin synthesis (*14, 15*). However, little is known about the mechanisms underlying the action of these agents in pancreatic islets. We have recently obtained evidence that one of the primary targets of these agents is the DNA of pancreatic islets (*4, 16–18*). That is, alloxan and streptozotocin induce DNA strand breaks and the damaged DNA stimulates nuclear poly(ADP-ribose) synthetase, which depletes the cellular pool of NAD, resulting in the depression of proinsulin synthesis. In this paper, we describe detailed results on which we base this suggestion. In addition, a possible implication of the DNA damage in the development of insulin-dependent diabetes and pancreatic B-cell tumors is also discussed.

### *Islet DNA Strand Breaks by Alloxan and Streptozotocin*

Figure 1 shows sedimentation profiles in an alkaline sucrose gradient of islet DNA. Islets isolated from rat pancreas were incubated with alloxan or streptozotocin for 5–20 min in Krebs-Ringer's bicarbonate medium. After incubation, islets were laid onto an alkaline 5–20% sucrose gradient and centrifuged. DNA from islets incubated for 10 min without the diabetogenic agents was recovered as a single peak

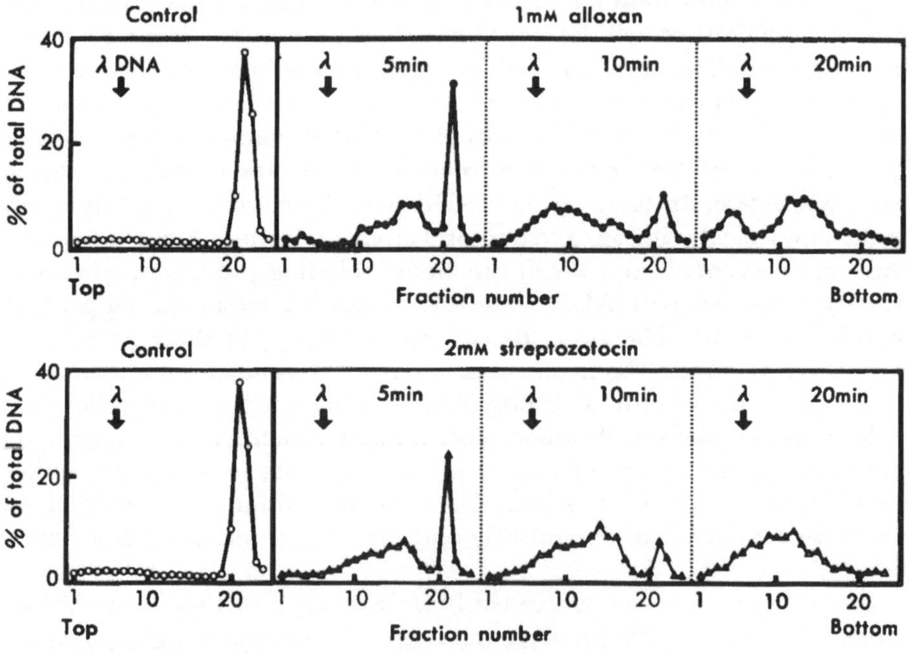


FIG. 1. Islet DNA strand breaks by alloxan and streptozotocin (*16*).

near the bottom of the gradient, the position at which undamaged DNA sediments. However, after only 5 min incubation with 1 mM of alloxan or 2 mM of streptozotocin, a considerable amount of DNA sedimented as a broad peak in the middle of the gradient with a concomitant decrease in undamaged DNA; after 10–20 min incubation, the DNA was almost completely fragmented. The effect of the two agents on islet DNA fragmentation was dose-dependent. These results indicate that alloxan and streptozotocin produce strand breaks in islet DNA.

*Poly(ADP-ribose) Synthetase Stimulation and Islet NAD Depletion*

In the eukaryotic cell nucleus, there is an enzyme which polymerizes the ADP-ribose moiety of NAD to form poly(ADP-ribose) (*19–21*). Recently, this enzyme, poly(ADP-ribose) synthetase, was purified to homogeneity from bovine thymus, and it was demonstrated that the enzyme is activated only when bound to fragmented or nicked DNA (*22, 23*). Therefore, we prepared a nuclear fraction from islets incubated in conditions causing the islet DNA fragmentation, and assayed poly(ADP-ribose) synthetase activity.

As shown in Fig. 2, both alloxan (1 mM) and streptozotocin (2 mM) induced a 2- to 3-fold increase in islet poly(ADP-ribose) synthetase activity, with a peak at 10 min.

NAD is the substrate of poly(ADP-ribose) synthetase. We therefore assayed the islet NAD level. Cellular NAD level was markedly reduced by either 1 mM of alloxan or 2 mM of streptozotocin within 20 min of incubation, and remained almost unaltered for 60 min (Fig. 2). There was striking temporal correlation between the decrease in level of islet NAD and the increase in islet poly(ADP-ribose) synthetase activity.

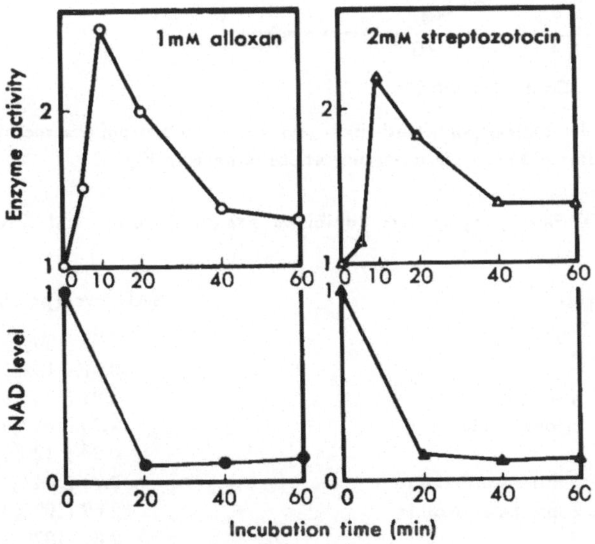


FIG. 2. Effect of alloxan and streptozotocin on islet nuclear poly(ADP-ribose) synthetase activity and islet NAD level (*16*).

These results indicate that both alloxan and streptozotocin cause DNA strand breaks which result in an increase in poly(ADP-ribose) synthetase activity, thereby depleting islet NAD level.

*Protection by Poly(ADP-ribose) Synthetase Inhibitors against Alloxan- and Streptozotocin-induced Decrease in Islet NAD Level and against the Inhibition of Proinsulin Synthesis*

Next, we examined the effect of poly(ADP-ribose) synthetase inhibitors on alloxan- and streptozotocin-induced NAD depletion. We have already shown that nicotinamide and picolinamide are potent inhibitors of islet nuclear poly(ADP-ribose) synthetase (*24*). In addition, Fig. 3 shows the inhibitory effect on the enzyme activity of benzamide, 3-aminobenzamide, 3-nitrobenzamide, 3-methoxybenzamide, theophylline, and 3-isobutyl-1-methylxanthine. The concentrations at which these compounds cause 50% inhibition of the enzyme activity are as follows: benzamide, 4.5 μM; 3-aminobenzamide, 6.5 μM; 3-nitrobenzamide, 18 μM; 3-methoxybenz-

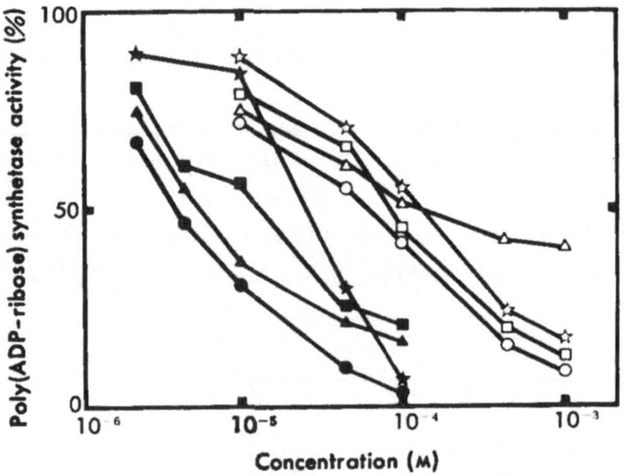


FIG. 3. Inhibition of islet nuclear poly(ADP-ribose) synthetase by benzamides, methylxanthines, nicotinamide, and picolinamide (*18*). Symbol indications same as in Fig. 4.

TABLE 1. Protection by Poly(ADP-ribose) Synthetase Inhibitors against Alloxan- and Streptozotocin-induced NAD depletion (*4*, *16*)

| Incubated with | NAD level (pmol/islet) |
|---|---|
| None | 2.65 (100%) |
| 1 mM alloxan | 0.51 ( 19%) |
| 1 mM alloxan and 2 mM nicotinamide | 2.81 (106%) |
| 1 mM alloxan and 2 mM picolinamide | 2.74 (103%) |
| 2 mM streptozotocin | 0.33 ( 12%) |
| 2 mM streptozotocin and 2 mM nicotinamide | 2.75 (104%) |
| 2 mM streptozotocin and 2 mM picolinamide | 2.69 (102%) |
| 2 mM nicotinamide | 2.84 (107%) |
| 2 mM picolinamide | 2.82 (106%) |

The numbers in parentheses give the percentage of the control without diabetogenic agents.

amide, 35 μM; theophylline, 65 μM; 3-isobutyl-1-methylxanthine, 120 μM; picolinamide, 95 μM; nicotinamide, 120 μM. Benzamides were found to be the most potent inhibitors of islet poly(ADP-ribose) synthetase. The inhibitory ability of methylxanthines was similar to that of nicotinamide and picolinamide.

Table 1 shows the effect of poly(ADP-ribose) synthetase inhibitors on alloxan- and streptozotocin-induced islet NAD depletion. The inhibitors used were nicotinamide and picolinamide. Alloxan (1 mM) decreased islet NAD level to 19% of the control level. Nicotinamide (2 mM) or picolinamide (2 mM) completely abolished the alloxan-induced decrease in islet NAD level. The same was true for the decrease

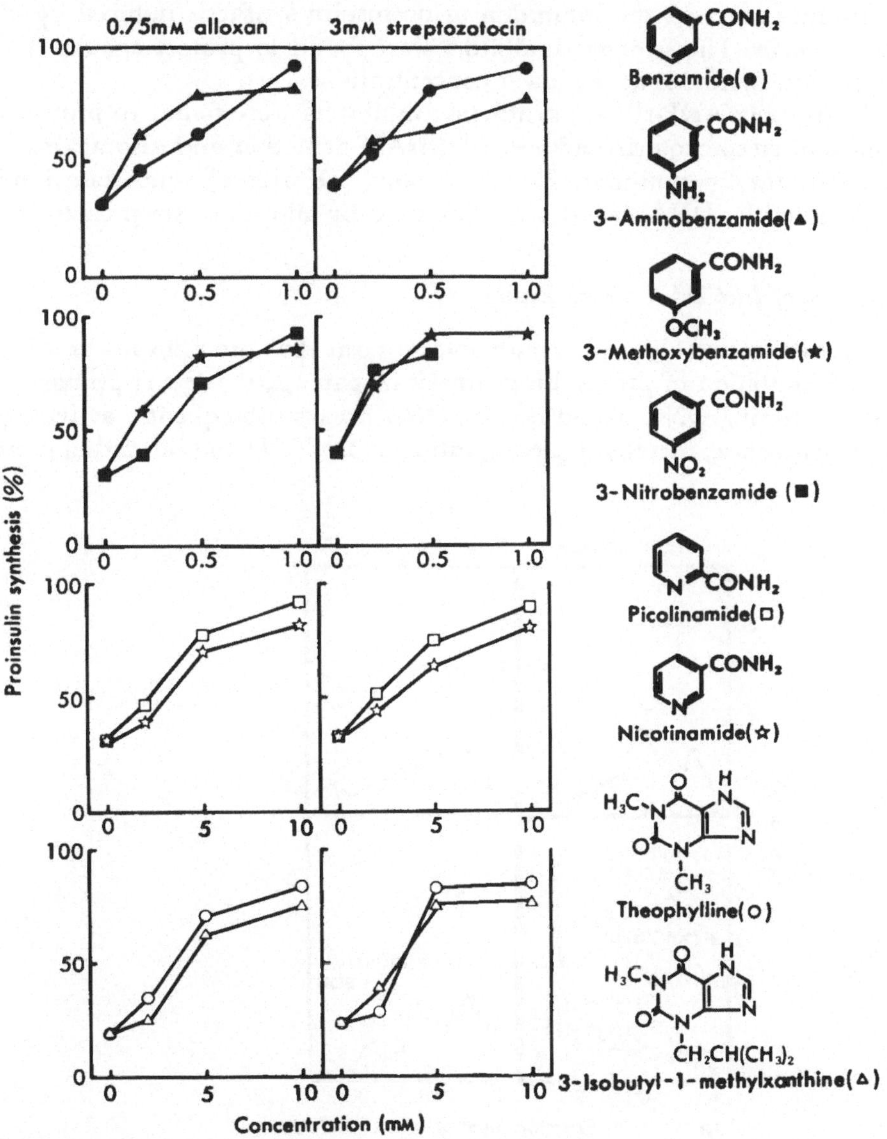


FIG. 4. Protection by poly(ADP-ribose) synthetase inhibitors against alloxan- and streptozotocin-induced depression of proinsulin synthesis (*18*).

in islet NAD level induced by streptozotocin. Addition of nicotinamide or picolinamide alone produced only a slight increase in islet NAD level.

The ability of islets to synthesize proinsulin is considered as a crucial marker for evaluation of diabetogenicity of alloxan and streptozotocin. Figure 4 shows results of experiments in which pancreatic islets were incubated in the presence of alloxan or streptozotocin with or without the addition of the islet poly(ADP-ribose) synthetase inhibitors. With 0.2, 0.5, or 1.0 mM of benzamide, alloxan-induced inhibition of proinsulin synthesis (32% of the control) was reversed to 48, 62, or 92%, respectively. Benzamide also protected against the streptozotocin-induced inhibition in a dose-dependent manner. Other islet poly(ADP-ribose) synthetase inhibitors also significantly reversed the inhibition of proinsulin synthesis induced by alloxan and streptozotocin. The stronger inhibitors were found to protect against the inhibition of proinsulin synthesis at the lower concentrations.

Thus, poly(ADP-ribose) synthetase inhibitors were found to protect against alloxan- and streptozotocin-induced islet NAD depletion and also against the inhibition of proinsulin synthesis. However, poly(ADP-ribose) synthetase inhibitors had no effect on islet DNA strand breaks induced by alloxan or streptozotocin (*18*).

### *Mechanism of Islet DNA Strand Breaks*

From these results, it is reasonable to posit that the alloxan- or streptozotocin-induced inhibition of proinsulin synthesis in pancreatic islets is primarily ascribable to their effecting DNA strand breaks. DNA breaks subsequently activate poly(ADP-ribose) synthetase, thereby depleting intracellular NAD and inhibiting proinsulin synthesis.

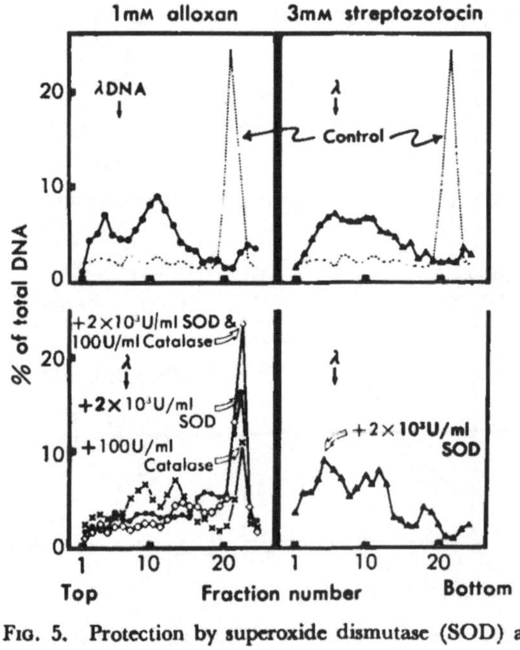


FIG. 5. Protection by superoxide dismutase (SOD) and catalase against alloxan-induced islet DNA strand breaks (*18*).

Recently, we also found that the biochemical events initiated by islet DNA strand breaks are actually induced *in vivo* by administration to rats of diabetogenic doses of alloxan and streptozotocin (*17*, *25*).

Next, we asked how these agents induce the DNA strand breaks. It has been suggested that reduction of alloxan to dialuric acid and reoxidation of dialuric acid to alloxan lead to the generation of radical oxygens such as superoxide, hydrogen peroxide and hydroxyl radical (*26*). Therefore, we examined the effect of superoxide dismutase and catalase, scavengers of radical oxygens, on islet DNA strand breaks induced by the diabetogenic agents.

As shown in Fig. 5, superoxide dismutase or catalase partially prevented the alloxan-induced DNA breaks. When both of these enzymes were present, almost complete prevention of the alloxan-induced DNA breaks was observed. This strongly suggests that it is hydroxyl radical rather than superoxide or hydrogen peroxide that attacks islet DNA (*18*). The scavenging enzymes also successfully protected against alloxan-induced inhibition of proinsulin synthesis. On the other hand, neither superoxide dismutase nor catalase affected the DNA breaks or inhibition of proinsulin synthesis induced by streptozotocin.

### *Basic Mechanism of Development of Insulin-dependent Diabetes Mellitus and Its Prevention*

We can now construct a basic model for the action of alloxan and streptozotocin in the induction of experimental diabetes (Fig. 6). The first step is the generation of hydroxyl radical by alloxan which attacks DNA to produce strand breaks. In the case of streptozotocin, the alkylating activity of this compound may be causally related to its ability to induce DNA strand breaks (*27*). Subsequently, the fragmented DNA activates poly(ADP-ribose) synthetase which depletes cellular NAD. Since NAD is the most abundant of cellular coenzymes and participates in many biological

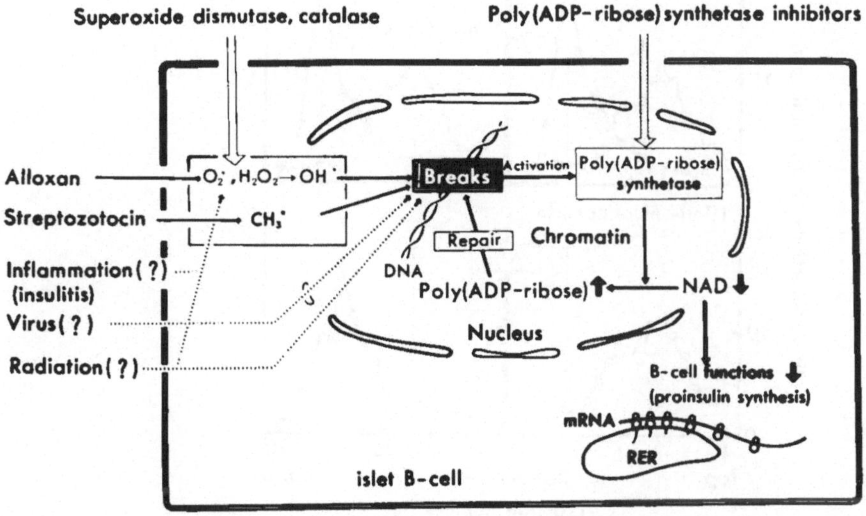


FIG. 6. Basic mechanism of development of insulin-dependent diabetes and its prevention (*4*, *18*).

reactions in mammalian cells, the reduction in intracellular NAD to such a non-physiological level by alloxan and streptozotocin may severely affect islet cell functions including proinsulin synthesis.

In addition, certain viruses and environmental agents such as radiation are thought to introduce islet DNA breaks to activate poly(ADP-ribose) synthetase and to be etiological agents of insulin-dependent diabetes. Furthermore, free radicals such as superoxide and hydroxyl radical generated during inflammation, "insulitis", may be expected to be etiological agents of insulin-dependent diabetes, because the outbreak of human insulin-dependent diabetes is often accompanied by infiltration of the islets with inflammatory cells (*28*) and such radical oxygens can be formed during inflammation (*29*). This, in turn, raises the possibility that insulin-dependent diabetes may be preventable or curable at least at a very early stage by inhibiting the occurrence of DNA strand breaks or the poly(ADP-ribose) synthetase (Fig. 6).

### *The Combined Administration to Rats of Streptozotocin and Alloxan with Poly(ADP-ribose) Synthetase Inhibitors Induces Islet B-cell Tumors*

So far, we have described the mechanism of the diabetogenic action of alloxan and streptozotocin. Streptozotocin also has an oncogenic action and it is well known that its administration combined with nicotinamide induces islet B-cell tumors (*30*). Next, we would like to present our experiments concerning the oncogenic action of the diabetogenic agents. Recently, it has been suggested that poly(ADP-ribosyl)-ation participates in the repair of damaged DNA and that inhibition of poly(ADP-

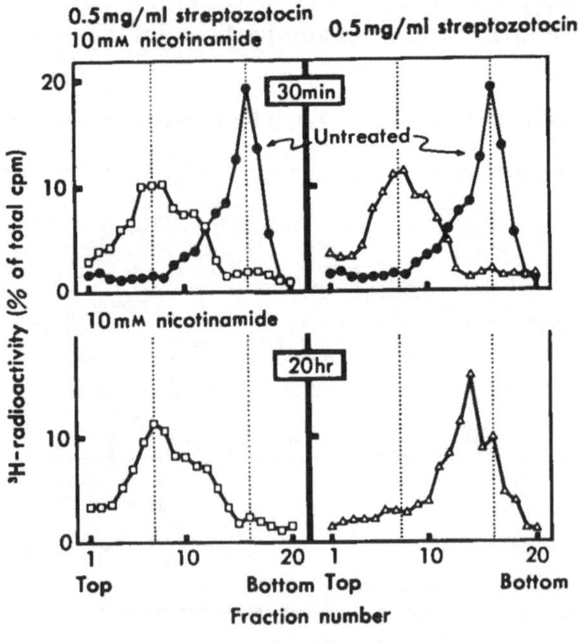


FIG. 7. Inhibition by nicotinamide of rejoining of streptozotocin-induced DNA strand breaks (*33*).

TABLE 2. Development of Islet B-cell Tumors in Rats Treated with Streptozotocin, Alloxan, and Poly-(ADP-ribose) Synthetase Inhibitors

| Treatment | Number of rats | | |
|---|---|---|---|
| | Treated | Survived for 1 year | With tumors |
| Streptozotocin-nicotinamide (50 mg/kg) (350 mg/kg) | 100 | 82 (100%) | 80 (98%) |
| Streptozotocin-picolinamide (50 mg/kg) (250 mg/kg) | 25 | 15 (100%) | 9 (60%) |
| Alloxan-nicotinamide (40 mg/kg) (350 mg/kg) | 60 | 38 (100%) | 10 (26%) |
| Alloxan-picolinamide (40 mg/kg) (250 mg/kg) | 36 | 20 (100%) | 5 (25%) |

Yamagami, T., Yamamoto, H., and Okamoto, H., in preparation.

ribosyl)ation blocks DNA repair (*31*, *32*). Using cultured hamster insulinoma cells, we investigated the effect of poly(ADP-ribose) synthetase inhibitors on the repair of alloxan- or streptozotocin-induced DNA strand breaks. In this experiment, DNA was broken down into small fragments within 30 min of incubation with streptozotocin, regardless of the presence, or absence of nicotinamide (Fig. 7). After 20-hr of incubation, DNA was repaired almost to its original size. However, as shown in Fig. 7, nicotinamide completely blocked this repair.

When administered alone, streptozotocin or alloxan efficiently induces diabetes in rats. However, in combination with nicotinamide or picolinamide, the diabetogenic action of these drugs was completely suppressed and after about one year B-cell tumors frequently developed in rats (Table 2).

## CONCLUDING REMARKS

As described above, alloxan and streptozotocin induce DNA breaks in islet B-cells. As shown in Fig. 8, in diabetes induction, fragmented DNA activates poly(ADP-ribose) synthetase resulting in the depletion of cellular NAD, and hence B-cell func-

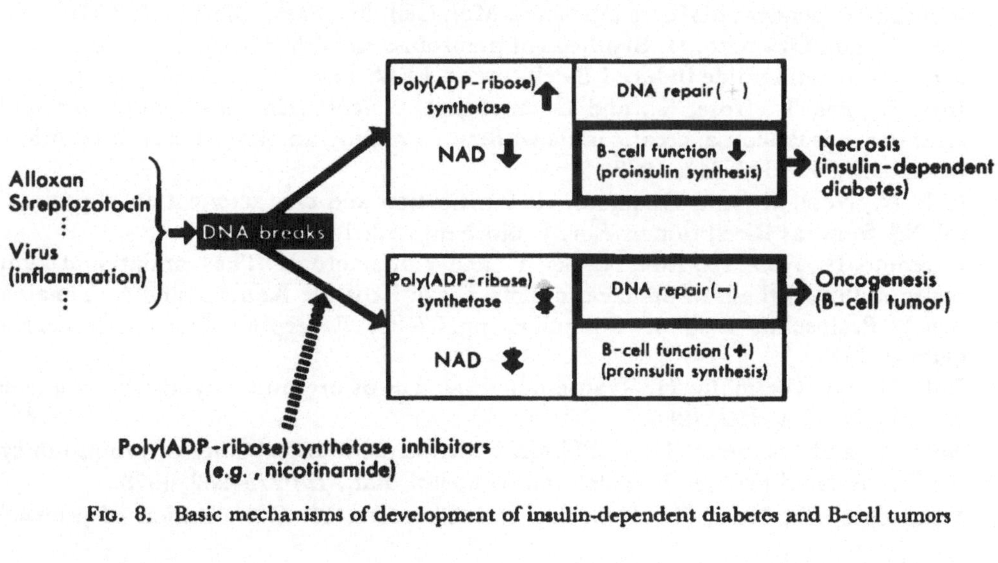


FIG. 8. Basic mechanisms of development of insulin-dependent diabetes and B-cell tumors

tions including proinsulin synthesis may be impaired leading to necrosis which eventually induces diabetes. This seems to be a "suicide response" of B-cells.

When the poly(ADP-ribose) synthetase inhibitor is present during this process, the cellular NAD level is maintained and, therefore, B-cell functions including proinsulin synthesis proceed normally. However, after a 1-year period, B-cells are found to be converted to the tumor cells with extraordinary high frequencies. The retardation of prompt DNA repair seems to be closely related to this phenomenon. Perhaps, incorrect repair of the damaged DNA may cause the activation of some oncogene(s) of B-cells.

This concept suggests that insulin-dependent diabetes and B-cell tumors are closely related with respect to their developmental processes. In other words, DNA breaks initiate two kinds of pathological state in B-cells, one is degeneratively and the other is oncogenically expressed.

## ACKNOWLEDGMENT

This work was supported in part by grants-in-aid for Cancer Research, for Scientific Research and for Special Project Research from the Ministry of Education, Science, and Culture, Japan.

## REFERENCES

1. Okamoto, H., Miyamoto, S., Mabuchi, H., Yoneyama, Y., and Takeda, R. Insulin-releasing effect of quinaldic acid and its relatives on isolated *Langerhans* islets. Biochem. Biophys. Res. Commun., *53*: 1297–1303, 1973.
2. Okamoto, H., Miyamoto, S., Mabuchi, H., and Takeda, R. Inhibitory effect on quinaldic acid on glucose-induced insulin release from isolated *Langerhans* islets of the rat. Biochem. Biophys. Res. Commun., *59*: 623–629, 1974.
3. Okamoto, H., Noto, Y., Miyamoto, S., Mabuchi, H., and Takeda, R. Inhibition by somatostatin of insulin release from isolated pancreatic islets. FEBS Lett., *54*: 103–105, 1975.
4. Okamoto, H. Regulation of proinsulin synthesis in pancreatic islets and a new aspect to insulin-dependent diabetes (Review). Mol. Cell. Biochem., *37*: 43–61, 1981.
5. Itoh, N. and Okamoto, H. Synthesis of preproinsulin with RNA preparation of streptozotocin-nicotinamide induced B-cell tumor. FEBS Lett., *80*: 111–114, 1977.
6. Itoh, N., Sei, T., Nose, K., and Okamoto, H. Glucose stimulation of the proinsulin synthesis in isolated pancreatic islets without increasing amount of proinsulin mRNA. FEBS Lett., *93*: 343–347, 1978.
7. Itoh, N., Nose, K., and Okamoto, H. Purification and characterization of proinsulin mRNA from rat B-cell tumor. Eur. J. Biochem., *97*: 1–9, 1979.
8. Okamoto, H., Nose, K., Itoh, N., Sei, T., and Yamamoto, H. The translational control of proinsulin synthesis in pancreatic islets. *In;* S. Baba, T. Kaneko, and N. Yanaihara (eds.), Proinsulin, Insulin, C-peptide, pp. 27–35, Excerpta Medica, Amsterdam/Oxford, 1979.
9. Itoh, N. and Okamoto, H. Translational control of proinsulin synthesis by glucose. Nature, *283*: 100–102, 1980.
10. Noto, Y. and Okamoto, H. Inhibition by kynurenine metabolites of proinsulin synthesis in isolated pancreatic islets. Acta Diabetol. Lat., *15*: 273–282, 1978.
11. Yamamoto, H., Nose, K., Itoh, N., and Okamoto, H. Quantitation of proinsulin

mRNA sequences in hamster insulinoma cells in culture by molecular hybridization. Experientia, *36*: 187–188, 1980.

12. Dulin, W. E. and Soret, M. G. Chemically and hormonally induced diabetes. *In;* B. W. Volk and K. F. Wellmann (eds.), The Diabetic Pancreas, pp. 425–465, Plenum Press, New York, 1977.
13. Cooperstein, S. J. and Watkins, D. Action of toxic drugs on islet cells. *In;* S. J. Cooperstein and D. Watkins (eds.), The Islets of *Langerhans*, pp. 387–425, Academic Press, New York, 1981.
14. Jain, K. and Logothetopoulos, J. Proinsulin biosynthesis by pancreatic islets of the rat and the study of alloxan cytotoxicity *in vitro*. Biochim. Biophys. Acta, *435*: 145–151, 1976.
15. Maldonato, A., Trueheart, P. A., Renold, A. E., and Sharp, G.W.G. Effects of streptozotocin *in vitro* on proinsulin biosynthesis, insulin release and ATP content of isolated rat islets of *Langerhans*. Diabetologia, *12*: 471–481, 1976.
16. Yamamoto, H., Uchigata, Y., and Okamoto, H. Streptozotocin and alloxan induce DNA strand breaks and poly(ADP-ribose) synthetase in pancreatic islets. Nature, *294*: 284–286, 1981.
17. Yamamoto, H., Uchigata, Y., and Okamoto, H. DNA strand breaks in pancreatic islets by *in vivo* administration of alloxan or streptozotocin. Biochem. Biophys. Res. Commun., *103*: 1014–1020, 1981.
18. Uchigata, Y., Yamamoto, H., Kawamura, A., and Okamoto, H. Protection by superoxide dismutase, catalase, and poly(ADP-ribose) synthetase inhibitors against alloxan- and streptozotocin-induced islet DNA strand breaks and against the inhibition of proinsulin synthesis. J. Biol. Chem., *257*: 6084–6088, 1982.
19. Sugimura, T. Poly(adenosine diphosphate ribose). Prog. Nucleic Acid Res. Mol. Biol., *13*: 127–151, 1973.
20. Hayaishi, O. and Ueda, K. Poly(ADP-ribose) and ADP-ribosylation of proteins. Annu. Rev. Biochem., *46*: 95–116, 1977.
21. Mandel, P., Okazaki, H., and Niedergang, C. Poly(adenosine diphosphate ribose). Prog. Nucleic Acid Res. Mol. Biol., *27*: 1–51, 1982.
22. Ohgushi, H., Yoshihara, K., and Kamiya, T. Bovine thymus poly(adenosine diphosphate ribose) polymerase — physical properties and binding to DNA. J. Biol. Chem., *255*: 6205–6211, 1980.
23. Benjamin, R. C. and Gill, D. M. Poly(ADP-ribose) synthesis *in vitro* programmed by damaged DNA. J. Biol. Chem., *255*: 10502–10508, 1980.
24. Yamamoto, H. and Okamoto, H. Protection by picolinamide, a novel inhibitor of poly(ADP-ribose) synthetase, against both streptozotocin-induced depression of proinsulin synthesis and reduction of NAD content in pancreatic islets. Biochem. Biophys. Res. Commun., *95*: 474–481, 1980.
25. Uchigata, Y., Yamamoto, H., Nagai, H., and Okamoto, H. Effect of poly(ADP-ribose) synthetase inhibitor administration to rats before and after injection of alloxan and streptozotocin on islet proinsulin synthesis. Diabetes, *32*: 316–318, 1983.
26. Cohen, G. and Heikkila, R. E. The generation of hydrogen peroxide, superoxide radical, and hydroxyl radical by 6-hydroxydopamine, dialuric acid, and related cytotoxic agents. J. Biol. Chem., *249*: 2447–2452, 1974.
27. Bennett, R. A. and Pegg, A. E. Alkylation of DNA in rat tissues following administration of streptozotocin. Cancer Res., *41*: 2786–2790, 1981.
28. Gepts, W. and LeCompte, P. M. The pancreatic islets in diabetes. Am. J. Med., *70*: 105–115, 1981.

29. Badwey, J. A. and Karnovsky, M. L. Active oxygen species and the functions of phagocytic leukocytes. Annu. Rev. Biochem., *49*: 695–726, 1980.
30. Rakieten, N., Gordon, B. S., Beaty, A., Cooney, D. A., Davis, R. D., and Schein, P. S. Pancreatic islet cell tumors produced by the combined action of streptozotocin and nicotinamide. Proc. Soc. Exp. Biol. Med., *137*: 280–283, 1971.
31. Durkacz, B. W., Omidiji, O., Gray, D. A., and Shall, S. (ADP-ribose)$_n$ participates in DNA excision repair. Nature, *283*: 593–596, 1980.
32. Gray, D. A., Durkacz, B. W., and Shall, S. Inhibitors of nuclear ADP-ribosyl transferase retard DNA repair after *N*-methyl-*N*-nitrosourea. FEBS Lett., *131*: 173–177, 1981.
33. Yamamoto, H. and Okamoto, H. Poly(ADP-ribose) synthetase inhibitors enhance streptozotocin-induced killing of insulinoma cells by inhibiting the repair of DNA strand breaks. FEBS Lett., *145*: 298–302, 1982.

ADP-RIBOSYLATION, DNA REPAIR AND CANCER, M. MIWA ET AL. (EDS.),
JAPAN SCI. SOC. PRESS, TOKYO/VNU SCIENCE PRESS, UTRECHT, PP. 309-319, 1983

# Possible Role of Poly(ADP-ribose) Polymerase on the Early Stage of Liver Carcinogenesis by Diethylnitrosamine in Rats

Yoichi KONISHI, Seiichi TAKAHASHI, Takeo OHNISHI,* Ayumi DENDA, Shuji MIKAMI, Yoko EMI, and Dai NAKAE

*Department of Oncological Pathology, Cancer Center, and Department of Biology,* Nara Medical College, Nara 634, Japan*

*Abstract:* In the present experiments, we studied the effect of poly(ADP-ribose) polymerase inhibitors on the early stage of liver carcinogenesis by diethylnitrosamine (DEN) in rat liver in order to clarify the biological role of this enzyme in cancer induction. We used 3-aminobenzamide(ABA), 5-methylnicotinamide(MNam), and thymidine as the inhibitors and measured the numbers and sizes of $\gamma$-glutamyltranspeptidase ($\gamma$-GTP) positive foci as a marker of initiated cell populations. When ABA was given within 4 hr after DEN treatment, it had almost the same effect as a partial hepatectomy and caused dose-dependent enhancement of the induction of $\gamma$-GTP positive foci. The administration of ABA at a dose of 600 mg/kg was effective to enhance the induction of the foci 1 day before to 1 day after 20 mg/kg DEN initiation. The enhancing effect of MNam and thymidine at a dose of 600 mg/kg was observed to the same extent as that of ABA. Based on these results the experiments were extended to the mechanisms of the enhancing effect of ABA. Liver cell necrosis was not detected by measuring serum GOT and GPT levels and histology after DEN and ABA administration. Further, the initiating and promoting activities of ABA in liver carcinogenesis were studied and ABA *per se* was not found to take part in either activity. These results indicate that poly(ADP-ribose) polymerase plays an important role in the early stage of liver carcinogenesis by DEN and provides a new avenue for studying the mechanisms of the initiation process in chemical carcinogenesis.

It is well accepted that there are at least two qualitatively different stages during chemical carcinogenesis, initiation, and promotion (*1*). Experimental models to study this two stage carcinogenesis in the liver have been developed (*2–9*). In these models, one can quantify the number of initiated cell population and study the promotion and progression of early preneoplastic and premalignant hepatocytes.

Poly(ADP-ribose) polymerase is a chromatin-bound enzyme which catalyzes the formation of homopolymer of repeating ADP-ribose unit from NAD (*10–12*); ADP-ribosylation of nuclear proteins has been reviewed (*13–15*). Inhibitors of poly(ADP-ribose) polymerase prevent the depletion of cellular NAD that is caused by

DNA damaging agents. These inhibitors retard the rejoining of DNA strand breaks caused by dimethyl sulphate (DMS) (*16*), potentiate the cytotoxicity of alkylating agents (*17*), and increase the frequency of sister chromatoid exchanges (*18*); ADP-ribosyl transferase may have a general function in the control of the breaking and rejoining of DNA (*19, 20*). These reports suggest that poly(ADP-ribose) polymerase may be relevant to carcinogenesis, however, no studies have shown that this enzyme is relevant to cancer induction. In the present experiment, we studied the effect of poly(ADP-ribose) polymerase inhibitors on the early stage of liver carcinogenesis and here we report the enhancing effect of the inhibitor on the induction of $\gamma$-glutamyltranspeptidase ($\gamma$-GTP) positive foci in the liver of rats treated with diethylnitrosamine (DEN).

*Experimental System*

Male Wistar rats or male Fisher 344 rats, 6–7 weeks old, approximately 165 g each, were used. Rats were housed in wire cages and water was given *ad libitum*. Experimental models used in the present experiment are shown in Fig. 1A and B. The model shown in Fig. 1A (model-1) was developed by Cayama *et al.* (*4*), and that shown in Fig. 1B (model-2) by Scherer and Emmelot (*5*). In experimental model-1, it is possible to quantify the number of initiated cell population which shows the $\gamma$-GTP positive phenotype (Fig. 2). DEN dissolved in 0.9% NaCl solution (4 mg/ml) was given intraperitoneally as an initiator. After 4 hr, the rats were submitted to partial hepatectomy (*21*) or received the inhibitors. 3-Aminobenzamide(ABA) dissolved in dimethylsulfoxide (DMSO: 360 mg/ml), 5-methylnicotinamide (MNam)

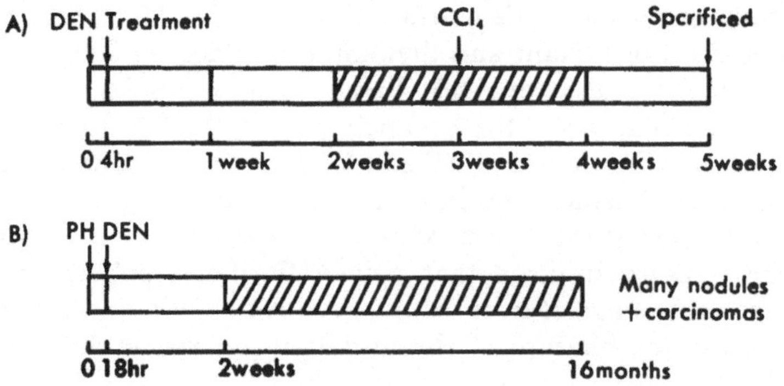


FIG. 1. A, schematic illustration of experimental model-1 for assay of $\gamma$-GTP positive foci in rat liver. DEN was given intraperitoneally as an initiator. After 4 hr rats were submitted to partial hepatectomy or received the inhibitors. Two weeks later, all animals were placed on a diet containing 0.02% 2-AAF, having a 'mito-inhibitory' effect, and 1 week after starting 2-AAF diet liver cell proliferation was stimulated with a single dose of $CCl_4$. After a total period of 2 weeks on 2-AAF diet, the rats were killed and liver samples were taken for histological and histochemical studies. Open bar, 1 week on basal diet; hatched bar; 1 week on basal diet+AAF. B, schematic illustration of experimental model-2 for assay of promotional activities of ABA. The rats were partially hepatectomized at 0 time followed in 18 hr by the administration of DEN. Two weeks later they were placed on the diet containing various doses of ABA. Animals were killed 16 months after partial hepatectomy and liver samples were taken for histological study. Open bar; basal diet; hatched bar; basal diet+ABA.

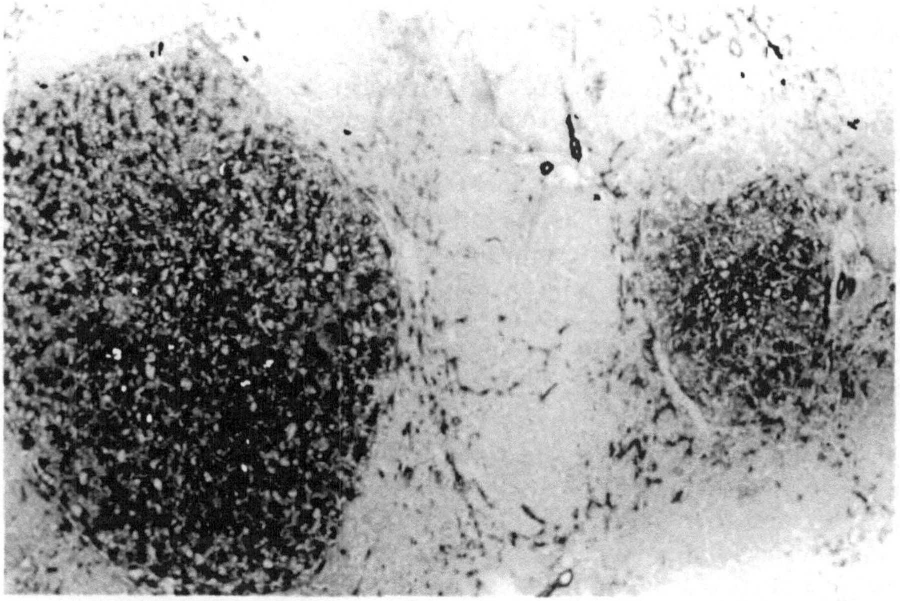

FIG. 2. $\gamma$-GTP positive foci seen in the liver of rats treated with 20 mg/kg DEN and 600 mg/kg ABA utilizing experimental model-1. $\gamma$-GTP stains, ×200.

dissolved in DMSO (480 mg/ml), and thymidine dissolved in 0.9% NaCl solution (50 mg/ml) were injected intraperitoneally.

In this process, DEN-induced DNA damage should be fixed by the cell proliferation induced by partial hepatectomy. In the present experiment, we gave the above inhibitors of poly(ADP-ribose) polymerase instead of partial hepatectomy. Two weeks later, the rats were placed on a commercial powder basal diet, Oriental MF, containing 0.02% 2-acetylaminofluorene(2-AAF). This short treatment of 2-AAF inhibits cell proliferation in non-initiated cells and does not produce tumors by itself. One week after starting the 2-AAF diet, proliferation was stimulated in initiated cells with a single dose of carbon tetrachloride ($CCl_4$: 2 ml/kg body weight at 1: 1 dilution with corn oil) administered by a gastric tube. In this protocol, cells initiated by DEN were selectively stimulated to proliferate while proliferation in non-initiated liver cells was prevented. After a period of 2 weeks on the 2-AAF diet, the rats were placed on a basal diet for 1 week and sacrificed. The liver was examined grossly and then 3 samples from different lobules were fixed first in ethanol at 4°C for histological and histochemical studies. Histochemically, $\gamma$-GTP was stained by the method of Rutenburg *et al.* (*22*). The number and diameter of $\gamma$-GTP positive foci were analyzed with an Imagelyzer model HTV-c 995 connected with a Desktop Computer System-45. In experimental model-2, it is possible to detect promoting activity of a test substance (*6*). DEN was given intraperitoneally 18 hr after partial hepatectomy. Rats were placed on a basal diet for 2 weeks after the operation and then on the diet containing various doses of ABA as indicated in Table 4. All rats were sacrificed after 16 months and the liver was examined grossly. Liver samples were fixed in 10% buffered formalin and processed for a routine histological examination.

*Enhanced γ-GTP Positive Foci Induction by the Inhibitors*

The effects of ABA, MNam, and thymidine on the induction of γ-GTP positive foci were studied. Part of the results were previously communicated (*23*). The results presented in Fig. 3 show that foci of resistant, phenotypically altered, γ-GTP positive hepatocytes were seen in rats treated with DEN+partial hepatectomy, DEN +DMSO, or DEN+ABA. This experiment was done in experimental model-1. DEN at a dose of 20 mg/kg body weight was given intraperitoneally. After 4 hr they were either submitted to partial hepatectomy, or received DMSO as a solvent of ABA or various doses of ABA. ABA enhanced the induction of γ-GTP positive foci at doses of more than 150 mg/kg. The average diameter of γ-GTP positive foci in the liver of rats treated with more than 300 mg/kg ABA was larger than that of those treated with DMSO(ABA dose is 0 mg/kg in Fig. 3), but similar to that of rats submitted to partial hepatectomy. To confirm the enhancing effect of ABA on γ-GTP positive foci, we studied the effect of a single fixed dose of ABA, 600 mg/kg, given 4 hr after 0, 2, 5, 10, and 20 mg/kg of DEN on the induction of γ-GTP positive foci. ABA did not enhance the induction of γ-GTP positive foci in the liver treated with 2 mg/kg or 5 mg/kg DEN but resulted in dose dependent induction of the foci with 10 mg/kg and 20 mg/kg DEN (*23*). We studied the timing of enhancing effect of ABA on the induction of γ-GTP positive foci in experimental model-1. In this experiment, DEN at a dose of 20 mg/kg was administered intraperitoneally. ABA at doses of 150 and 600 mg/kg was given 1 day before DEN, simultaneously with DEN, and 1, 3, 7, and 14 days after DEN. Clear results with 600 mg/kg ABA were obtained and are shown in Fig. 4. The open circle shows the ratio of the number of foci per square cm and the closed circle the ratio of total area of foci per square cm of the liver of rats treated with 600 mg/kg ABA against those of rats treated with DMSO. The administration of ABA was effective to enhance the induction of γ-GTP positive foci 1 day before DEN, simultaneously with DEN, and 1 day after DEN initiation.

Next, we studied the effect of other poly(ADP-ribose) polymerase inhibitors

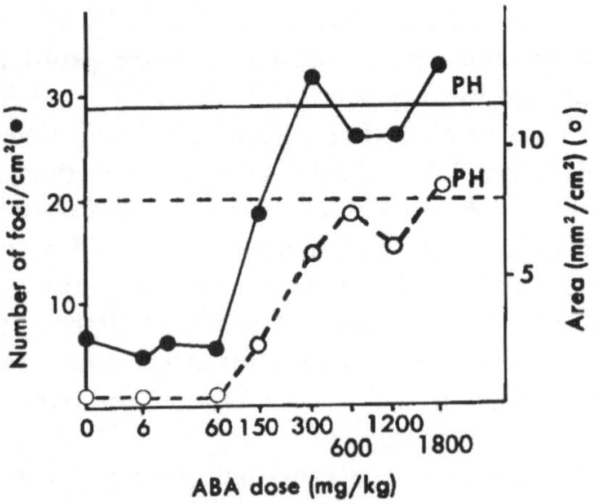


FIG. 3. Enhancing effect of ABA on the induction of γ-GTP positive foci in rat liver initiated by DEN. PH, partial hepatectomy.

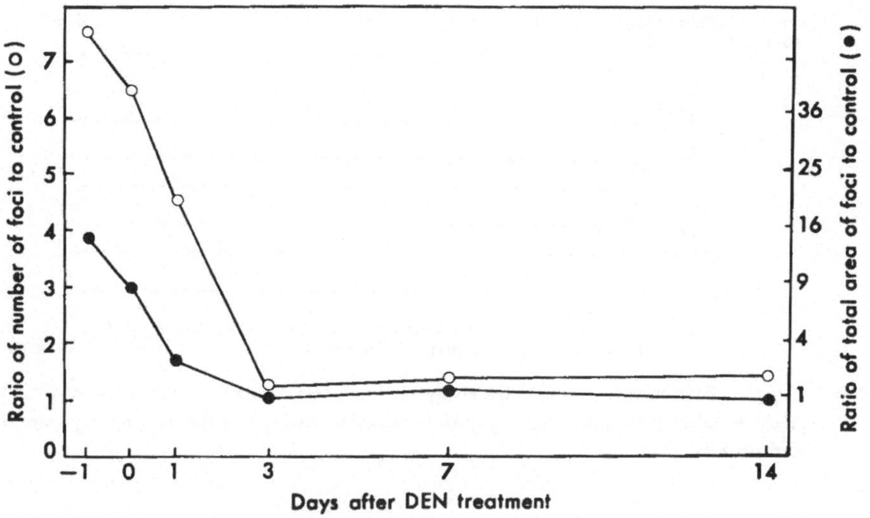


FIG. 4. Effective time of ABA administration which enhanced induction of $\gamma$-GTP positive foci in rat liver initiated by DEN. ABA dose was 600 mg/kg and each point represents of 5–6 rats.

TABLE 1. Enhanced Induction of $\gamma$-GTP Positive Foci by Poly(ADP-ribose) Polymerase Inhibitors in Rat Liver Initiated by DEN

| Groups | Doses of DEN (mg/kg) | Treatments | | Effective number of rats | Number of $\gamma$-GTP positive foci (foci/cm²) | Average diameter of foci |
|---|---|---|---|---|---|---|
| | | Inhibitors of poly(ADP-ribose) polymerase | Dose (mg/kg) | | | |
| 1 | 20 | DMSO | — | 5 | 1.8 | 404 |
| 2 | 20 | ABA | 600 | 7 | 8.3[a,b] | 577 |
| 3 | 0 | ABA | 600 | 7 | 1.2 | 421 |
| 4 | 20 | MNam | 600 | 6 | 9.4[a,b] | 566 |
| 5 | 0 | MNam | 600 | 6 | 0.7 | 409 |
| 6 | 20 | Thymidine | 600 | 5 | 6.9[a,b] | 552 |
| 7 | 0 | Thymidine | 600 | 5 | 0.6 | 492 |

[a] $p<0.01$ compared with group 1. [b] $p<0.001$ compared between groups 2 and 3, 4 and 5, and 6 and 7.

such as MNam and thymidine on the induction of $\gamma$-GTP positive foci in experimental model-1. In this experiment, DEN at a dose of 20 mg/kg was given intraperitoneally. After 4 hr the rats received ABA, MNam, or thymidine at doses indicated in Table 1 instead of partial hepatectomy. As shown, the numbers of $\gamma$-GTP positive foci were increased in groups 2, 4, and 6 which received 600 mg/kg ABA, MNam, or thymidine. The enhancing effects of MNam and thymidine were the same as those of ABA at this dose.

### *Liver Cell Necrosis Measured by Serum Enzymes and Histology after DEN and ABA Administration*

It is generally known that cell proliferation is required for the fixation process

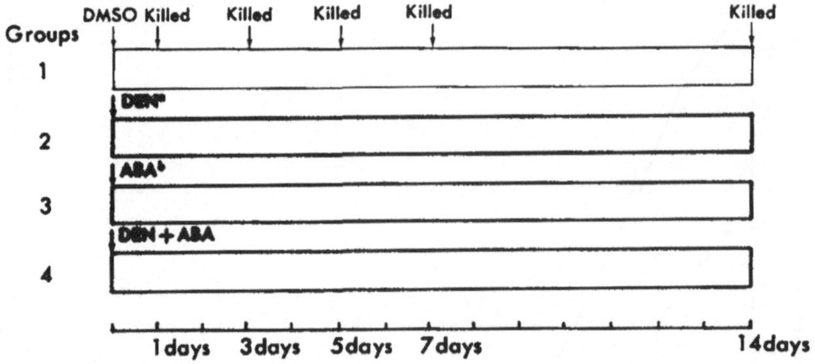


FIG. 5. Experimental design for study on liver cell necrosis in rats treated with DEN and/or ABA. [a] DEN 20 mg/kg was injected intraperitoneally. [b] ABA 600 mg/kg was injected intraperitoneally.

TABLE 2. Serum GOT and GPT Not Elevated in Rats Treated with ABA and/or DEN

| Groups | Treatments | Doses | Serum enzymes | Units/ml (days after treatments)[a,b] | | | | |
|---|---|---|---|---|---|---|---|---|
| | | | | 1 | 3 | 5 | 7 | 14 |
| 1 | DMSO | 2.5 ml/kg | GOT | 157±30 | 98± 8 | 118±13 | 120±21 | 102± 6 |
| | | | GPT | 30± 7 | 29± 3 | 29± 3 | 21± 1 | 24±12 |
| 2 | DEN | 20 mg/kg | GOT | 120±21 | 111±10 | 80± 7 | 170±44 | 91± 1 |
| | | | GPT | 42±34 | 27±13 | 30± 4 | 33± 6 | 45±11 |
| 3 | ABA | 600 mg/kg | GOT | 113±25 | 111± 8 | 90± 5 | 142±45 | 98± 1 |
| | | | GPT | 28± 3 | 30± 6 | 20± 2 | 34± 9 | 48± 4 |
| 4 | DEN+ABA | 20 mg/kg+600 mg/kg | GOT | 171±52 | 118±19 | 96±16 | 106± 4 | 87±20 |
| | | | GPT | 52±14 | 38±15 | 28± 8 | 33± 9 | 43±16 |

[a] Values of serum GOT and GPT do not significantly differ between groups. [b] Mean±S.D.

of DNA damage by a carcinogen (*4*). The present results indicate that poly(ADP-ribose) polymerase inhibitors enhanced the induction of $\gamma$-GTP positive, phenotype foci to the same extent as a partial hepatectomy. In this experiment, we studied liver cell necrosis after administration of DEN and/or ABA in rats since cell proliferation might occur after liver cell necrosis induced by DEN and/or ABA. The experimental design is shown in Fig. 5. Group 1 was the control receiving 2.5 ml/kg DMSO; group 2 received 20 mg/kg DEN; group 3 received 600 mg/kg ABA; and group 4 received 20 mg/kg DEN and 600 mg/kg ABA simultaneously. All injections were made intraperitoneally. The rats were sacrificed by taking blood from the aorta under ether anesthesia 1, 3, 5, 7, and 14 days after the initial treatments. Liver samples were taken for histological studies and serum GOT and GPT levels were measured from blood samples by SMAC autoanalyzer (Table 2) . The values of serum GOT and GPT levels did not significantly differ and were not changed as a function of time after the treatments. Histological examination also revealed no liver cell necrosis in rats of any group.

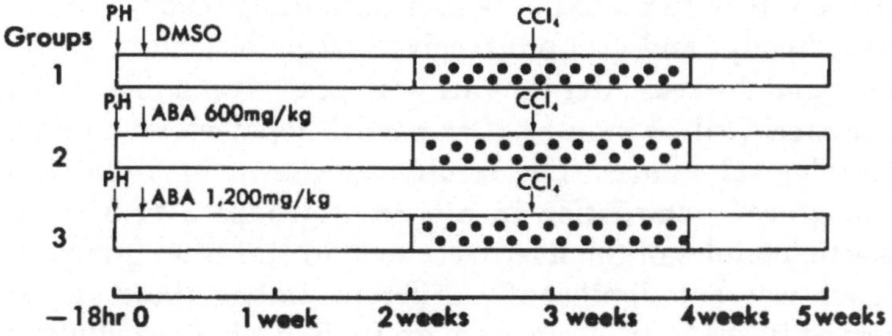


FIG. 6. Experimental design for initiating activity of ABA in rat liver. Open bars; basal diet; hatched bars; basal diet containing 0.02% AAF. PH, partial hepatectomy.

TABLE 3. No Initiating Activity of ABA for Liver Carcinogenesis

| Groups | Treatments | Doses | Effective number of animals | Number of γ-GTP positive foci (foci/cm²)[a] | Average diameter of foci[b] |
|---|---|---|---|---|---|
| 1 | DMSO | 2.5 ml/kg | 5 | 2.7±1.0[c] | 391±141 |
| 2 | ABA | 600 mg/kg | 6 | 1.0±0.8 | 391±153 |
| 3 | ABA | 1,200 mg/kg | 4 | 1.3±0.8 | 476±118 |

[a,b] No significant differences between groups. [c] Mean±S.D.

TABLE 4. No Promoting Activity of ABA for Liver Carcinogenesis

| Group | Treatment | ABA concentration in diet (%) | Period of ABA administration (months) | Number of rats | | Body weights (g) | | Incidence of hyperplastic nodules (%) |
|---|---|---|---|---|---|---|---|---|
| | | | | Initial | Effective | Initial | Final | |
| 1 | PH+DEN | 0 | 16 | 18 | 12 | 155 | 555 | 9 (75) |
| 2 | PH+DEN | 0.05 | 16 | 18 | 12 | 164 | 549 | 10 (83) |
| 3 | PH+DEN | 0.1 | 16 | 19 | 13 | 170 | 568 | 10 (77) |
| 4 | PH+DEN | 0.2 | 16 | 21 | 12 | 151 | 544 | 8 (67) |
| 5 | PH+Saline | 0.2 | 16 | 18 | 12 | 154 | 532 | 0 (0) |

PH, partial hepatectomy. DEN was given intraperitoneally at a dose of 20 mg/kg body weight.

### *ABA Has No Initiating or Promoting Activities for Liver Carcinogenesis*

Initiating activity of ABA was studied in the experimental design shown in Fig. 6. The concept of this experimental design is the same as that of experimental model-1, however, this design is more sensitive for testing the initiating activity of the test chemical (ABA) since ABA was given as an initiator 18 hr after partial hepatectomy when liver cell DNA synthesis was at maximum (*6*). Group 1 received DMSO, group 2 received 600 mg/kg ABA, and group 3 received 1,200 mg/kg ABA (Table 3). The numbers of γ-GTP positive foci were not increased and the average diameter of the foci was not enlarged in groups 2 and 3 as compared to group 1. The promoting activity of ABA was studied in experimental model-2. Group 1 received 20 mg/kg DEN 18 hr after partial hepatectomy and 2 weeks later the rats were placed on a powder basal diet; groups 2, 3, and 4 received 20 mg/kg DEN 18 hr after partial hepatectomy and 2 weeks later the rats were placed on a diet containing 0.05% ABA, 0.1% ABA, or 0.2% ABA, respectively; and group 5 received saline instead of DEN 18 hr after partial hepatectomy and 2 weeks later the rats were placed on a diet containing 0.2% ABA. All rats were sacrificed 16 months after partial hepatectomy and liver samples were taken for histological studies. The results are shown in Table 4. Final body weights showed no growth retardation in rats from groups 2 to 5 as compared to group 1. Hyperplastic nodules of the liver were seen in rats from groups 1 to 4, however, their incidence was not significantly different among those groups. No such nodules were observed in the liver of rats from group 5. No hepatocellular carcinomas were observed in rats from groups 1 to 5.

## DISCUSSION

It is evident from the present results that the inhibitors of poly(ADP-ribose) polymerase enhanced the induction of $\gamma$-GTP positive foci in rat liver initiated by DEN. It was reported that increased activity of poly(ADP-ribose) polymerase and decrease in cellular NAD content were observed in cells treated with N-methyl-N-nitrosourea or $\gamma$-rays (*24, 25*). The activity of this enzyme also increases in human lymphoblastoid cell lines after UV-irradiation and the response is abolished or delayed in xeroderma pigmentosum cells (*26*), which are unable to excise pyrimidine dimers and are associated with a high risk of skin cancer in humans. The enzyme activity also increases in permeabilized cells incubated with either endogenous (*27*) or exogenous endonucleases (*28*) *in vitro* reaction using nicked DNA (*29*). These findings support the idea that stimulation of the enzyme is due to DNA strand-breakages induced by chemical carcinogens. Recently, Althaus *et al.* (*19*) reported that a rapid increase in ADP-ribosyltransferase activity was observed in primary rat hepatocyte monolayer culture. Inhibition of ADP-ribosyltransferase activity with pyrimidine derivatives was accompanied by an enhanced DNA repair synthesis in response to the direct-acting carcinogen, methyl methanesulfonate, or UV irradiation. It has also been reported that ABA inhibits excision repair of DNA damage caused by chemicals in a human cell (*26*). The present results might be due to either inhibition of poly(ADP-ribosyl)ation or excision repair.

Initiation of chemical carcinogenesis is reported to require cell proliferation (*4*). In the present experiment, we could not detect any liver cell necrosis measured by serum enzyme and histology. However, further studies are required on the effect of the inhibitors on DNA synthesis in rat liver. Ishikawa *et al.* (*30*) suggested that carcinogen-induced DNA damage, if not repaired before cell proliferation, is intimately related to the initiation-fixation process of carcinogenesis. In the present experiment, ABA was effective to enhance the induction of $\gamma$-GTP positive foci from 1 day before to 1 day after DEN administration. DEN-induced DNA damage might be intimately related to the initiation-fixation process without cell proliferation. In this context, further experiment is needed to explain the mechanisms of the initiation-fixation process of DEN-induced DNA damage in rat liver.

It was reported that $\gamma$-GTP, glucose-6-phosphatase, and canalicular ATPase can be utilized to identify altered foci, however, not all foci exhibited the same phenotype in three different enzyme assays (*6*). In the present experiment, only $\gamma$-GTP was utilized to identify altered foci since previous reports (*31–33*) showed that γ-GTP positive foci induced by DEN can develop into to carcinomas. Further characterization of the γ-GTP positive foci seen in the present experiment utilizing different enzyme markers is needed. Established promoters for liver carcinogenesis in rats have been reported (*33, 34*). Studies on the progression of DEN-induced γ-GTP positive foci as seen in the present experiment by giving promoters are in progress.

ABA had no initiating or promoting activities for liver carcinogenesis initiated by DEN and enhanced the induction of γ-GTP positive foci in rat liver, probably without cell necrosis. These results provide a new avenue not only for studying the biological function of poly(ADP-ribose) polymerase but also the initiation-fixation process in chemical carcinogenesis.

ACKNOWLEDGMENT

This work was supported by Grants (57015087 and 57010069) from the Ministry of Education, Science and Culture of Japan and by a contribution from the late Fumiko Horikawa.

REFERENCES

1. Berenblum, I. Some implications of the two-stage mechanism of carcinogenesis. An. Med. Perugia, *48*: 822–828, 1957.
2. Solt, D. and Farber, E. New principle for the analysis of chemical carcinogenesis. Nature, *263*: 701–703, 1976.
3. Solt, D., Medline, A., and Farber, E. Rapid emergence of carcinogen-induced hyperplastic lesions in a new model for the sequential analysis of live carcinogenesis. Am. J. Pathol., *88*: 595–618, 1977.
4. Cayama, E., Tsuda, H., Sarma, D.S.R., and Farber, E. Initiation of chemical carcinogenesis requires cell proliferation. Nature, *275*: 60–62, 1978.
5. Scherer, E. and Emmelot, P. Kinetics of induction and growth of precancerous liver-cell foci, and liver tumor formation by diethylnitrosamine in the rat. Eur. Cancer, *11*: 689–701, 1975.
6. Pitot, H., Barsness, L., and Kitagawa, T. Stages in the process of hepatocarcinogenesis in rat liver. *In;* T. J. Slaga, A. Sivak, and R. K. Boutwell (eds.), Carcinogenesis — A Comprehensive Survey, Vol. 2, pp. 433–442, Raven Press, New York, 1978.
7. Ito, N., Tatematsu, M., Nakanishi, K., Hasegawa, R., Takano, T., Imaida, K., and Ogiso, T. The effect of various chemicals on the development of hyperplastic liver nodules in hepatectomized rats treated with N-nitrosodiethylamine or N-2-fluorenylacetamide. Gann, *71*: 832–842, 1980.
8. Columbano, A., Ledda, G. M., Rao, P. M., Rajalakschmi, S., and Sarma, D.S.R. Dietary orotic acid, a new selective growth stimulus for carcinogen altered hepatocytes in rat. Cancer Lett., *16*: 191–196, 1982.
9. Shinozuka, H., Takahashi, S., Lombardi, B., and Abanobi, S. E. Effects of varying fat content of a choline-devoid diet on promotion of the emergence of γ-glutamyltranspeptidase positive foci in the liver of carcinogen-treated rats. Cancer Lett., *16*: 43–50, 1982.
10. Chambon, P., Weil, P. D., Doly, J., Strosser, M., and Mandel, P. On the formation of a noval adenylic compound by enzymatic extracts of liver nuclei. Biochem. Biophys. Res. Commun., *25*: 638–643, 1966.

11. Hasegawa, S., Fujimura, S., Shimizu, Y., and Sugimura, T. The polymerization of adenosine 5'-diphosphate-ribose moiety of NAD by nuclear enzyme. II. Properties of the reaction. Biochim. Biophys. Acta, *148*: 369–376, 1967.
12. Nishizuka, Y., Ueda, K., Nakazawa, K., and Hayaishi, O. Studies on the polymer of adenosine-diphosphate-ribose. I. Enzymatic formation from nicotinamide adenine dinucleotide in mammalian nuclei. J. Biol. Chem., *242*: 3164–3171, 1967.
13. Hilz, H. and Stone, P. Poly(ADP-ribose) and ADP-ribosylation of proteins. Rev. Physiol. Biochem. Pharmacol., *76*: 1–58, 1976.
14. Hayaishi, O. and Ueda, K. Poly(ADP-ribose) and ADP-ribosylation of proteins. Annu. Rev. Biochem., *46*: 95–116, 1977.
15. Purnell, M. R., Stone, P. R., and Wish, W.J.D. ADP-ribosylation of nuclear protein. Biochem. Soc. Trans., *8*: 215–227, 1980.
16. Durkacz, B. W., Omidiji, O., Gray, D. A., and Shall, S. (ADP-ribose)$_n$ participates in DNA excision repair. Nature, *283*: 593–596, 1980.
17. Jacobson, M. K., Levi, V., Juarez-Salinas, H., Barton, R. A., and Jacobson E. L. Effect of carcinogenic N-alkyl-N-nitroso compound on nicotinamide adenine dinucleotide metabolism. Cancer Res., *40*: 1719–1802, 1980.
18. Utakoji, T., Hosoda, K., Umezawa, K., Sugimura, M., Matsushima, T., Miwa, M., and Sugimura, T. Induction of sister chromatoid exchanges by nicotinamide in Chinese hamster lung fibroblasts and human lymphoblastoid cells. Biochem Biophys. Res. Commun., *90*: 1147–1152, 1979.
19. Althaus, F. R., Lawrence, S. D., Sattler, G. L., and Pitot, H. C. ADP-ribosyltransferase activity in cultured hepatocytes. Interactions with DNA repair. J. Biol. Chem., *257*: 5528–5535, 1982.
20. Miwa, M., Kanai, M., Kondo, T., Hoshino, H., Ishihara, K., and Sugimura, T. Inhibitors of poly(ADP-ribose)polymerase enhance unscheduled DNA synthesis in human peripheral lymphocytes. Biochem. Biophys. Res. Commun., *97*: 1313–1316, 1980.
21. Higgins, G. M. and Anderson, R. M. Experimental pathology of the liver restoration of the liver of the white rat following partial surgical removal. Arch. Pathol., *12*: 186–202, 1931.
22. Rutenburg, A. H., Kim, H., Fischbeim, J. W., Hankers, J. S., Wasserkrug, H. L., and Seligman, A. M. Histochemical and ultrastructural demonstration of $\gamma$-glutamyl-transpeptidase activity. J. Histochem. Cytochem., *17*: 517–526, 1969.
23. Takahashi, S., Ohnishi, T., Denda, A., and Konishi, Y. Enhancing effect of 3-aminobenzamide on induction of $\gamma$-glutamyl transpeptidase positive foci in rat liver. Chem-Biol. Interact., *39*: 363–368, 1982.
24. Davies, M. I., Shall, S., and Skidmore, C. J. Poly(adenosine diphosphate ribose) polymerase and deoxyribonucleic acid damage. Biochem. Soc. Trans., *5*: 949–950, 1977.
25. Davies, M. I., Halldorsson, H., Nduka, N., Shall, S., and Skidmore, C. J. The involvement of poly(adenosine diphosphate-ribose) in deoxyribonucleic acid repair. Biochem. Soc. Trans., *6*: 1056–1057, 1978.
26. Durkacz, B. W., Nduka, N., Omidiji, O., Shall, S., and Zia'ee, A. A. ADP-ribose and DNA repair. *In;* M. E. Smulson and T. Sugimura (eds.), Novel ADP-ribosylations of Regulatory Enzymes and Proteins, pp. 207–216, Elsevier/North-Holland, Amsterdam, 1980.
27. Halldorsson, H., Gray, D. A., and Shall, S. Poly(ADP-ribose)polymerase activity in nucleotide permeable cells. FEBS Lett., *85*: 349–352, 1978.

28. Berger, N. A., Weber, G., and Kaichi, A. S. Characterization and comparison of poly(adenosine diphosphoribose) synthesis and DNA synthesis in nucleotide-permeable cells. Biochim. Biophys. Acta, *519*: 87–104, 1978.
29. Ohgushi, H., Yoshihara, K., and Kamiya, T. Bovine thymus poly(adenosine diphosphate ribose)polymerase. J. Biol. Chem., *255*: 6205–6211, 1980.
30. Ishikawa, T., Takayama, S., and Kitagawa, T. Correlation between time of partial hepatectomy after a single treatment with diethylnitrosamine and induction of adenosinetriphosphate-deficient island in rat liver. Cancer Res., *40*: 4261–4264, 1980.
31. Farber, E. and Cameron, R. The sequential analysis of cancer development. Adv. Cancer Res., *31*: 125–226, 1980.
32. Takahashi, S., Lombardi, B., and Shinozuka, H. Progression of carcinogen-induced foci of $\gamma$-glutamyltranspeptidase positive hepatocytes to hepatomas in rats fed a choline-devoid diet. Int. J. Cancer, *29*: 445–450, 1982.
33. Shinozuka, H., Lombardi, B., and Abanobi, S. A comparative study of the efficacy of four barbiturates as promoters of the development of $\gamma$-glutamyltranspeptidase-positive foci in the liver of carcinogen treated rats. Carcinogenesis, *3*: 1017–1020, 1982.
34. Kitagawa, T. and Sugano, H. Enhancing effect of phenobarbital on the development of enzyme-altered islands and hepatocellular carcinomas initiated by 3′-methyl-4-dimethylaminoazobenzene or diethylnitrosamine. Gann, *69*: 679–687, 1978.

ADP-RIBOSYLATION, DNA REPAIR AND CANCER, M. MIWA ET AL. (EDS.), JAPAN SCI. SOC. PRESS, TOKYO/VNU SCIENCE PRESS, UTRECHT, PP. 321-331, 1983

# Consequences of Altered Poly(ADP-ribose) Metabolism on Hepatocyte Functions *In Vitro*

Felix R. ALTHAUS* and Henry C. PITOT

*McArdle Laboratory for Cancer Research, University of Wisconsin, Madison, Wisconsin 53706, U.S.A.*

*Abstract:* Cultured hepatocytes derived from adult rats were used to determine the consequences of altered poly(ADP-ribose)-metabolism on DNA repair and the expression of hepatocellular functions *in vitro*. Inhibition of poly(ADP-ribose)-biosynthesis affected the process of DNA repair in response to carcinogens at more than one single step. The expression of two fetal enzymes, gamma-glutamyl transpeptidase and the K-(111) type isozyme of pyruvate kinase, was inhibited in hepatocytes in which poly(ADP-ribose)-biosynthesis was suppressed with 3-aminobenzamide or other inhibitors. Our data show that cellular functions as diverse as DNA repair and gene expression are affected by the same experimental manipulations of poly(ADP-ribose)-metabolism in the same cell. This suggests that poly(ADP-ribose)-modification of chromatin proteins affects a mechanism which is involved in the regulation of more than one chromatin function.

Poly(ADP-ribose)-modification of chromatin proteins has been implicated in cellular functions as diverse as differentiation, DNA replication, and DNA repair (*1*, *2*). The most conclusive evidence for an involvement of poly(ADP-ribose) in any of these processes relates to its role in DNA repair (*3*–*5*). An involvement of ADP-ribosylation in cellular differentiation has been inferred from the observation of a temporal relationship between alterations in nuclear ADP-ribosyltransferase (E. C. 2.4.99) activity and engagement of cells in terminal differentiation (*6*–*11*). However, poly(ADP-ribose) might primarily regulate the proliferative activity of cells undergoing terminal differentiation. It has, therefore, not been possible to ascribe definitively the expression of individual cellular functions during differentiation to changes in ADP-ribosylation activity (*1*). The recent introduction by Purnell and Whish (*12*) of 3-aminobenzamide, the most selective inhibitor of ADP-ribosyltransferase known at this time, has made it possible to determine directly the potential involvement of this enzyme in a particular biological process.

* Present address: University of Zurich, Institute of Pharmacology and Biochemistry, CH-8091 Zurich, Switzerland.

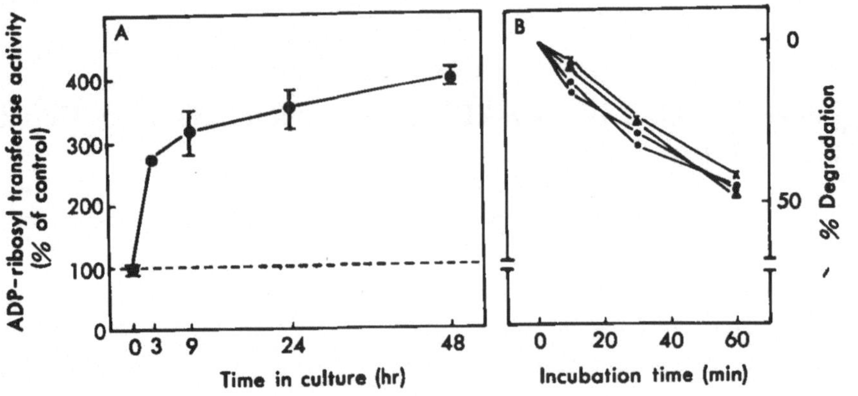


FIG. 1. ADP-ribosyltransferase activity (A) and poly(ADP-ribose) degrading activity (B) of hepatocytes at different times after initiation of cultures. Both activities were determined in permeabilized hepatocytes (for experimental details see ref. *14*). A. The ADP-ribosyltransferase activity is expressed as percent of control (=100%), which represents the activity of freshly isolated hepatocytes ("0 hr in culture"). Control value: 80±12 dpm/μg DNA/hr (mean±S.D., *n*=3). B. Poly(ADP-ribose) degradation in freshly isolated hepatocytes (●) and hepatocytes after 9 hr (○), 24 hr (▲), and 48 hr (×) in primary monolayer culture. (Data taken from ref. *14*).

We have used this and other inhibitors of ADP-ribosyltransferase to determine the consequences of altered poly(ADP-ribose)-metabolism on selected functions of cultured hepatocytes derived from adult rats. The present report demonstrates that the process of DNA repair as well as the expression of two fetal enzymes by these adult hepatocytes is sensitive to experimental manipulation of poly(ADP-ribose)-biosynthesis.

### *Relationship between poly(ADP-ribose)-metabolism and DNA-repair*

When freshly isolated hepatocytes derived from adult rats are established in culture, they undergo a number of morphological and functional changes that reflect their adaptation to the *in vitro* environment (for review see *13*). We have investigated whether or not poly(ADP-ribose)-metabolism would undergo such adaptive changes within the first 48 hr of culture (*14*). The data in Fig. 1 demonstrate that the activity of the nuclear enzyme ADP-ribosyltransferase rapidly increases within a few hours after freshly isolated hepatocytes have been plated at high seeding density directly on the plastic surface of tissue culture dishes (Fig. 1A). This increase in ADP-ribosyltransferase activity is paralleled by a rapid disappearance of intracellular $NAD^+$ as has been reported by others (*15*).

In the past few years, a number of inhibitors of ADP-ribosyltransferase activity have been described (*1*, *12*). Among these inhibitors, 3-aminobenzamide has particularly useful properties since it is a competitive inhibitor of the enzyme with a relatively low $K_i$, and has very little cytotoxic effects and probably little metabolic involvement in other biochemical reactions (*1*, *4*, *12*). Figure 2A shows the inhibitory action of 3-aminobenzamide and two structural analogues, *i.e.*, 2-amino- and 4-aminobenzamide, on ADP-ribosyltransferase activity in permeabilized hepatocytes. The results in Fig. 2A suggest a requirement for the amino group to be in the 3-posi-

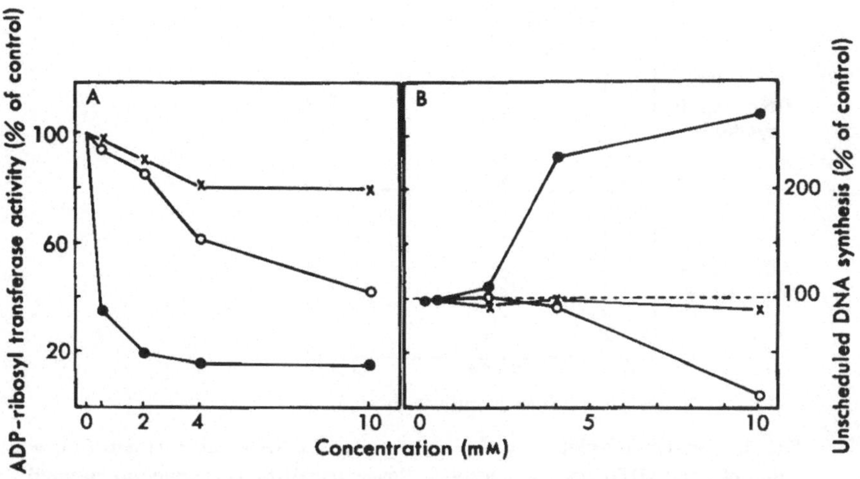


FIG. 2. Effects of 2-aminobenzamide (2-aba, ○), 3-aba (●), and 4-aba (×) on the ADP-ribosyltransferase activity (A) and on DNA repair synthesis (B) of cultured hepatocytes. ADP-ribosyltransferase activity (A) was determined after 27 hr in culture as described previously (*14*). The dashed horizontal line (B, 100% value) represents maximum stimulation of DNA repair synthesis in response to treatment with methyl methanesulfonate ($5 \times 10^{-4}$ M) alone. DNA repair synthesis was measured after a 48 hr pretreatment with the various aminobenzamides or dimethylsulfoxide (DMSO) (1%) which served as a solvent control. Further experimental details are described in ref. *14*.

tion in order to be effective as an inhibitor of this activity. The partial inhibition by 2-aminobenzamide occurred at concentrations that were cytotoxic to intact hepatocytes indicating that such inhibition may be secondary to cytotoxicity.

Interestingly, the specific inhibition of poly(ADP-ribose)-biosynthesis in cultured hepatocytes had distinct effects on the DNA repair function of these cells subsequent to treatment with chemical or physical carcinogens. Hepatocytes maintained for 48 hr in the continuous presence of the ADP-ribosyltransferase inhibitor, 3-aminobenzamide, exhibited a markedly increased capacity for DNA repair synthesis while treatment with the non-inhibitory analogues, 4-aminobenzamide or 3-aminobenzoic acid (data not shown for the latter), did not affect DNA repair synthesis following DNA damage by UV irradiation or the direct alkylating agent, methyl methane sulfonate. Similar results were obtained with ADP-ribosyltransferase inhibitors of different chemical structure, such as nicotinamide, theophylline, or thymidine (*14*). On the basis of these and other data (*14*), we concluded that DNA repair synthesis subsequent to DNA damage by a number of carcinogens is sensitive to the manipulation of poly(ADP-ribose)-biosynthesis.

This "ADP-ribose-sensitivity" of DNA repair synthesis was absent during the first few hours of culture and only gradually developed thereafter. Figure 3 summarizes this development in a schematic manner. DNA repair synthesis in response to the alkylating agent, N-methyl-N'-nitro-N-nitrosoguanidine (MNNG), or UV light (254 nm) was not enhanced by inhibitors of poly(ADP-ribose)-biosynthesis during the first 5 hr of culture but started to increase when poly(ADP-ribose)-biosynthesis was continuously suppressed over a period of 24 or 48 hr. Separate experiments

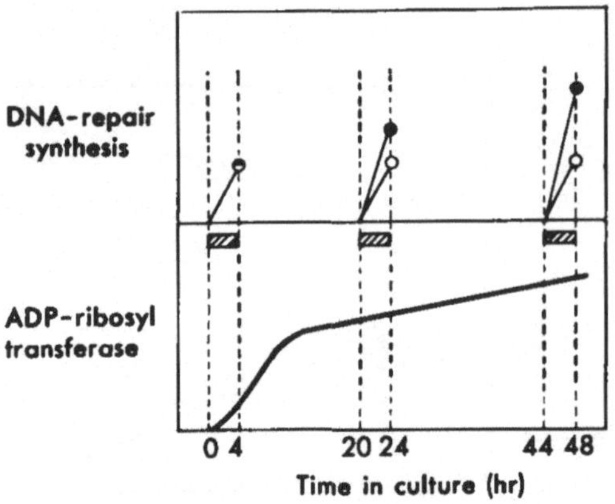


FIG. 3. Gradual development of the sensitivity of DNA repair synthesis (upper panel) to inhibition of poly(ADP-ribose)-biosynthesis (lower panel) by benzamide in cultured hepatocytes as a function of time (schematic summary). Hepatocytes were maintained in the continuous presence of benzamide (●) for 4, 24, or 48 hr and allowed to repair DNA damage inflicted by an identical UV (48 J/$M^2$) or MNNG ($10^{-4}$ M) treatment for an equal time period (shaded areas). With increasing culture time, a gradual enhancement of repair synthesis was observed in benzamide-treated (●) cells compared with untreated controls (○). The lower panel demonstrates the time relationship of this phenomenon to the spontaneous rise in poly(ADP-ribose)-biosynthesis in these cells.

indicated that for this enhancement of DNA repair synthesis, suppression of the initial rise in poly(ADP-ribose)-biosynthesis early after plating of hepatocytes (Fig. 1A) was critical, since a 48 hr treatment with 3-aminobenzamide after day 2 of culture only caused a marginal enhancement of DNA repair synthesis (data not shown). This suggested to us that enhancement of DNA repair synthesis did not result simply from a temporary inhibition of the nuclear ADP-ribosyltransferase activity, but required the prevention of the poly(ADP-ribose)-modification of a nuclear acceptor protein(s) that spontaneously occurs during the initial adaptation of hepatocytes to culture conditions (Althaus *et al.*, unpublished observations).

Concomitant with an enhancement of DNA repair synthesis, we observed that inhibition of poly(ADP-ribose) formation was associated with a decreased sensitivity of unscheduled DNA synthesis to inhibition by aphidicolin (Fig. 4). Since aphidicolin is assumed to probe selectively for the participation of DNA polymerase $\alpha$ in DNA repair (*16*), we tentatively conclude that experimental manipulation of poly(ADP-ribose)-metabolism in cultured hepatocytes differentially alters the contributions of DNA polymerase $\alpha$ and DNA polymerase $\beta$ in the repair of DNA damage. It was interesting to note that repair synthesis following DNA damage by the carcinogen, 4-nitroquinoline-1-oxide, was similarly affected in this regard (Fig. 4), although we had previously observed no enhancement of this synthesis by ADP-ribosyltransferase inhibitors (*14*) which enhance repair of DNA damage inflicted by the carcinogen, MNNG.

Another step in the repair of DNA damage, namely, strand rejoining, appeared to be sensitive to experimental manipulation of poly(ADP-ribose)-metabolism in cultured hepatocytes (Fig. 5). We observed that the removal of strand breaks caused by the alkylating agent, methyl methanesulfonate, was accelerated in the presence

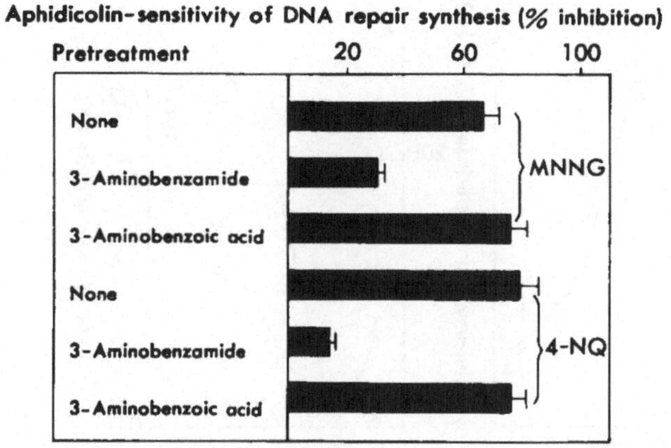


FIG. 4. Effect of the ADP-ribosylation inhibitor 3-aminobenzamide (10 mM) and its non-inhibitory analogue, 3-aminobenzoic acid (10 mM), on the sensitivity of DNA repair synthesis to inhibition by aphidicolin. Hepatocytes were pretreated for 44 hr as indicated, and aphidicolin (10 $\mu$g/ml) was added to the medium 30 min prior to carcinogen treatment (4 hr, MNNG, $10^{-4}$ M or 4-NQ, $10^{-4}$ M). DNA repair synthesis occurring during this carcinogen treatment was measured as described (*14*), and the inhibition of repair synthesis by aphidicolin was expressed as percent of control (= no aphidicolin).

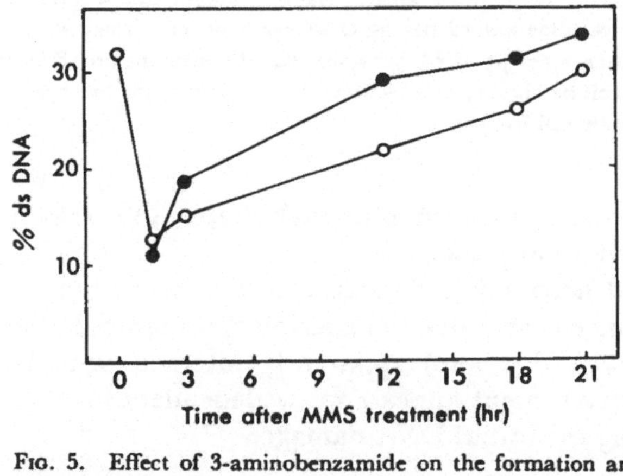


FIG. 5. Effect of 3-aminobenzamide on the formation and removal of DNA strand breaks in hepatocytes cultured for 48 hr and exposed to $4 \times 10^{-4}$ M methyl methanesulfonate (MMS) for 30 min. At different times after MMS-treatment, the percent double stranded DNA (dsDNA) remaining after a 50 min incubation under alkaline conditions was determined as described by Birnboim and Jevcak (*26*). Under these conditions, the unwinding velocity of DNA is directly proportional to the number of strand breaks present in DNA. ● hepatocytes maintained in the presence of 3-aminobenzamide; ○ control cells (solvent control = 1% DMSO in medium). For more details see refs. *14* and *26*.

of inhibitors of ADP-ribosyltransferase while the number of strand breaks formed under these conditions was unaffected by this treatment. These findings have been confirmed by Bohr and Klenow (*17*), who used normal human lymphocytes for their investigations. However, these results are in apparent contradiction to the findings of Shall and coworkers (*4*), who clearly demonstrated an inhibition of strand rejoin-

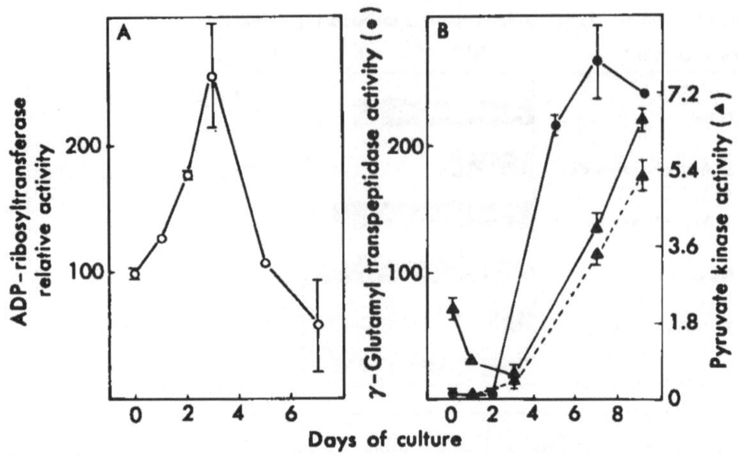


FIG. 6. Alterations in the activities of ADP-ribosyltransferase, gamma-glutamyltranspeptidase, and pyruvate kinase of adult rat hepatocytes cultured on collagen gel-nylon meshes. A. Hepatocytes were cultured on collagen gel-nylon meshes as previously described (*18*). At the times indicated, hepatocytes were recovered from the substratum by short collagenase treatment (*18*) and ADP-ribosyltransferase activity was measured as described (*14*). All activities are expressed relative to the activity of freshly isolated hepatocytes. B. The activities of gamma-glutamyltranspeptidase (● ●) and pyruvate kinase (▲—▲) were determined as described (*19*). ▲---▲ fetal (K-type) isozyme activity of pyruvate kinase. Gamma-glutamyltranspeptidase activity is expressed as nmol of *p*-nitroaniline formed per $\mu$g DNA per hour, and pyruvate kinase activity reflects $\Delta$ absorbance at 340 nm per $\mu$g DNA per hour. Bars indicate means$\pm$S.D. from 3 to 5 measurements in separate cell batches; symbols without bars represent the averaged values from 2 determinations in 2 separate cell batches.

ing in L1210 cells following treatment with dimethylsulfate. The reason for this discrepancy of results is not obvious at present.

Our studies using cultured hepatocytes demonstrate that more than one single step in the repair of DNA damage is sensitive to experimental manipulation of poly-(ADP-ribose)-biosynthesis. Poly(ADP-ribose) apparently interacts with late stages in the repair process and its involvement appears to be dependent on the chemical nature of the carcinogen causing the initial DNA damage.

### *Consequences of Altered Poly(ADP-ribose) Metabolism on the Expression of Hepatocellular Functions In Vitro*

For the following studies hepatocytes derived from adult rats were cultured on collagen gel-nylon meshes as previously described (*18*). This substratum allows for the maintenance of these cells for periods of up to 3 weeks. We found that in rat hepatocytes cultured on collagen gel-nylon meshes, ADP-ribosyltransferase activity increases to a maximum after 3 days of culture (Fig. 6A) and subsequently decreases to the low levels seen in freshly isolated hepatocytes (*19*). Coincident with the decrease in ADP-ribosyltransferase activity, the activities of gamma-glutamyltranspeptidase and the fetal (K-type) isozyme of pyruvate kinase sharply increased (Fig. 6B). These two enzyme activities are normally high in fetal hepatocytes *in vivo* but disap-

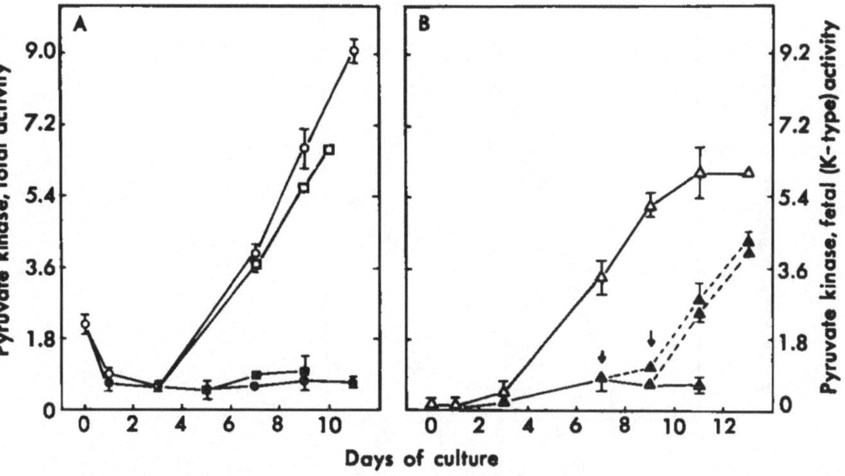


Fig. 7. Effects of the ADP-ribosyltransferase inhibitors, 3-aminobenzamide (10 mM) and nicotinamide (25 mM), and the non-inhibitory analogue, 3-aminobenzoic acid (10 mM), on the expression of pyruvate kinase activity of hepatocytes cultured on collagen gel-nylon meshes. A. Total pyruvate kinase activity of hepatocytes maintained in the presence of nicotinamide (●), 3-aminobenzamide (■), 3-aminobenzoic acid (□), and in control cultures (○). B. Activity of the fetal (K-type) isozyme of pyruvate kinase in untreated control (△—△) and nicotinamide-treated (▲—▲) hepatocyte cultures. ▲---▲ fetal isozyme activity after withdrawal (arrows) of nicotinamide from the culture medium. Bars indicate the mean±S.D. obtained from measurements in 3 to 5 separate cell preparations; symbols without bars represent the averaged values from 2 determinations in 2 separate cell batches. None of the ADP-ribosyltransferase inhibitors had any direct effect on pyruvate kinase activity *in vitro*. (Data taken from ref. *19*).

pear shortly after birth (*20*, *21*). Previous studies of the fetal (K-type) isozyme of pyruvate kinase in cultured hepatocytes have clearly shown that this expression of fetal isozyme activity is the result of induction of the isozyme protein (*21*). We observed that the induction of the fetal pyruvate kinase isozyme was totally inhibited in hepatocyte cultures treated with 3-aminobenzamide or nicotinamide (Fig. 7) indicating that the induction of the fetal pyruvate kinase isozyme and the alterations in poly(ADP-ribose)-biosynthesis (Fig. 6) are causally related (*19*). The specificity of action of the ADP-ribosyltransferase inhibitors was indicated by the inability of the non-inhibitory analogue, 3-aminobenzoic acid (1 to 10 mM), to prevent the expression of the fetal pyruvate kinase isozyme (Fig. 7A). A 70 to 95% inhibition of ADP-ribosyltransferase activity was observed when hepatocytes were maintained for 2 to 7 days in medium containing 10 mM 3-aminobenzamide or 25 mM nicotinamide. This inhibition was reversible as was the repression of the fetal pyruvate kinase isozyme (Fig. 7B).

Similar observations were made on the expression of gamma-glutamyltranspeptidase, which was also suppressed by inhibitors of ADP-ribosyltransferase (Fig. 8). This suggested to us that ADP-ribosylation might have a coordinative role in the expression of some hepatocellular functions *in vitro*. That suppression of poly(ADP-ribose)-biosynthesis did not cause a general perturbation of gene expression in these cells was suggested by the following observations (Fig. 9). The expression of $\alpha$-feto-

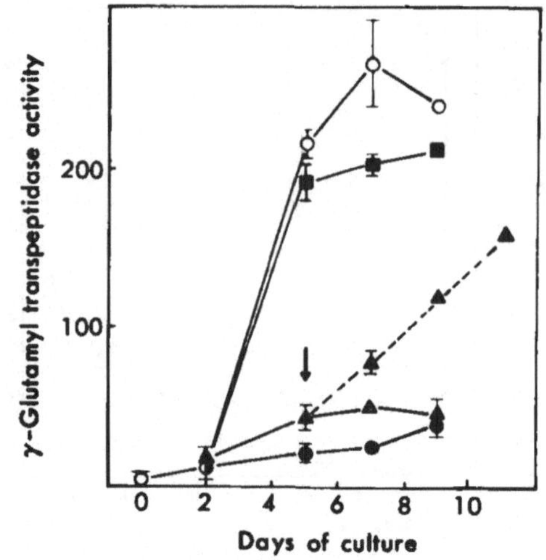


FIG. 8. Effect of nicotinamide (25 mM), 3-aminobenzamide (10 mM), and 3-aminobenzoic acid (10 mM) on the expression of gamma-glutamyltranspeptidase activity by hepatocytes cultured on collagen gel-nylon meshes. ○—○ enzyme activities of untreated control cultures; ●—● 3-aminobenzamide; ■—■ 3-aminobenzoic acid; ▲—▲ nicotinamide-treated hepatocyte cultures; ▲---▲ enzyme activity after withdrawal (arrow) of nicotinamide from the culture medium. Inhibition of gamma-glutamyltranspeptidase expression was also observed with other ADP-ribosylation inhibitors, such as isonicotinamide, benzamide, and theophylline. Mean ± S.D., 3 to 5 separate measurements; symbols without bars represent the averaged values from 2 determinations in 2 separate cell batches. None of the ADP-ribosyltransferase inhibitors had any direct effect on gamma-glutamyltranspeptidase activity *in vitro*. (Data taken from ref. *19*).

protein (Fig. 9A) and albumin (Fig. 9B) was unaltered in cells where the initial rise in ADP-ribosyltransferase activity was suppressed with the same inhibitors.

The inhibition of gamma-glutamyltranspeptidase expression required the presence of ADP-ribosylation inhibitors during the first 5 days of culture. Treatment of hepatocytes after day 5 of culture, *i.e.*, when ADP-ribosyltransferase activity had returned to low levels (Fig. 6A), caused only marginal inhibition (approximately 10%) of this expression. A histochemical study revealed that gamma-glutamyltranspeptidase was uniformly suppressed in all hepatocytes of a culture dish, indicating that the actions of the ADP-ribosyltransferase inhibitors were not selective for a small cell population within the liver (data not shown). In addition, the action of these inhibitors on gamma-glutamyltranspeptidase expression was reversible (Fig. 8).

Nicotinamide has been reported to inhibit general protein synthesis during the first 24 hr of hepatocyte culture (*22*). Although we could confirm this observation, we found that both nicotinamide (25 mM) and 3-aminobenzamide (10 mM) stimulated [$^3$H]leucine incorporation into total hepatocellular proteins after this initial culture period. Therefore, we conclude that the effects of these two ADP-ribosyltransferase inhibitors on the expression of pyruvate kinase and gamma-glutamyltranspeptidase are not the result of a general inhibition of protein synthesis.

Similarly, previous studies from our laboratory had demonstrated the lack of dependence of the expression of fetal functions by cultured hepatocytes on DNA synthesis (*18*), which increases after 2 to 3 days in culture in the absence of measur-

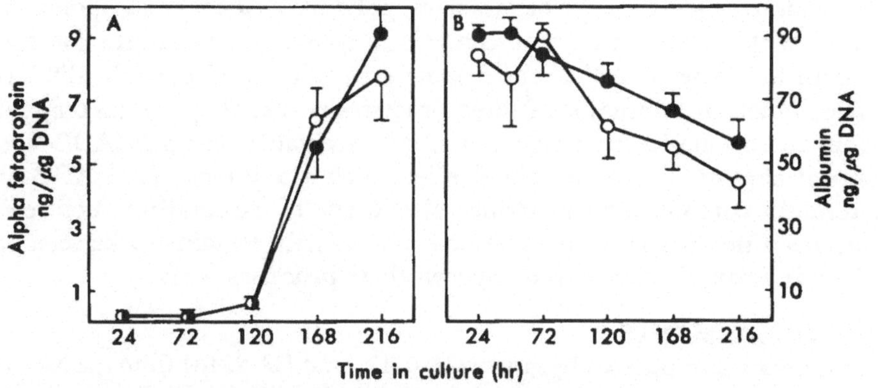


FIG. 9. Effect of 3-aminobenzamide on the expression of alpha-fetoprotein (A) and albumin (B) by cultured hepatocytes. Hepatocytes were maintained in the absence (○) or presence (●) of 3-aminobenzamide (10 mM) in the culture medium, and alpha-fetoprotein and albumin production were determined as described elsewhere (*18*).

able cell replication. Although this increase in DNA synthesis was slightly inhibited by nicotinamide (25 mM), continuous treatment of hepatocyte cultures for up to 9 days with 3-aminobenzamide did not affect this activity (data not shown). Thus the action of ADP-ribosyltransferase inhibitors on pyruvate kinase expression was not related to effects on DNA synthesis by these hepatocytes.

Increasing evidence suggests that ADP-ribosyltransferase activity is largely dormant in the cell nucleus unless factors inducing DNA damage are manifest (*1*, *23–25*). Presumably DNA damage, especially in the form of strand breaks, then initiates the activation of ADP-ribosyltransferase. In order to test this presumed correlation between DNA damage and ADP-ribosylation, we examined hepatocellular DNA for the presence of DNA strand breaks in relation to changes in ADP-ribosylation. Using a slight modification (*14*) of the sensitive technique by Birnboim and Jevcak (*26*), we were unable to find an increase in the number of DNA strand breaks prior to or concomitant with changes in ADP-ribosyltransferase activity (Fig. 6), although we observed a gradual but slow increase in the number of DNA strand breaks with increasing time of culture. This increase was unaffected by 3-aminobenzamide or nicotinamide (data not plotted). These results suggest that the transient and spontaneous alterations in ADP-ribosyltransferase activity seen in primary cultures of hepatocytes (Fig. 6) are not causally related to DNA fragmentation. These data are in agreement with the observation by Jackowski and Kun (*6*), who found age-dependent variations in the ADP-ribosyltransferase activity of cardiocyte nuclei in the absence of measurable DNA fragmentation.

## CONCLUSIONS

Our results demonstrate the consequences of altered poly(ADP-ribose)-metabolism on two diverse functions within the same cell, *i.e.*, DNA repair and expression of some fetal functions by hepatocytes *in vitro*. The mechanism(s) of poly(ADP-ribose)-involvement in these two processes remain(s) to be elucidated. However, our results seem to suggest that in hepatocytes, poly(ADP-ribose) acts upon a "regulatory cascade" within chromatin that controls a common mechanism for such diverse chromatin functions as DNA repair, gene expression, and possibly DNA replication (*3*). A common denominator of these processes is that they are accompanied by alterations in chromatin structure, and it is conceivable that poly(ADP-ribose)-modification of chromatin proteins could affect such alterations. That ADP-ribosylation of chromatin components can induce alterations of chromatin structure *in vitro* has already been demonstrated by Poirier *et al.* (*27*). It remains to be seen whether or not there is a causal relationship between these processes *in vivo*.

ACKNOWLEDGMENTS

This work was supported by grants CA-07175 and CA-22484 from the National Cancer Institute. F. R. Althaus is a recipient of a Public Health Service International Research Fellowship (5 F05 TWO 2756-2).

REFERENCES

1. Purnell, M. R., Stone, P. R., and Whish, W.J.D. ADP-ribosylation of nuclear proteins. Biochem. Soc. Trans., *8*: 215–227, 1980.
2. Smulson, M. E. and Sugimura, T. (eds.). Developments in Cell Biology, Vol. 6, Elsevier/North-Holland, Amsterdam, 1980.
3. Lindahl, T. Worthy of a detour. Nature, *198*: 424–425, 1982.
4. Durkacz, B. W., Omidiji, O., Gray, D. A., and Shall, S. (ADP-ribose)$_n$ participates in DNA excision repair. Nature, *283*: 593–596, 1980.
5. Creissen, D. and Shall, S. Regulation of DNA ligase activity by poly(ADP-ribose). Nature, *296*: 271–272, 1982.
6. Jackowski, G. and Kun, E. Age-dependent variation of rates of polyadenosine-diphosphoribose synthesis by cardiocyte nuclei and the lack of correlation of enzymatic activity with macromolecular size distribution of DNA. J. Biol. Chem., *256*: 3667–3670, 1981.
7. Porteous, J. W., Furneaux, H. M., Pearson, C. K., Lake, C. M., and Morrison, A. Poly(adenosine diphosphate ribose) synthetase activity in nuclei of dividing and of non-dividing but differentiating intestinal epithelial cells. Biochem. J., *180*: 455–463, 1979.
8. Savard, P., Aubin, R., Whish, W.J.D., Lord, A., and Poirier, G. Poly(adenosine diphosphoribose) polymerase activity and adenosine diphosphate ribosylation of proteins during pancreatic degeneration and regeneration. Cancer Res., *41*: 1417–1421, 1981.
9. Pekala, P. H., Lane, M. D., Watkins, P. A., and Moss, J. On the mechanism of preadipocyte differentiation. J. Biol. Chem., *256*: 4871–4876, 1981.
10. Bredehorst, R., Wilckens, K., Adamietz, P., Steinhagen-Thiessen, E., and Hilz, H. Mono(ADP-ribosyl)ation and poly(ADP-ribosyl)ation of proteins in developing liver and hepatomas. Eur. J. Biochem., *120*: 267–274, 1981.
11. Gartemann, A., Bredehorst, R., Wielckens, K., Stratling, W.H.H., and Hilz, H. Mono- and poly-ADP-ribosylation of proteins in mouse kidney after castration and testosterone treatment. Biochem. J., *198*: 37–44, 1981.
12. Purnell, M. R. and Whish, W.J.D. Novel inhibitors of poly(ADP-ribose) synthetase. Biochem. J., *185*: 775–777, 1980.
13. Althaus, F. R. Hepatocyte cultures as a research tool in pharmacology and toxicology.

*In;* Y. Ruckebusch (ed.), I.N.R.A. Colloquia, MTP Press, Lancaster, 1983, in press.

14. Althaus, F. R., Lawrence, S. D., Sattler, G. L., and Pitot, H. C. ADP-ribosyltransferase activity in cultured hepatocytes: Interactions with DNA repair. J. Biol. Chem., *257*: 5528–5535, 1982.
15. Paine, A. J., Hockin, L. J., and Legg, R. F. Relationship between the ability of nicotinamide to maintain nicotinamide-adenine dinucleotide in rat liver cell culture and its effect on cytochrome P450. Biochem. J., *184*: 461–463, 1979.
16. Huberman, J. A. New views of the biochemistry of eukaryotic DNA replication revealed by aphidicolin, an unusual inhibitor of DNA polymerase $\alpha$. Cell, *23*: 647–648, 1981.
17. Bohr, V. and Klenow, H. 3-Aminobenzamide stimulates unscheduled DNA synthesis and rejoining of strand breaks in human lymphocytes. Biochem. Biophys. Res. Commun., *102*: 1254–1261, 1981.
18. Sirica, A. E., Richards, W., Tsukada, Y., Sattler, C. A., and Pitot, H. C. Fetal phenotypic expression by adult rat hepatocytes on collagen gel-nylon meshes. Proc. Natl. Acad. Sci. U.S., *76*: 283–287, 1979.
19. Althaus, F. R., Lawrence, S. D., He, Y. Z., Sattler, G. L., Tsukada, Y., and Pitot, H. C. Effects of altered (ADP-ribose)$_n$-metabolism on expression of fetal functions by adult hepatocytes. Nature, *300*: 366–368, 1982.
20. Pitot, H. C. and Sirica, A. E. Hepatocarcinogenesis as a problem in developmental biology. *In;* R. G. McKinnell, M. A. DiBerardino, M. Blumenfeld, and R. D. Bergad (eds.), Differentiation and Neoplasia, pp. 241–250, Springer-Verlag, Berlin, 1980.
21. Leffert, H., Moran, T., Sell, S., Skelly, H., Ibsen, K., Mueller, M., and Arias, I. Growth state-dependent phenotypes of adult hepatocytes in primary monolayer culture. Proc. Natl. Acad. Sci. U.S., *75*: 1834–1838, 1978.
22. Villa, P., Hockin, L. J., and Paine, A. J. The relationship between the ability of pyridine and substituted pyridines to maintain cytochrome P450 and inhibit protein synthesis in rat hepatocyte cultures. Biochem. Pharmacol., *29*: 1773–1777, 1980.
23. Tanaka, Y., Hashida, T., Yoshihara, H., and Yoshihara, K. Bovine thymus poly(ADP-ribose) polymerase histone-dependent and $Mg^{2+}$-dependent reaction. J. Biol. Chem., *254*: 12433–12438, 1979.
24. Benjamin, R. C. and Gill, D. M. ADP-ribosylation in mammalian cell ghosts. J. Biol. Chem., *255*: 10493–10501, 1980.
25. Cohen, J. J., Catino, D. M., Petzold, S. J., and Berger, N. A. Activation of poly (adenosine diphosphate ribose) polymerase by SV40 minichromosomes: effects of DNA damage and histone H1. Biochemistry, *21*: 4931–4940, 1982.
26. Birnboim, H. C. and Jevcak, J. J. Fluorometric method for rapid detection of DNA strand breaks in human white blood cells produced by low doses of radiation. Cancer Res., *41*: 1889–1892, 1981.
27. Poirier, G. C., de Murcia, G., Jongstra-Bilen, J., Niedergang, C., and Mandel, P. Poly(ADP-ribosyl)ation of polynucleosomes causes relaxation of chromatin structure. Proc. Natl. Acad. Sci. U.S., *79*: 3423–3427, 1982.

# Author Index

# Subject Index